AGGREGATE RESOURCES
A GLOBAL PERSPECTIVE

Aggregate Resources

A *global perspective*

Edited by
PETER T. BOBROWSKY
BC Geological Survey Branch, Victoria, BC, Canada

A.A.BALKEMA/ROTTERDAM/BROOKFIELD/1998

Cover photograph: Late Pleistocene section on the Finlay River of British Columbia, Canada consisting of horizontally stratified, well-sorted pebble gravel, underlying Late Wisconsinan diamicton. Glaciofluvial gravels like this are common throughout glaciated regions of North America and provide an excellent source of natural aggregate.

Published by
A.A. Balkema, P.O. Box 1675, 3000 BR Rotterdam, Netherlands
Fax: +31.10.4135947; E-mail: balkema@balkema.nl; Internet site: http://www.balkema.nl

A.A. Balkema Publishers, Old Post Road, Brookfield, VT 05036-9704, USA
Fax: 802.276.3837; E-mail: info@ashgate.com

ISBN 90 5410 675 1

Contents

VI *Contents*

Aggregate resources in global perspective

PETER T. BOBROWSKY
BC Geological Survey Branch, Victoria, BC, Canada

1 INTRODUCTION

Sand, gravel, crushed stone, natural aggregate, and construction materials, are but a few of the terms used to describe one important item in the physical environment of society. We take for granted the thousands of kilometres of gravel roads and paved highways which traverse the landscape around the globe. Whether we travel by foot, horse and carriage or high speed automobile, the need for road networks is never fully satisfied. Indeed, we are a society fully dependent on ease of mobility and expect to reach the farthest of locations via some defined route both conveniently and quickly. Today, existing transportation arteries are constantly being upgraded and newly created roads are reaching frontier terrains at an incredible rate. The process of constructing transportation routes is not peculiar to present day society and can be formally traced back to the engineered roads of the Incas or more primitive prehistoric paths and tracks which extend additional millennia. In concert with this observation of aggregate as a key element in transportation, we also recognize the need for aggregate resources in the construction of homes, buildings and other structures such as bridges. Again, it is easy for some to overlook the essential need for aggregate in simple home construction. As far as the average home buyer and highway commuters are concerned, the role of aggregate in day-to-day life is not that relevant (Figs 1 to 4).

The important role of aggregate in society will not diminish in the future. Available sources of aggregate are constantly under threat from competing landuse options, sterilization and simple over-exploitation. Aggregate resources rarely occur in convenient locations and the costs involved in locating and transporting aggregates from source area to use are constantly increasing. This prominent role in society is best exemplified by the attention aggregates have gained in the published arena. During the last few years several texts have appeared on the subject and as public and government awareness expands, more publications are likely to appear.

In 1991, a landmark publication appeared in North America entitled *The Aggregate Handbook* (Barksdale 1991). Published by the National Stone Association, this seminal work provided a comprehensive guidebook to the technical aspects of aggregate resources. The intent of the book was to 'guide the growth of the industry and the proper application of its products in engineering design, construction and other

Figure 1. An example of small scale technology for aggregate processing at a local scale in Nepal (photo courtesy of D. VanDine).

Figure 2. View of hand screening in processing of aggregate in Nepal (photo courtesy of D. VanDine)

Figure 3. View of extensive aggregate operation in competition for local landuse near Vancouver, Canada (photo by P. Bobrowsky).

Figure 4. High technology aggregate processing contrasting with that illustrated in Nepalese examples (photo by P. Bobrowsky).

uses'. The text remains an essential reference manual to the technical aspects of the aggregate industry. In Europe, a comparable text was published in 1985 and reprinted by the Geological Society in 1992, under the title *Aggregates: sand, gravel and crushed rock aggregates for construction purposes* (Smith & Collis 1993). With an obvious emphasis on the United Kingdom, the text is also an essential reference manual to the technical components of the aggregate industry. Lastly, as a product of the 2nd International Aggregates Symposium held in 1990, an edited volume entitled *Aggregates – Raw Materials' Giant* was published in 1994 (Luttig 1994). Less technical and more academic in content, the text is a very useful introduction to European perspectives on aggregate in relation to technologies, the environment and landuse issues important at the turn of the decade.

The significance of aggregate remains a global issue. Building on the regional and technical success of the aforementioned texts, the purpose of this volume is to present, under a single cover, the many diverse aspects of aggregates including industry production and demand, geologic occurrence, mapping, technical advances, economic implications, legislative management, landuse decisions as well as others. To meet this objective successfully, I consider it imperative that contributions and case studies from a number of countries be included. Hence, the papers in this book provide data from Australia, Belgium, Canada, Lebanon, the Netherlands, Norway, South Africa, United Kingdom and the United States. The book will appeal to all involved with aggregate resources: Geologists, producers, construction engineers, technicians, developers, land-use planners, legislators, academics and the public consumer, especially since all of us are in some manner either directly dependent or indirectly affected by this resource.

2 STRUCTURE

Given the breadth of topics covered in the 27 papers presented in this volume a coherent organization that will be suited to every reader is clearly untenable. The papers are broadly organized under general themes following legislation, economics, geology, landuse, techniques, and case studies, but it will be apparent that most papers overlap several of these categories.

The first collection of papers deal with production, legislation and the signficance of aggregate and their deposits. The lead paper by Vagt and Irvine is an overview of the Canadian aggregate market. It includes both discussion and statistics on differences in provincial production, quantities and values shipped, uses of sand and gravel, trends in construction as well as the impact of recycling and substitution. Although centred on a single country, Vagt and Irvine's paper illustrates the dynamic temporal and spatial trends characteristically experienced by the aggregate industry in many places around the world. Beeby's paper on California provides an informative historical review of the development of a successful aggregate management plan, policy and legislation for the most actively growing state in the USA. He shows the importance of California's Surface Mining and Reclamation Act of 1975 (SMARA) in shaping the future of aggregate resource management. His paper is a good case study in the successful implementation of legislation. Stewart's paper deals with the aggregate industry in Ontario, the largest in Canada and thus a complement to the preceding paper by Beeby. He examines how the provincial government of Ontario has responded with time to the increasing pressures for necessary regulation of aggregate extraction by discussing the role that the physical, social and economic environment has had on influencing aggregate resource management. The paper contains a fairly detailed history of legislative changes and their effects over a period of a few decades. Knight's paper on conservation provides an innovative twist to the importance of legislation and land management. He describes a new perspective on aggregate resources as a landscape entity which is considerate of conservation principles. The United Kingdom's National Nature Reserves and Sites of Special Scientific Interest are intimately tied to the countries aggregate resource base.

The second series of papers focus on the geological aspects of aggregate, but rely heavily on case studies to exemplify their points. Florsheim and colleagues examine the geomorphological effects of in channel gravel extraction in the Russian River of California in terms of the evolution of channel morphology. They integrate aerial photographs, maps, and cross sections to show a series of time transgressive changes along 13 km of river channel which has responded to intensive aggregate extraction. Edwards outlines the types of geological deposits hosting exploitable sand and gravel in Alberta. The primary deposits include certain preglacial sediments, glaciofluvial, alluvial, glaciolacustrine, eolian and colluvial accumulations. He then explains the exploration characteristics which were used for developing predictive conceptual models. Kondolf's paper is a practical examination of the qualitative and quantitative environmental impacts of instream mining on river channels and floodplains. His discussion topics include transient effects, channel incision, sediment starvation and knickpoint migration, undermining of structures, groundwater effects, channel instability, flood control, bed coarsening and several other items. The paper is well documented with case studies. A paper by Dunlevey and Stephens follows in which they examine the current usage, availability and quality of coarse aggregate in South Africa. The authors discuss in detail regional lithologies and their properties which is useful to the construction aggregate industry in the region. Bogemans presents a paper that stresses the importance of primary Quaternary geologic data in landuse planning (i.e. field data). Her paper outlines basic profile-type maps and thematic applied maps. The author documents the premise with case study examples from north Belgium in the Flemish Valley and the Northern Campine area. The final paper in this group is one by Thoms which deals with modern process studies. Noting the importance of alluvial sand and gravel deposits in eastern Australia, which are routinely exploited with instream mining, Thoms presents new data on field studies which elucidate processes of sand accumulation. This practical exercise has far reaching implications for modeling aggregate deposit accumulations in many areas outside of Australia.

The third series of papers centre on landuse issues associated with aggregate resources. The first paper by Baker examines the application of an integrated resource management principle to planning and regulating aggregate resources. He focuses on two Canadian provinces, Ontario and British Columbia, reviewing current legislation and flaws in the practice of aggregate management. There is a good discussion on the importance of sound policy to accommodate multiple interest groups and competing landuse strategies. The next paper by Blackett and Tripp provides a good discussion on multiple landuse conflict issues at a local scale in southern Utah, USA. Here rapid urban expansion is the prime element affecting the poorly mapped aggregate resources. De Jong and de Mulder present an examination of the nature and distribution of clay, sand and gravel in the Netherlands including discussion on production demands for construction materials, followed by a review of the policy in place by government and industry to regulate, manage and use the three construction materials. Poulin and Martin's paper is a good review of basic landuse planning principles and planning for aggregate. They analyze the economics of recycled aggregate within the framework of multiple landuse issues. The last paper in this section is by Bobrowsky and Manson who examine the importance of qualitative and quantitative modeling of aggregate deposits for landuse planning. They illustrate how inventory, area and volume data can be manipulated for predictive purposes useful for future landuse planning.

The fourth collection of papers revolve around technical aspects of aggregate resources. The first paper by Bliss is a superb review of the methods and reasons why aggregate resources are qualitatively and quantitatively assessed. He discusses classic quality evaluation tests such as soundness, absorption, fineness, as well as resistance to statistical modeling for quantitative evaluation. This is followed by an excellent paper by Langer and Knepper who review basic field and lab testing methods. Their paper includes a discussion of the types of aggregate deposits, as well as aggregate assessments. It is a good introduction for the field geologist interested in aggregates. Jol and colleagues provide a lengthy explanation of the method and utility of ground penetrating radar (GPR) as an effective exploration tool for aggregate studies. Using a series of examples, the authors illustrate how GPR can be used to map the depth and lateral extent of sand and gravel deposits including braid river/delta deposits, fan-foreset delta deposits and catastrophic flood accumulations. This is followed by an in-depth discussion by Verhoeff and van de Wall of the application of petrography in durability assessment of rock aggregate. Issues addressed include mineralogy, texture and weathering, laboratory testing, methodology and mechanical behavior.

The text concludes with a series of papers loosely considered to be case study examples. Rowan and Kitetu begin by identifying the magnitude and significance of environmental effects of sand harvesting from Kenyan rivers. Their quantitative evaluation provides a framework for policy makers and environmental planners to cope with long term conservation and resource sustainability. Gonggrijp examines the important role of aggregates in the Netherlands including the impact of mining on the landscape. He focuses on the geological significance of outcrops resulting from mining, the need for restoration and reclamation of pits and quarries, outcrop stability, arrangement and management. He concludes with a series of Dutch case study examples. Edwards provides the summary statistics and interpretations resulting from a detailed aggregate survey in the province of Alberta. From the data he describes patterns of production, transport distances, as well as geologic, environmental and legislative conditions including environmental concerns which affect production. Several noteworthy and relevant observations are presented. Khawlie provides a good introduction to the escalating problems of construction material management in Lebanon. Problems range from high population density and rapidly increasing demands to unchecked use of explosives, health issues, geological instability and environmental degradation. The author stresses the need for better databases, securing technical and management issues and implementing legislative enforcement. Hora's paper deals with a high population, high density coastal urban setting in western North America. The author reviews changes in production and demand, types and distribution of deposits, quarried and crushed aggregate and future challenges for the city of Vancouver. Neeb introduces the Norwegian Gravel and Hard Rock Aggregate Database which is an inventory and description of the 8790 aggregate deposits currently recorded in the country. He discusses testing methods and standards, types of bedrock suitable for crushing, uses of the database, super-quarries and production – consumption statistics unique to Norway. Kelly and colleagues review the legislation and policy surrounding aggregate resources in the province of Ontario, Canada. They discuss in detail a very useful methodology currently used in the province to develop aggregate potential maps. The final paper in the volume is by Mossa and Autin who provide an in-depth overview of the history and technology of aggregate mining in the state of Louisiana, USA. Their paper includes a detailed discussion of the geology of deposits, the complexities of dealing with aggregate primarily located below the water table, and most importantly the environmental and economic impacts of aggregate mining.

REFERENCES

Barksdale, R.D. (ed.), 1991. *The Aggregate Handbook*. National Stone Association, Washington, DC.
Luttig, G.W. (ed.), 1994. *Aggregates – Raw Materials' Giant*. Erlangen, Germany.
Smith, M.R. & Collis, L. (eds), 1993. *Aggregates; sand, gravel and crushed rock aggregates for construction purposes*. Bath, UK: Geological Society Special Publication 9.

ACKNOWLEDGMENTS

I am extremely grateful to the contributors of this volume for responding in such a professional manner to the deadlines imposed at the start of the effort. The success of this book rests entirely on the quality of their contributions. The support of the British Columbia Geological Survey Branch is appreciated and was most significant to ensuring a successful completion of the project. Similarly, my thanks to the IUGS Commission on Geological Sciences for Environmental Planning which are recognized as co-sponsors of the book. The Commission has provided me with an international level of appreciation for environmental geology problems. Considerable credit goes to the many professionals who carefully reviewed the numerous manuscripts originally submitted for inclusion in this book. Each manuscript was reviewed by me and a minimum of two other independent scientists including: F. Bachhuber, D. Baker, P. Barnett, D. Beeby, R. Blackett, G. Brooks, B. Broster, N. Catto, J. Clague, E. de Mulder, J. Dunlevey, D. Edwards, P. Egginton, J. Florsheim, B. Grant, S. Hicock, D. Hora, D. Huntley, L. Jackson, H. Jol, P. Karrow, R. Kelly, J. Knight, M. Kondolf, B. Langer, J. Lewis, D. Liverman, J. Logan, G. Manson, T. Morris, J. Mossa, P.-R. Neeb, E. Nielsen, R. Poulin, T. Pronk, M. Roberts, M. Roed, J. Rowan, B. Schreiner, M. Sharp, C. Smith, D. Smith, I. Spooner, R. Stea, J. Teller, M. Thoms, O. Vagt, D. VanDine, B. Ward, and B. Whelan. My thanks to all those listed above for their time and effort. At the production end of the book, I appreciate the support of the staff at A.A. Balkema for seeing the project to quality completion. Special thanks in this regard to Ms. Monique Verdonk and her incredible patience. I am extremely grateful to my wife Theresa for her tolerance, love and support. The difficulties of dealing with a geologist fascinated with 'dirt' are extensive. Honorable mention to Oreana and Killian, keepers of the global peace.

March 1997
P.T. Bobrowsky
Victoria, British Columbia

Construction aggregates in Canada – An overview

G.O. VAGT & R.D. IRVINE
Minerals and Metals Sector, Natural Resources Canada, Ottawa, Ontario

1 INTRODUCTION

Access to allow orderly development and rehabilitation of lands endowed with construction aggregates needed for building and engineering construction has become an important concern in many regions of Canada.

In 1994, the production value of non-fuel minerals in Canada was about $14.5 billion, based on preliminary figures. This output consists of two large fields: metals production valued at $9.4 billion and industrial minerals production, comprising non-metals and structural materials, valued at about $5.1 billion, as shown in Table 1 (after Pilsworth & Kokkinos 1994). Similarly, the sources of most minerals-related data on shipments and uses is Natural Resources Canada as well as Statistics Canada, as noted on the relevant figures.

As shown, the industrial minerals field accounts for about 35% of the value of Canada's non-fuel minerals industry. This importance is reflected at the provincial level. For example, industrial minerals account for 27% of the value of Ontario's non-fuels production (Anon. 1994a) and for 39% of Quebec's non-fuels production (Anon. 1994b).

Based on the current reporting system for provincial and federal government accounts, the subfield of structural materials, as this relates to the term *field* defined in general terms (Bates 1994), includes all natural construction aggregates, cement, lime, heavy clay products mainly derived from shale or clay, and other rock (or stone) that, considered together, are used mainly for construction purposes. The volumes and values of these materials are generally reported as shipments fob pit or quarry, by the relevant companies, or more precisely, by establishments reporting as specific operating units, as cited under the heading 'Interpretation' (Anon. 1993b).

In 1994, the reported annual shipments of natural construction aggregates (mainly sand and gravel and crushed stone) were valued at $1.2 billion, or nearly half of the value of all structural materials produced, as shown. About 73% of all crushed stone production in Canada is dependent on reserves of limestone and dolostone, followed by granite including traprock, sandstone including quartzite, shale including slate, and marble (Vagt 1994a). The relative importance of limestone resources is also found in United States markets (Tepordei 1994).

9

Table 1. Canada, non-fuel minerals industry production, by value, 1994. Sources: Natural Resources Canada; Statistics Canada.

	(C$ billions)*
1. Metals	$ 9.4
2. Industrial minerals	
Nonmetals	$ 2.6
Structural materials	$ 2.5
Total	$14.5

*Production excludes crude petroleum, natural gas, natural gas by-products and coal. Note: Figures are preliminary.

There are more than 2450 companies with an estimated 9700 direct employees in the construction aggregates sector, based on available 1992 reports and estimates (Anon. 1994c). This sector accounts for the largest proportion of the non-fuel minerals industry, as measured by volume of shipments destined for final consumption. The impact of this output, and consumption, on resource availability is magnified even more if limestone raw materials for producing cement and lime are included, (Anon. 1994a).

This overview addresses only briefly related subjects including aggregates qualities and specifications, recycling and re-use, international trade, and, transportation modes and costs. References to lightweight aggregates, as well as to ultra-lightweight minerals that are used in construction are noted elsewhere (Vagt 1994b). At the national level, the Canada Centre for Mineral and Energy Technology (CANMET) conducts scientific and technological activities related to cement, aggregates and the durability of concrete for infrastructure i.e., building and engineering construction (Malhotra 1987). Major research projects include: 1. Development of high performance concrete for offshore structures, 2. Use of flyash, slag, silica fume and other pozzolanic materials as a replacement for cement in concrete, and 3. The role of supplementary cementing materials for controlling alkali-aggregate reactions.

Issues related to environmental and permitting procedures that may involve both provincial and federal government jurisdictions are not considered in this overview. A good general reference concerning numerous related technical, economic and environmental subjects is 'The Aggregate Handbook' (Barksdale 1991).

2 CONSTRUCTION AGGREGATES

The diverse nature of companies and businesses involved in producing sand and gravel and crushed stone presents challenges in efforts to capture complete production and consumption data. Based on some reporting methods used in the past, the recorded output of these construction aggregates from all sources may, in some cases, be understated by as much as 25%. The province of Ontario (the largest producer and consumer of construction aggregates in Canada) has developed a comprehensive system for reporting its total output of aggregates. Specifically, Ontario includes designated licensed areas, crown land permits, private lands, and wayside permits as part of its reporting process (OMNR 1995).

Crushed stone is produced mainly for road-surfacing materials, concrete aggregates, asphalt aggregates, railroad ballast, and numerous other uses. A miscellaneous stone category, which accounts for less than 1% of all construction aggregates produced, includes rubble and riprap, roofing granules, and special materials for other uses such as manufacture of artificial building stone. (A detailed breakdown of uses for sand and gravel is described under a separate heading).

Shipments of all construction aggregates combined, of sand and gravel shown separately, as well as expenditures on construction, are portrayed regionally on a cumulative basis as shown in most of the figures. Aggregates production and shipments for the Northwest Territories and Yukon Territory are included with those in the Prairies region, shown as *Prairies and North*, in Figures 1 and 2.

2.1 *Quantities shipped*

Shipments of all aggregates combined, including in detail, natural gravel, sand and crushed gravel, and crushed stone and miscellaneous stone, are based on quantities reported by operators of sand and gravel pits, dredges or quarries (Fig. 1).

Variations in regional shipments of all construction aggregates, and analysis by the Minerals and Metals Sector (Irvine 1994) both suggest that demand is very much affected by the general economic conditions. This linkage can be seen in the growth in aggregates production, especially in Ontario during the expansion of the mid-1980s and in the subsequent decline toward the end of the 1980s as the onset of the recession. Netting out these economic swings, Canadian shipments of construction aggregates have been relatively stable, totaling about 320 million tonnes a year in both 1974 and 1993. Considering the entire period from 1971-1993 and Ontario in

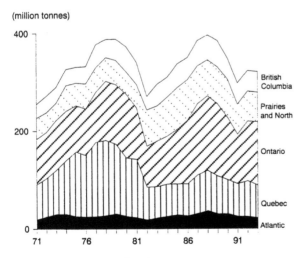

(million tonnes)

Source: Natural Resources Canada.
Note: Includes sand and gravel, crushed stone, and miscellaneous stone.

Figure 1. Canada, Shipments of construction aggregates (by volume) by region, 1971-1993.

(1986$ millions)

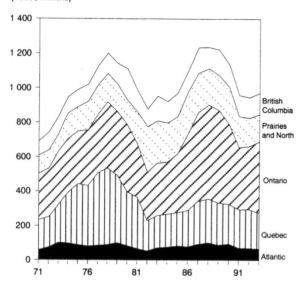

Source: Natural Resources Canada.
Note: Includes sand and gravel, crushed stone, and miscellaneous stone.

Figure 2. Canada, Shipments of construction aggregates (by value) by region, 1971-1993.

particular (because this province accounts for the largest proportion of Canada's aggregates output), the peak of the construction boom in 1988-1989 is well delineated. Over the same period of time, the relative contribution of each region to total Canadian shipments of construction aggregates has remained relatively stable.

Canada has historically been the world's leading per capita producer and consumer of construction aggregates, on the basis of analysis that considers all sources of aggregates production (Burton 1993). About 16 tonnes of aggregates per capita is produced and consumed annually (1988 and 1989 data). In comparison, aggregates consumption in three leading Scandinavian countries has been estimated to be about 11 tonnes per capita. More recently, consumption in Canada has declined to about 11 tonnes per capita, based on depressed levels of shipments in 1993.

2.2 *Values shipped*

The value of Canadian shipments of all construction aggregates combined was $1.2 billion in 1993, based on final statistics, or $972 million, in 1986 dollars adjusted for inflation (Fig. 2). As shown, the real adjusted annual value of construction aggregates produced over the 20 year period from 1974-1993 inclusive, has been about $950 billion, after considering the economic swings.

3 SAND AND GRAVEL

Sand and gravel deposits are widespread generally as a result of glaciation, and large producers have established plants as conveniently as possible to major consuming centres. These aggregate operations are usually associated with other activities such as ready-mix concrete or asphalt plants. They are also usually complemented by many small producers which serve local markets seasonally or on demand. Some relatively large operations may operate intermittently, serving as suppliers to heavy construction companies when required. Provincial highways departments may operate regional quarries to supply roadbed material for both repair work and new projects.

3.1 *Quantities shipped*

Shipments of sand and gravel amount to about 250 million tonnes per year, or about 73% of the total volume of all construction aggregates shipped, based on 1991-1993 data inclusive (Fig. 3). As shown, there has not been a recovery to the relatively high levels of production of sand and gravel in 1988-89.

3.2 *Values shipped*

The value of shipments of sand and gravel was $764 million in 1993, or about $600 million in 1986 dollars adjusted for inflation (Fig. 4). These values account for 60-65% of the total value of all construction aggregates shipped, and, in real dollar terms as shown, have remained about the same during the 20 year interval from 1974 to 1993 inclusive.

(million tonnes)

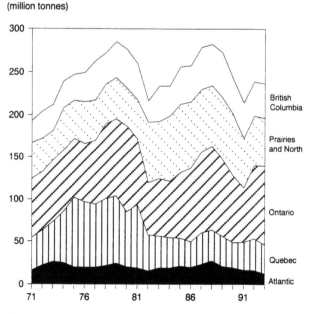

Sources: Natural Resources Canada; Statistics Canada.

Figure 3. Canada, Shipments of sand and gravel (by volume) by region, 1971-1993.

(1986$ millions)

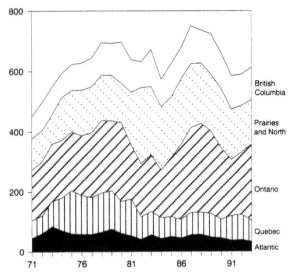

Figure 4. Canada, Shipments of sand and gravel (by value) by region, 1971-1993.

Sources: Natural Resources Canada; Statistics Canada.

4 TRADE IN CONSTRUCTION AGGREGATES

Trade in construction aggregates is small relative to the total volume of aggregates consumed in Canada. However, where transportation costs allow, cross-border shipments are important in some populated regions. In 1994, exports amounted to a little more than two million tonnes valued at $17.6 million (Fig. 5). Ontario accounted for more than half of this amount; the Atlantic region and British Columbia accounted for most of the remainder. Imports of construction aggregates in 1994 amounted to about 630,000 tonnes (Fig. 6). British Columbia accounted for nearly two-thirds of this trade, and Ontario, Quebec and the Atlantic region accounted for most of the remainder.

International bulk shipping of aggregates has been feasible in some areas. In British Columbia, large-volume ocean transportation facilities have been used for many years to supply high-quality aggregates or high-calcium limestone for higher value-added uses. For example, limestone producers on Texada Island, situated about 100 km northwest of Vancouver in the Strait of Georgia, supply raw material to cement and lime producers on the lower mainland and in the State of Washington (Anon. 1995). In Nova Scotia, granite aggregates have been shipped at regular intervals since 1986 from a coastal quarry. Also, analysis at this time suggested that as prices of local aggregates increased in major urban markets along the eastern coastal United States, sources in Canada's Maritime provinces could become more favourable (ADI 1986). More recently, studies have suggested that trends toward urbanization in the US Eastern Seaboard, together with growing local concerns associated with mineral extraction, present opportunities for producers in Atlantic Canada and elsewhere (Poulin & Vagt 1993).

**Exports
(HS 2517.10)**

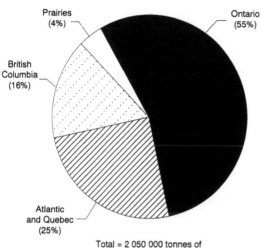

Total = 2 050 000 tonnes of
material used for aggregates.

Source: Statistics Canada.
Note: Nomenclature refers to international Harmonized System
principles for commodity classification.

Figure 5. Canada, Exports of construction aggregates (by volume) by region, 1994.

**Imports
(HS 2517.10)**

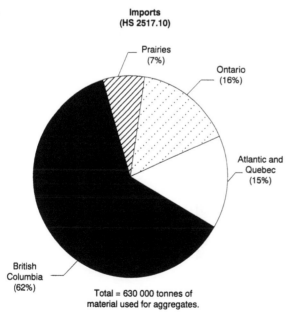

Total = 630 000 tonnes of
material used for aggregates.

Source: Statistics Canada.
Note: Nomenclature refers to international Harmonized System
principles for commodity classification.

Figure 6. Canada, Imports of construction aggregates (by volume) by region, 1994.

Canadian imports of construction aggregates in 1994 amounted to nearly 630,000 tonnes (Fig. 6). British Columbia accounted for nearly two-thirds of this traffic, and Ontario and the Atlantic region accounted for most of the remainder.

5 USES OF SAND AND GRAVEL

Uses for sand and gravel in Canada, based on the available data for 1993, are as follows: road bed and surfacing, 60%; concrete aggregate, 12%; asphalt aggregate, 7%; fill material, 7%; and, other uses (including railroad ballast, ice control, mortar sand, backfill, and other miscellaneous uses) 14% (Fig. 7). Information is available from the Minerals and Metals Sector concerning the uses of high purity industrial silica sands in Canada (Boucher 1993).

6 TRENDS IN BUILDING AND ENGINEERING CONSTRUCTION

Expenditures for building and engineering construction influence to a substantial degree demand for aggregates. The cumulative total value of these two types of expenditures, which include both new and repair construction, was valued at about $94.4 billion in Canada in 1993 (Anon. 1991-1993). In terms of 1986 dollars, adjusted for inflation, this value is equivalent to about $78 billion (Fig. 8).

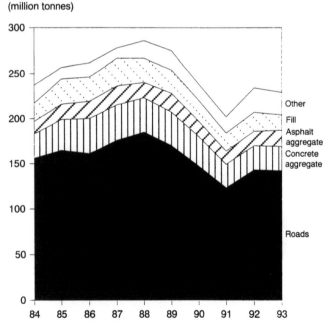

Source: Natural Resources Canada.
Note: Other includes railroad ballast, mortar sand and backfill for mines.

Figure 7. Canada, Reported uses for sand and gravel, 1984-1993.

(1986$ billions)

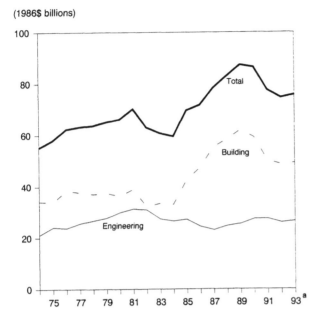

Source: Statistics Canada catalogue no. 64-201, Annual,
Construction in Canada (includes new and repair work).
[a] Intended expenditures only.
Note: Includes total value of actual expenditures for new and repair
work purchased.

Figure 8. Canada, Value of all
construction by type, 1974-1993.

During the period 1974-1993, residential and non-residential building construction combined, accounted for about two-thirds of the value of all construction expenditures. In the same period, engineering construction accounted for about one-third of construction expenditures, as the trend lines in Figure 8 show. From the national perspective, the value of engineering construction, including both new and repair construction adjusted for inflation, has remained at about the same level during the period. (It should be noted that new construction, historically, has accounted for 80%-90% of all construction. Repair or renovation has accounted for the remainder.)

Based on new reporting systems, data on construction expenditures now highlight actual capital expenditures that include major renovations (Anon. 1993a). The value of all expenditures on construction by region, during the 20 year period 1974-1993, adjusted for inflation, has been relatively stable (Fig. 9).

As shown in Figure 9, intended expenditures on all construction amounted to nearly $76 billion in 1993. Also as shown, expenditures for all construction during the period demonstrate an upward trend. (It should be noted that related expenditures in the Yukon Territory and Northwest Territories combined, normally account for less than 5% of regional expenditures, in this case allocated to British Columbia).

(1986$ billions)

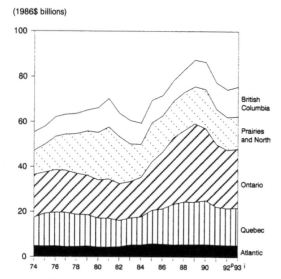

Source: Statistics Canada, Catalogue 64-201, Annual, Construction
in Canada (includes new and repair work).
Note: Figures for 1992 are preliminary; figures for 1993 are
intentions.

Figure 9. Canada, Value of all construction by region, 1974-1993.

7   TRENDS AND PROJECTIONS IN NEW CONSTRUCTION

Considering the value of total new construction only, adjusted for inflation, there has
been modest growth since 1992 as the economy has moved out of recession (Fig.
10). This scenario, including a forecast to 1996, is based on information from
Southam Construction Information Services (Southam 1994-1996). As noted, the
upward trend in the real value of construction that began in 1992, albeit modest, is
expected to continue.

Much of the underlying strength in the inflation-adjusted value of construction,
again according to the above-mentioned scenario by Southam, is expected to continue to be based on new engineering construction, which in the Prairies region can
account for more than 30% of all new construction (Fig. 11). In detail, this construction mainly relates to oil- and gas-related development.

8   REGIONAL TRENDS IN UNIT VALUES OF SAND AND GRAVEL

The trends from 1974 to 1993 in average real unit values of sand and gravel, on a
regional basis, have been relatively flat (Fig. 12). However, year-to-year differences
in these values may change as much as 25%. As shown, there have been trends toward similar unit values during the recessionary periods of 1982-1985 and 1990-
1992. These periods have been marked by relatively lower expenditures for construction, as shown in Figure 9.

(1986$ billions)

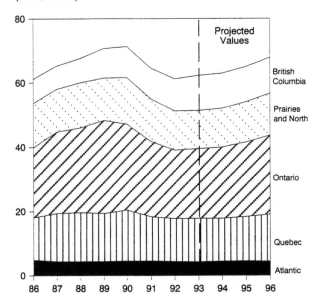

Source: Southam Construction Information Services Annual Forecast, 1994-96.

Figure 10. Canada, Value of new construction by region, 1986-96.

(1986$ billions)

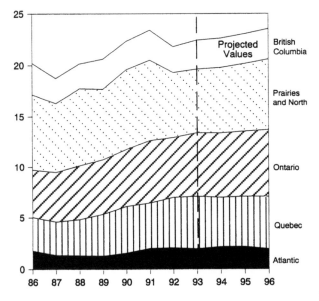

Source: Southam Construction Information Services Annual Forecast, 1994-96.

Figure 11. Canada, Value of new engineering construction by region, 1986-96.

9 INTERPRETATION

The average real fob pit or quarry values associated with the regional shipments of all aggregates, have not increased during the 20 year period from 1974 to 1993, inclusive, based on the data presented in Figures 2 and 4. Also, as shown in Figure 12, the average unit values of sand and gravel have tended to be flat, after discounting the effects of peak periods of aggregates demand and construction spending.

One reason for the relatively flat trend in the average real fob prices of aggregates has been horizontal integration through the merging of firms that has taken place during the past 20 years. In other words, consolidated ownership has resulted in marketing and financial efficiencies (Anon. 1992a). The trend toward fewer primary establishments operating in the aggregates business may be confirmed from a review of some of the available statistics (Anon. 1993b).

A second reason is that, during this 20 year period, there has been a trend toward larger scale operations and the use of larger equipment. From this, it may be assumed that production costs have benefited from economies of scale at relevant pits and quarries in Canada, as has been demonstrated elsewhere (Bronitsky 1973).

In comparison to the relatively flat trend in the real fob values of shipments of aggregates in Canada, real expenditures for building and engineering construction have increased by an average compound rate of about 1.7% per year during the period 1974-1993, as reflected in Figure 9. Viewed in this context, the importance of relatively low-valued aggregates, as a major materials input for building and engineering construction, is understated in terms of their value to society.

Moreover, it may be argued that for planning purposes, the value of construction

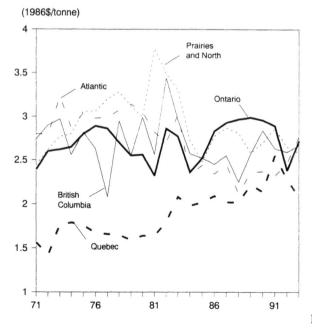

Source: Natural Resources Canada.

Figure 12. Canada, Trends in unit value of sand and gravel, 1971-93.

aggregates, considered as a diminishing resource in several regions in Canada, would be represented more accurately if viewed in terms of their final in-place value. In other words, attendant high transportation and handling costs should be considered integral to the value of reserves and resources of aggregates, particularly at sites strategically located relative to urban areas.

According to a recent study of aggregates in Ontario, relating to markets for more than 90% of the province's total aggregates output, the average annual costs of transportation for the years 1986 to 1990 were estimated to have been approximately $500 million. Transportation distances were mainly in the range of 20-50 kilometers (Anon. 1992b). These high transportation costs were roughly equivalent to the average annual fob value of construction aggregates i.e., $555 million, during the equivalent 5-year period, as shown in Figure 2. It is therefore expected that the average on-site or in-place values of aggregates delivered to most urban areas throughout North America, are essentially double the reported fob pit or quarry values. In this regard, it is expected that the values of delivered aggregates in the United States are at least double their fob values because transportation distances are frequently from 45 to 80 kilometers (Langer & Glanzman 1993).

The importance of ensuring accessible reserves of relatively low cost aggregates, including limestones and dolostones in particular, cannot be overstated because related uses in Canada extend to the chemicals, metallurgical, and the dimensional stone industries. One of the most comprehensive studies of limestones in North America, from the perspective of their importance in industrial sectors in addition to construction, was commissioned by the Ontario Ministry of Natural Resources (Derry et al. 1989). More recently, high-calcium limestones have played a key role in new environmental control and remediation technologies. For example, limestone uses for flue-gas desulphurization and for fluidized bed coal-fired boilers have become important in Atlantic Canada (MacDonald & Boehner 1994). In this context, high-purity limestones, and some of their dolomitic varieties, have been increasingly referred to as 'industrial antacid', or 'green gold'.

10 POSSIBLE OFFSHORE SOURCES OF AGGREGATES

As existing land-based sources of aggregates become depleted, there is growing potential for economically viable marine dredging of sand and gravel in Canada. Offshore sand and gravel resources in Canada have been utilized to meet special job requirements in the Beaufort Sea, the Prince Rupert area, and at the Roberts Bank port facility near Vancouver. In Atlantic Canada, it has been established that there is a good possibility of defining sufficient quantities of sand and gravel for marine dredging (Prime 1988). General issues relating to offshore mineral resources, with references to restrictions on beach mining, have been summarized more recently (Hale 1990).

11 RECYCLING AND SUBSTITUTION

The recycling of concrete and other construction materials is expected to increase because of limitations on the use of landfill sites and the increased use of mater.als management practices previously limited to Europe. Although regulations have generally not been established concerning the environmentally sound use of secondary materials from construction and demolition wastes, it has been documented that the use of these materials contributes to aggregates conservation. In Ontario, materials of most interest include: old asphalt, old concrete, blast furnace slag, steel slag, nickel and copper slags, crushed brick, and fly ash and bottom ash (OMNR 1992). Recycled asphalt pavement (RAP) and reclaimed crushed Portland cement concrete, as aggregates, have been used for many years and are included in the Ontario provincial specifications (Senior et al. 1994)

Despite the trend towards greater recovery and reuse of materials, it is considered that suitable substitutes are not available in large enough quantities to moderate the trend towards diminishing accessible reserves of primary mineral resources.

12 PROJECTED SHIPMENTS OF CONSTRUCTION AGGREGATES

Economic variables including Canadian Gross Domestic Product (GDP), housing starts, and one- and five-year mortgage rates were selected to test the association with shipments of construction aggregates. As shown in Figure 13, this association for the years 1980-1993 inclusive has been confirmed to be very close. The T-

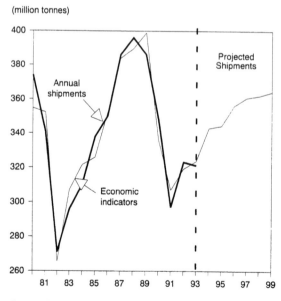

Source: Natural Resources Canada.
Note: The economic indicators used are: Canada's Gross Domestic Product, housing starts, and one- and five-year mortgage rates.

Figure 13. Canada, Projected shipments of construction aggregates, 1980-99.

statistics were very significant for each of the variables tested, and the adjusted R^2 of the regression equation was 0.89. With this information, shipments of construction aggregates were projected to 1999, using a 1994 five-year economic forecast of the Conference Board of Canada (CBC), also as indicated.

As shown, two major points emerge. First, there is in fact a very high statistical association between shipments of aggregates on the one hand, and housing starts and one- and five-year mortgage rates on the other. Second, shipments will increase as the economy recovers from recession. However, applying the Conference Board economic forecast and assuming that all other factors remain equal, this recovery may be relatively modest in comparison with the peaks of earlier economic cycles. In other words, it appears that the highs will not be as high in the 1990s as they were in the 1980s.

13 SUMMARY AND CONCLUSIONS

Total Canadian shipments of primary construction aggregates, namely, sand and gravel, crushed stone, and miscellaneous stone, amount to about 320 million tonnes per year valued at $1.2 billion, or 8% of value of the non-fuel minerals industry.

In terms of volumes of shipments destined for final consumption, these primary construction aggregates account for the largest proportion of the non-fuel minerals industry. The impact of this production and consumption on resource planning and availability is magnified even more if limestone raw materials for producing cement and lime are included.

Data confirm that a very high base level of demand for construction aggregates exists in Canada and that the average annual value of shipments of these primary materials (fob pit or quarry) during the period 1974-93, has been about $1 billion per year in 1986 dollars, adjusted for inflation. The importance of construction aggregates, in terms of volumes and values in the country's five main regions, has remained about the same during the period from 1974 to 1993. However, during peak years of construction expenditures, often led by Ontario, the most populous province, per capita production and consumption of aggregates reaches 16 tonnes, possibly the highest in the Western World based on the data available.

Compared to the essentially flat trend in the real fob values of shipments of aggregates in Canada, real expenditures for building and engineering construction have increased by an average compound rate of about 1.7% per year. Viewed in the context of resource planning and valuation, the importance of relatively low-valued aggregates, as a major materials input for building and engineering construction, appears to be understated in terms of their value to society.

Moreover, it may be argued that the importance of aggregates, which now are clearly recognized as a diminishing resource in several regions of Canada, would be represented more accurately if viewed in terms of their final in-place values i.e., values that account for relatively high transportation and handling costs.

The demand for aggregates is mainly local or regional and influenced to a major degree by trends in domestic construction. However, where low-cost resources are not available or land is not available for multi-use or sequential use, international sources of supply may be important.

A thorough understanding of the chemical and physical characteristics of aggregates e.g. alkali reactivity, is particularly important when shipments are destined for non-local use, or for export for use under jurisdictions having new or less familiar codes or specifications.

There are two types of alkali-aggregate reaction that occur in Canada; these are alkali-silica reaction, and, alkali-carbonate reaction. Comprehensive documentation is provided in two recent references (Langley et al. 1993, DeMerchant et al. 1995).

Although regulations have generally not been established concerning the environmentally sound use of secondary materials from construction and demolition wastes, the use of these materials contributes to aggregates conservation. However, it is considered that substitutes are simply not available in volumes that would moderate the trend towards diminishing reserves and resources of primary construction materials in several regions of Canada.

Canada's relatively high per capita consumption of construction aggregates will continue because of new capital expenditures and the ongoing need for repair and replacement of transportation infrastructure. With this scenario, and considering that in many areas there has been a long-term trend away from bulk movements by rail and water toward the more localized trucking mode, it is becoming more important to encourage multiple or sequential land-use management, with proper environmental considerations.

Based on a review of the data series, and of some of the methods used to collect the data, the annual levels of production and shipments of aggregates in Canada may have been understated in terms of quantity and value. This possibility has encouraged efforts by authorities to try to improve data collection and reporting, both for the purposes of long-term, integrated resource management and for the respective provincial and national accounts.

ACKNOWLEDGEMENTS

The authors wish to express their thanks to several employees of Natural Resources Canada, namely: Greig Birchfield, Lorraine Dugas, Heather Miller, Ron Mosher, and Francine Sarazin, for the mineral statistics gathering, compilation and formatting; Frank Penton for the regression analysis of the economic data and projections, and; Jean Bureau for coordinating the preparation of the final version of all files prior to submission for publication.

REFERENCES

ADI Ltd. 1986. Study of markets for aggregate materials accessible from Nova Scotia and New Brunswick, open file report 86-9, Nova Scotia Department of Natural Resources, and, New Brunswick Department of Natural Resources and Energy, respectively.

Anon. 1991-1993. *Construction in Canada*. Statistics Canada, Catalogue No. 64-201 Annual, Ottawa, Canada, various pages.

Anon. 1992a. *Aggregate resources of southern Ontario: A state of the resource study*, Chpt. 6 5.1, pp. 6.5-6.6. Prepared by Planning Initiatives Ltd. and Associates, for the Ontario Ministry of Natural Resources, Toronto, Ontario.

Anon. 1992b. *Aggregate resources of southern Ontario: A state of the resource study*, Chpt. 7.2.2., pp. 7.4. Prepared by Planning Initiatives Ltd. and Associates, for the Ontario Ministry of Natural Resources, Toronto, Ontario.

Anon. 1993a. *Capital expenditures by type of asset*. Statistics Canada, Catalogue No. 61-223, Ottawa, Canada, various pages.

Anon. 1993b. *Quarries and sand pits*. Statistics Canada, Catalogue No. 26-225, prepared under the direction of Natural Resources Canada, Mining Sector, Ottawa, Canada, pp. 21-25.

Anon. 1994a. *Canada's mineral production*. Statistics Canada, Catalogue No.26-202, prepared under the direction of Natural Resources Canada, Mining Sector, Ottawa, Canada, pp. 7-19.

Anon. 1994b. L'industrie minérale du Québec: Bilan et faits saillants; Perspectives (1995). Ministère des Ressources naturelles, Services de la statistique et de l'économie minérale, Québec, pp. 4.

Anon. 1994c. *Market trends for industrial minerals*. Mining Sector, Mineral and Metal Commodities Branch, Natural Resources Canada, pp. 19.

Anon. 1995. *Focus on industrial minerals*. Ministry of Energy, Mines and Petroleum Resources, and BC Trade, Victoria, British Columbia 2(2): 1.

Barksdale R.D. (ed.) 1991. *The Aggregate Handbook*, National Stone Association, Washington, D.C., various pages. Georgia Institute of Technology.

Bates, R.L. 1994. Overview of the industrial minerals. In *Industrial Minerals and Rocks, 6th edition*, Society for Mining, Metallurgy, and Exploration, Inc., Littleton, Colorado. Edited by D.D. Carr. Indiana Geological Survey.

Boucher, M.A. 1993. Silica. *Canadian Minerals Yearbook: Review and Outlook*, Natural Resources Canada, Mineral and Metal Commodities Branch, Ottawa, Canada, pp 43.1-43.8.

Bronitsky, L. 1973. The economics of construction mineral aggregate: unpublished Ph. D. dissertation, Rensselaer Polytechnic Institute, Troy, New York, 333 pp.

Burton, B. 1993. Aggregate resources: A Canadian perspective, special report, *Canadian Aggregates* 8(2): 5-9.

CBC 1994. Canadian outlook, Autumn economic forecast, Conference Board of Canada, Ottawa, Canada 10(1): Table 1.

Derry, Michener, Booth & Wahl 1989. *Limestone industries of Ontario, volume I – Geology, properties and economics*. Ontario Ministry of Natural Resources, Land Management Branch, Toronto, Ontario, 158 pp.

DeMerchant, Fournier & Malhotra 1995. *Alkali-aggregate reactivity in New Brunswick*. Natural Resources Canada, and New Brunswick Department of Transportation, CANMET, Ottawa, Canada, 44 pp.

Hale, P.B. 1990. 'Mineral Resources,' Chapter 13, Geology of the continental margin of eastern Canada, M.J. Keen & G.L. Williams (eds), Geological Survey of Canada, *Geology of Canada*, No. 2, pp.721-741 (also Geological Society of America, *The Geology of North America*, v. I-1).

Irvine, R.D. 1994. Canadian Industrial Minerals: Prospects to 2000. Proceedings of Industrial Minerals '94. Sixth Annual Canadian Conference on Markets for Industrial Minerals, Toronto, Ontario, October, 1994, a submission of 14 pp.

Langer & Glanzman 1993. *Natural aggregate: Building America's future*, US Geological Survey Circular 1110, pp. 21.

Langley, Fournier & Malhotra 1993. Alkali-aggregate reactivity in Nova Scotia. Natural Resources Canada, and Nova Scotia Department of Transportation and Communications, CANMET, Ottawa, Canada, 36 pp.

MacDonald, R.H. & Boehner, R.C. 1994. Limestone in Nova Scotia: Green gold. *Proceedings of the 30th Forum on the Geology of Industrial Minerals*. New Brunswick Department of Natural Resources and Energy, and Nova Scotia Department of Natural Resources, Miscellaneous Report 16, pp. 105-120.

Malhotra, V.M. (ed.) 1987. *Supplementary cementing materials for Concrete*. Canada Centre for Mineral and Energy Technology (CANMET), Energy, Mines and Resources Canada, Communication Group – Publishing, Ottawa, Canada. CANMET.

OMNR 1992. *Mineral aggregate conservation: Reuse and recycling.* John Emery Geotechnical Engineering Limited, prepared for the Ontario Ministry of Natural Resources.

OMNR 1995. *Mineral aggregates in Ontario: Overview and statistical update – 1993.* Ministry of Natural Resources, Resource Stewardship and Development Branch, Toronto, Ontario, numerous pages.

Pilsworth, D. & Kokkinos, K. 1994. General Review. *Canadian Minerals Yearbook: Review and Outlook.* Natural Resources Canada, Mining Sector, Ottawa, Canada, pp. 1.1-1.14.

Poulin, R. & Vagt, O. 1993. *Aggregates markets of the US eastern Seaboard.* Internal publication, in cooperation with Natural Resources Canada, Mining Sector, Ottawa, Canada, 16 pages.

Prime, G. 1988. *Aggregates in Nova Scotia.* N.S. Department of Mines and Energy, Halifax, Nova Scotia, Information Circular No. 20, pp. 4.

Senior, S.A., Szoke, S.I. & Rogers, C.A. 1994. Ontario's experience with reclaimed materials for use as aggregates. *International Road Federation/Transportation Association of Canada Conference Proceedings, Volume 6,* pp. A31-A55.

Southam, publ. 1994-1996. *Annual Construction Forecast,* 88 pp. *Published by* Southam Construction Information Services, Toronto, Ontario.

Tepordei, V.V. 1994. Crushed stone. *Mineral Industry Surveys, Annual Review,* United States Bureau of Mines, pp. 1-3.

Vagt, G.O. 1994a. Stone. *Canadian Minerals Yearbook: Review and outlook.* Natural Resources Canada, Mineral and Metal Commodities Branch, Ottawa, Canada, pp. 47.1-47.19.

Vagt, G.O. 1994b. Mineral Aggregates. *Canadian Minerals Yearbook: Review and outlook.* Natural Resources Canada, Mineral and Metal Commodities Branch, Ottawa, Canada, pp. 32.1-32.14.

Successful integration of aggregate data in land-use planning: A California case study

DAVID J. BEEBY
California Department of Conservation, Division of Mines and Geology, Sacramento, USA

1 INTRODUCTION

This paper describes the efforts of California's government to deal with the competition for land between aggregate mining and urbanization. It focuses on a 20-year old law called SMARA. That act is used to identify and protect valuable mineral resources from other land uses and to assure that reclamation follows mining. This paper addresses only the land-use competition side of SMARA. The act has been quite successful although there have been some problems. Case studies, which are included here, explain how the act has been used and modified. California's experience in using this act can be used to guide other governments who must address conflicts between mining and other uses for the land.

California's Surface Mining and Reclamation Act of 1975 (SMARA) is one of the earliest attempts by government to deal with problems caused by the acute land-use competition between aggregate mining and urbanization. The background of SMARA and why it was passed will be presented, as well as a discussion of what is being done in California to identify and protect valuable mineral resources from preclusive land uses.

Land-use competition between mining and other uses is inevitable but need not be contentious. It is inevitable because of the nature of the aggregate commodity itself, which must be produced locally to remain economically viable. It need not be contentious if adequate planning, based on objective and accurate data, is accomplished by elected decision makers.

Aggregate is a low unit-cost, bulky commodity that is expensive to transport. Unit trains or marine transport of aggregate are uncommon in California, and according to annual production statistics from the US Bureau of Mines, more than 90% of all aggregate in California is transported by trucks on public roads and highways. Minimum haulage charges established by the California Public Utilities Commission average between 7 and 10 cents per tonne per km, plus time charges and fuel surcharges. Rates vary throughout the State, but as a generalization, to haul aggregate by truck more than about 50 km doubles its cost to the consumer. Private contractors can pass this higher cost on to the consumer. What about aggregate used by the government?

Historically, about half of all aggregate used in California is used by the government – Federal, State, county, city, and local – in public works projects. Public roads, dams, freeways, bridges, buildings, and airports consume huge amounts of aggregate. Increases in cost of this construction can result in three actions: 1. Ignore needed repair work to the urban infrastructure, 2. Diminish government service in other areas and transfer those funds to cover needed infrastructure repairs, or 3. Impose higher taxes. None of these actions is politically or socially acceptable, and none are likely to get local government officials reelected. Fortunately, the need for making these difficult economic choices can be prevented by maintaining local construction aggregate supplies.

Dealing only with the economic cost of hauling aggregate long distances still leaves a serious environmental cost unaddressed. More truck miles means more trucks on the road at any given time, causing serious problems with traffic congestion (requiring more roads to be built or existing roads to be widened) and greater road wear. To put this in perspective, construction projects within the Los Angeles Basin consume about 54 million tonnes of aggregate annually. Assuming that 90% of this aggregate is transported to the job site by truck and each truck is loaded to the maximum legal limit of 22 tonnes, this tonnage equates to 2,250,000 truckloads, or almost 6,200 truckloads per day. Doubling the distance from the mine to the consumer doubles the amount of time these trucks are in the traffic flow of an already crowded transportation network. More truck-miles means greater fuel use, more air pollution, more tire wear, more noise and dust, and more road wear. From either an economic or an environmental perspective, it is in the best interests of the citizens of any government jurisdiction to have locally available mines for construction aggregate.

2 BACKGROUND

To put the California experience into proper perspective and assess its applicability to other jurisdictions, four facts are important to an understanding of the State's efforts to protect mineral resources.

First, California has always been a mining state, with a rich mining heritage dating from the gold rush of 1849. That heritage continues to the present, and according to the California Department of Conservation's Office of Mine Reclamation, today there are over 1300 permitted mines in California, with over 900 active in 1995 (see Fig. 1). These mines are widely distributed, with mining occurring in 57 of 58 counties (CDMG Minefile 1996).

Mineral production in California has evolved considerably since our gold rush days. In 1860, 70% of the production was attributed to a single commodity – Gold, and only 4% was due to fuel minerals (oil, gas, and coal). Today those figures have reversed and oil and natural gas now dominate mineral production (Fig. 2). More significant to the mining industry, however, is the ten-fold growth of the industrial minerals (including aggregate) market over that period as the California population has grown and the urban infrastructure has matured. According to the former US Bureau of Mines, California now leads the United States in the production of industrial minerals, both in number of different mineral commodities (28) and in dollar value

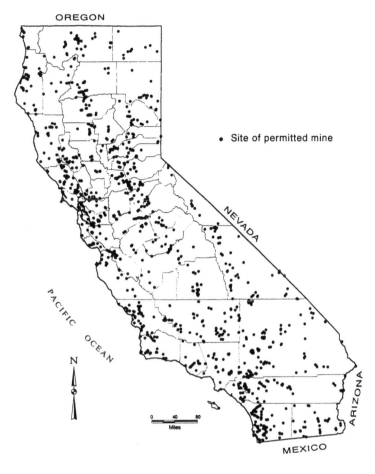

Figure 1. Location of the 1315 permitted mines in California (source: Department of Conservation, Division of Mines and Geology).

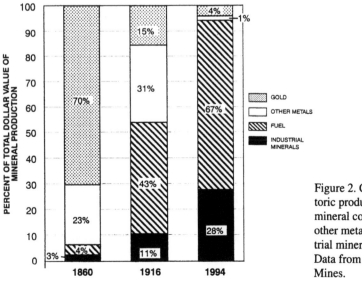

Figure 2. Comparison of historic production of selected mineral commodities (gold, other metals, fuel and industrial minerals) in California. Data from US Bureau of Mines.

($2.1 billion). Excluding the fuel minerals, industrial minerals now make up 86% of the total nonfuel mineral production, which contributes an annual mineral production of $2.5 billion to California's economy.

The variety and relative value of mineral production shows that California, which began statehood through mining, continues to be a leading mining state. Today, however, the construction commodities – Portland cement, sand and gravel, and crushed stone dominate the mining scene instead of gold. Large open-pit mines dominate the industry, with very few remaining underground operations. Large corporate mining now controls the future of California's overall mineral production, but small intermittently active operations remain common in the construction aggregate industry (Tooker & Beeby 1991).

Second, huge tonnages of construction aggregate are used in California. From some 750 active gravel pits and quarries, California produced 121 million tonnes of construction aggregate in 1994, worth $731 million dollars (US Bureau of Mines 1995). Almost two-thirds of this total is from sand and gravel, with the rest from crushed stone (Fig. 3). Construction aggregate is the most important mineral commodity produced in California, not only because of value, but because it is essential to the maintenance of societal infrastructure. For the past several decades, California has lead the US in sand and gravel production (USBM 1995).

Third, the gold rush also started a tradition of independent thinkers in the State who mistrusted centralized State or Federal control in land-use decision making. This was such a fundamental desire of California pioneers that the original California constitution, written in 1850, assures the citizenry of local control of land use and land use permits. Mining permits have always been approved at the local level – usually by one of the 58 counties, but also by any of 470 incorporated cities. At present 133 different lead agencies have active mines within their jurisdictions.

Fourth, California's population is large (more than 33 million), highly urbanized, well educated, politically aware, and environmentally conscious. This large population is the engine that fuels the huge demand for construction aggregate in the State.

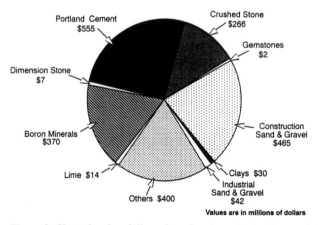

Figure 3. Chart showing dollar value of total production for nonfuel industrial minerals produced in California in 1994 (total 2.2 billion). Values are in millions of dollars. Data from the US Bureau of Mines.

Mining is widespread and common, particularly pits and quarries where construction aggregate is mined, but ironically in light of the proximity of mining to urban centers, the general population knows almost nothing about modern mines, the mining industry, or the importance of mineral commodities in their daily lives.

These four facts began combining to make mine permitting more difficult in the early-1960's as urban sprawl came into increasing conflict with traditional alluvial sand and gravel mining near metropolitan market regions.

3 THE SURFACE MINING AND RECLAMATION ACT OF 1975

Ironically the aggregate industry had been caught in the backwash of its own success. Urbanization, which is dependent upon construction aggregate for both building and maintenance, created land-use pressures that often caused premature closure of mines on the urban fringe. However, in an industry where costs are dominated by transportation distances, pits and quarries need to be situated as close to their urban markets as possible. Aggregate pits that were originally situated in rural areas decades earlier were being surrounded by residential subdivisions and groups of hostile neighbors began to appear with increasing frequency at use-permit renewal hearings.

To protect itself, the aggregate industry lobbied the California legislature for help to develop a regional mineral deposit database that would enable local planners to direct new subdivisions away from remaining aggregate deposits. In the early 1960's, a select blue-ribbon task force was set up by the California Senate Natural Resources Committee to study the problem and make recommendations. The process was heavily politicized as issues of local versus state control, development versus environment, sensitivity to local needs, proprietary data, economic competition, and personal property rights were debated and addressed. Differences between the numerous construction aggregate associations and the 'hard rock' mining associations were reconciled, and a united front was eventually presented to the legislature. After numerous impasses and 12-years of effort by the California aggregate industry, spearheaded by Don Reining of the Southern California Rock Products Association, perseverance was rewarded. The Surface Mining and Reclamation Act (SMARA) was passed into law in 1975, the first law of its kind in the US (DMG 1995). The first article of SMARA states 'The legislature hereby finds and declares that the extraction of minerals is essential to the continued economic well-being of the State and to the needs of the society, and that the reclamation of mined lands is necessary to prevent or minimize adverse effects on the environment and to protect the public health and safety'. This landmark statement was, at the time of its passage, unique in its recognition of the importance of mineral resources, giving them an equal footing with other natural resources.

In addition to providing for the protection of aggregate (and other mineral resources) from development incompatible with mining, SMARA established and defined criteria for the reclamation of surface mines, provided for reclamation compliance monitoring, and established requirements for financial assurances for completion of reclamation. These reclamation aspects of SMARA have been considerably more volatile than the mineral resource issues and are beyond the scope of this paper.

SMARA was amended in 1980 to double the mineral land classification program and allow for expansion into non-urban regions of the State in response to increased Federal mineral withdrawals and proposed new roads and/or wilderness areas, National Parks, and National Monuments. Since 1984, the act has been amended 14 more times to deal with mined-land reclamation issues.

The legislature wisely provided funding for the newly created program through the use of a portion of the Federal lease dollars collected from California mines operating on Federal lands within the State. The $1.1 million annually, which was originally allocated, enabled the fledgling program to develop initially with some immunity from budget cuts. This was increased to $2 million in 1980, and in 1990, authority to charge fees was provided to fund a new reclamation compliance unit. In recent years, money allocated to fund the SMARA mineral land classification program has averaged about $1.2 million per year, split equally between returned Federal lease dollars and State General Fund dollars.

4 SMARA IMPLEMENTATION

Relative to land-use planning, implementation of SMARA is a shared responsibility between the California Department of Conservation's Division of Mines and Geology (DMG), the State Mining and Geology Board (SMGB), and the lead agencies (local governments) of California. Only when all three of these partners take their responsibilities seriously does the SMARA process work effectively.

4.1 *Role of the Division of Mines and Geology*

Under the policy guidance of the Director of the California Department of Conservation and the SMGB, the State Geologist, who is Chief of DMG, is mandated to classify specified lands within the State on the basis of mineral content. The classification is mandated to be done 'on the basis solely of geologic factors, and without regard to existing land use and land ownership.' Giving the State responsibility for the preparation of an accurate, objective, quantified mineral-resource data base reduces the ability of special interest groups from influencing land classification. This assures that all valuable mineral deposits recognized by DMG economic geologists will be identified. It also assures that worthless ground will be identified as such, even if it is being fraudulently promoted as having mineral value. State-wide oversight assures regional and state-wide consistency in the information provided to local government, which is especially important when mineral deposits cross jurisdictional boundaries or are of special interest to several governmental bodies. Finally, state-wide standardization of the classification process eliminates unfair advantages between neighboring mining companies in adjacent jurisdictions, giving all identified mineral properties the same level of protection under the law. It is important to note that the process of identifying potential mineral deposits does not require that any properties be permitted for mining. It simply provides decision makers with a valid non-biased data base against which various land use permits, including mining permits, can be evaluated, and requires them to recognize such data and use it in their land-use decisions.

The California Mineral Land Classification System, developed in 1983, formalized the scientific basis and the economic geologic criteria used to evaluate individual mineral deposits in the field. Details of its development and application can be found in Loyd et al. (1994), but in summary, it is based upon a matrix modified after the USBM/USGS (1980) McKelvy diagram, where increasing economic value is plotted against increasing degree of knowledge about a mineral deposit to determine varying degrees of economic potential (Fig. 4). Six categories of Mineral Resource Zones (MRZs) are defined based upon this diagram. The highest categories of MRZ for construction aggregate are MRZ-2a and MRZ-2b. Areas classified MRZ-2a are underlain by economically significant measured or indicated aggregate resources, quantified by drilling and laboratory testing. Because of the amount of data needed to justify this classification, MRZ-2a areas are almost always currently permitted for mining, and are therefore defined to be aggregate reserves in SMARA terminology. Areas classified MRZ-2b are underlain by economically significant inferred aggregate resources. Generally MRZ-2b areas lie in the same geologic unit known to contain the reserves, but which has not been as extensively drilled or tested for aggregate quality.

The MRZs are the map units used in the classification maps. This makes it unnecessary for the non-geologist to use the more complex geologic maps from which the MRZs are derived. The MRZs convey to the land-use planner information regarding the type and quality of mineral resources within their jurisdiction. These MRZs have proven quite effective in the identification of potential economic deposits of construction aggregate, and convey that information to non-geologists in a plain and straightforward manner. Planners have found the information in the MRZs easy to convey to their elected officials who have ultimate land-use authority.

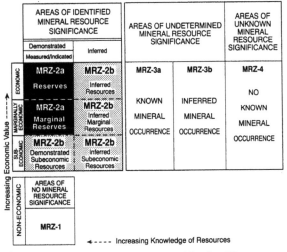

Figure 4. California Mineral land Classification Diagram presenting the relationship between mineral resource occurrence and economic significance. The horizontal axis of the diagram represents the degree of knowledge about mineral occurrence, while the vertical axis portrays economic characteristics of mineral deposits (grade and size). Modified from a mineral resource classification scheme developed by the US Bureau of Mines and the US Geological Survey (1980).

The SMARA Mineral Land Classification Process has been previously described (Beeby 1988) so will only be summarized here. The user should be aware that all SMARA Mineral Land Classification Reports do not follow the same model and have evolved through time. In the 20 years since the program was started, policy, style and project management have evolved, and different classification techniques have been tried and retained or discarded. The classification process has varied through time to accommodate different needs and to deliver better products faster and more economically. For example, reports designed to provide data to influence potential Federal wilderness withdrawals target remote rural parts of the State. In these areas, aggregate is too far from any market to have economic significance, so is not classified. Those reports are therefore dominated by deposits with metallic mineral potential.

For a time the SMGB directed DMG to classify all lands within entire jurisdictions for all mineral commodities in a single pass. This soon proved too cumbersome and wasteful of staff resources since many commodities were unlikely to ever become economic and many areas were not threatened by land uses incompatible with mining.

In the current classification approach, the geologist works with the local government planners, the local mining industry, and the SMGB to selectively classify only high potential mineral commodities in regions of anticipated growth or known mineralization.

The reader interested in more detail on these differences, or in details of technique or policy, is referred to any of the region-specific studies cited in the bibliography. Since this volume specifically deals with construction aggregate, the classification methodology used most successfully for that commodity will be the focus.

Once a specific area is targeted for mineral land classification the following steps have been followed for the most complete aggregate studies:

1. Local government is consulted before the study begins, and is consulted and kept informed throughout the study.

2. Active mines are visited. Local aggregate deposit parameters are gathered for percent waste, in-situ density, local stratigraphy, etc.

3. The market region (called a 'production-consumption region') of the producing mines is identified.

4. Appropriate geologic mapping covering the production-consumption region is collected, compiled, and reconciled.

5. Existing subsurface data, usually from water-well logs and foundation borings, is collected, analyzed for particle-size distribution, and plotted.

6. Historic mine production records are evaluated both for mineral product quality (highest use) and tonnage produced. This data is extrapolated to other portions of the same geologic units, as appropriate.

7. Aggregate quality test data from current and former producing pits and quarries or exploratory test sites is gathered and analyzed. Company exploration data is utilized if available.

8. The above is synthesized and extrapolated into MRZ categories and is drafted onto classification maps.

9. Areas which are already urbanized (defined as having a housing density of more than one home per 4 hectares) are economically unavailable for mining. Spe-

cific areas where mining would be socially unacceptable, such as developed parks, golf courses, or cemeteries are equally unavailable for mining. Such areas are excluded from MRZ-2b areas. The aggregate-bearing lands that are left after this screening process are called 'resource sectors' or more simply, 'sectors'. They often include agricultural and forest lands, low density residential land, or open space land-uses.

10. Tonnage within the resource sectors is quantified, and compared to regional need, which is based upon 20-years historic consumption.

11. Data and conclusions are informally discussed with local government planners, and refined if necessary.

12. Report and maps are reviewed and accepted by the State Geologist and the State Mining and Geology Board (SMGB) after public SMGB meetings.

13. The report and maps are formally transmitted from the SMGB to local government, starting a 12-month period during which local government must incorporate the report and maps into it's general plan.

14. The report is presented to local government planners for their use, and is formally presented to local decision makers at public meetings (usually City Council or Board of Supervisors meetings).

15. The report and maps are published by DMG and made available for sale to interested parties.

Typical mineral land classification maps are compiled at a scale of 1:24,000 to meet planners needs, but are printed at a scale of 1:48,000 to keep costs down. They are intended to focus attention on mineral deposits classified MRZ-2a or MRZ-2b for construction aggregate, including the primary production districts and all active mines within the study region.

These maps show what exists without regard to land use or ownership. Had they been available to an earlier generation of planners, rich aggregate deposits currently preempted from mining because of unplanned urbanization might still be available. It is of little value to planners to dwell on what might have been, so the focus of the SMARA maps is on what remains. However, not all of the remaining non-urbanized aggregate deposits are socially or politically acceptable targets for future mining. This means that local choices for mineral extraction are even more limited. But, for the first time, the choices are clearly in front of the local decision makers.

One of the beauties of SMARA is that it focuses attention of planners on the relatively small portion (generally less than 2%) of their jurisdiction known to contain aggregate resources (areas classified as MRZ-2a and 2b), and the even smaller (generally less than 0.5%) containing aggregate reserves (areas classified MRZ-2a and controlled by aggregate producers holding valid mining permits for them). This use of these restrictive and very specific definitions serves three purposes. First, it gives planners an accurate menu of known potential deposits to plan for future needs, without including speculative deposits. Second, the definitions let report users know exactly what the basis for classification has been. This is particularly valuable for the mineral exploration community, but also gives all users confidence in report quality and consistency from one area to another. Third, the McKelvy definitions are widely used by the economic geologic, legal, and banking professions, and are thus defensible in disputes or potential lawsuits.

The reports also address what those future needs are likely to be, through comparison of historic aggregate consumption rates, projected population growth, and quantification of existing reserves. The overall collective depletion date of existing permitted reserves is presented to planners and elected officials, giving them for the first time a sense of the urgency (or lack of urgency) about the longevity of their local aggregate inventory.

4.2 *Role of the State Mining and Geology Board*

The State Mining and Geology Board (SMGB) is comprised of a group of nine individuals representing specifically defined areas of expertise, who are appointed by the Governor. In conjunction with the Department of Conservation, they set the policies under which the DMG operates. In regard to SMARA, the SMGB develops and approves policies and procedures for it's implementation (California State Mining and Geology Board 1983), establishes priorities for areas to be classified by the State Geologist, and formally accepts and distributes the completed classification reports to lead agencies. The SMGB conducts its business in six open public meetings per year prior to any decision making. Additional formal public hearings are held by the SMGB as needed, and occasional legal appeals regarding reclamation plans or compliance issues are conducted.

The SMGB acts as a liaison between the State and local government or citizens groups, assuring that the legal requirements of SMARA are adhered to and that all parties have an opportunity to make their positions known. SMGB reviews and approves mining ordinances used by local jurisdictions, and publishes an annual list of all aggregate mines in the State that are in compliance with SMARA (called the 'AB 3098 List' after its enabling legislation in the California Assembly). Since Caltrans (California's highway department) cannot legally buy aggregate from companies not on this list, it is a powerful incentive toward assuring compliance with the reclamation elements of SMARA. The SMGB also has two additional authorities under SMARA. First, they can hear and accept petitions from property owners to have their land classified by the State Geologist.

This has usually happened when a mining company is the property owner who feels that an upcoming use-permit hearing may be influenced by the fact that the land has been classified, and the presence of a mineral deposit has been substantiated by the State. A property classified by petition has the same legal status as a property classified in the normal method based upon state-wide priority. Second, the SMGB can take a legal action called 'Designation', whereby a property already classified MRZ-2 by the State Geologist is recognized as being of regional or state-wide significance. In the designation process the SMGB holds public hearings in the local jurisdictions affected by the classification reports. The SMGB then prepares a generic environmental impact report which addresses the impact of mineral land designation. After a series of public hearings, the SMGB designates certain classified mineral deposits as being Regionally Significant, giving them added legal protection. High-quality aggregate resources in twelve metropolitan regions of the State have been identified and designated by the SMGB. The designation process has happened infrequently in recent years so that DMG fiscal resources can be directed to complete additional classification in new jurisdictions.

4.3 *Role of lead agencies*

Local government is mandated by SMARA to recognize the classification data developed by the State and establish appropriate mineral management polices to be incorporated into their general plans within one year of receipt of data from the State Geologist. Under SMARA, local government retains all land-use decision making authority relative to the granting of mining permits, but because local government generally lacks the technical staff expertise or regional perspective to evaluate the often conflicting testimony given by the pro- and anti-mining interests, State prepared Mineral Land Classification reports give a stable base of information from which informed land-use decisions can be made. In some cases, DMG geologists are called upon by local government in mine-permit hearings to testify about aggregate resources under consideration. This ability to provide a non-special interest expert to assist decision makers with technical issues is a major strength of the SMARA process. SMARA geologists testify at these hearings about four times per year. Almost every decision where State presence was requested was in favor of aggregate resource protection. Mine-permit decisions will probably always remain controversial. However, with the locally quantified resource inventories provided under SMARA, these decisions can be based on objective data, balanced against a perspective of long-term local resource needs rather than on emotion.

5 SMARA PROGRAM RESULTS

After two small pilot studies (Rapp 1975, Evans et al. 1979) were completed to resolve and refine policy and technical issues, the first Mineral Land Classification Reports were released in 1978. A steady stream has followed, and today 58 classification reports and 33 classification petitions have been completed and published, more than half of which deal exclusively with construction aggregate (Fig. 5). About 138,000 km^2 of California have been mapped under the mandate of SMARA. Within this area, almost 6400 million tonnes of aggregate reserves and 68,000 million tonnes of aggregate resources have been classified. Almost 1100 km^2 containing almost 32,000 million tonnes of high quality (Portland-cement-concrete grade) construction aggregate have been designated as being regionally significant. For the past 10 years, an average of three classification reports and two classification petitions per year have been completed.

A typical classification report now takes about two years to complete and costs between \$150,000 to \$200,000 to produce, publish, and distribute. A typical classification petition, covering only a single property, can be completed in as little as one month and costs about \$10,000 to finish. The cost of a classification petition is billed to the company that requested it.

The primary user groups of Urban SMARA classification reports are lead agencies, closely followed by the aggregate industry. However, a diverse variety of other users has emerged as report availability has become better known. Users include bankers, other government agencies, mineral appraisers, lawyers, realtors, geologic consultants, investors, landowners, and students. Out-of-state and international mining companies have frequently used the reports to evaluate entry into the California aggregate market.

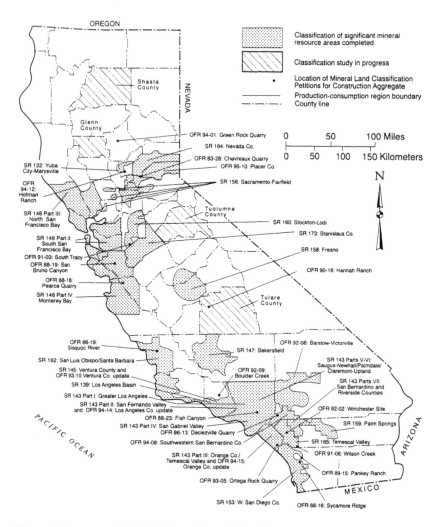

Figure 5. Progress of SMARA Mineral Land Classification for construction aggregate in California. This map shows locations of Division of Mines and Geology Special Reports (SR) and Open-File Reports (OFR) covering areas classified for construction aggregate resources (source: Department of Conservation, Division of Mines and Geology).

The aggregate classification reports have generally been well received by lead agencies, and the presence of a data base, where none previously existed, has resulted in land-use decisions based on better information. Lead agencies and individual aggregate producers, initially somewhat suspicious of the State program, are now requesting accelerated classification of their regions. The State Mineral Land Classification efforts under SMARA appear to be working well in California. In 1994 the classification program received excellent marks from both the mining industry and local government in a mandated effectiveness study independently conducted by the University of California. In that report it is stated 'Classification is almost universally regarded as an example of government playing an appropriate and competent role – success is mainly limited by funding.' (O'Hare et al. 1995).

Case studies: Absolute successes of the SMARA Classification Program relative to mineral conservation are difficult to measure with confidence in the short term because of the long-term nature of the SMARA goals. Final judgment of program success must come from future generations of Californians. However, numerous anecdotal examples of the effectiveness of the program include the speeding of the permitting process, longer life span on renewed use-permits, discovery of new deposits, attracting new industry to the State, the opening of new mines, and requests from local government for more or faster classification within their jurisdictions (Fig.6).

CASE STUDIES

1. San Joaquin River – Fresno County
2. Ventura County
3. Sacramento/Fairfield Area
4. Mather Air Force Base
5. Yuba City – Marysville Area
6. Apperson Ridge – Alameda County
7. Russian River Area – Sonoma County
8. Hospital Creek Alluvial Fan – San Joaquin/Stanislaus counties
9. Sisquoc River Petition – Santa Barbara County
10. Incompatible Adjacent Land Use – Los Angeles County
11. Incompatible Adjacent Land Use – Contra Costa County
12. Mining Use-Permit Extension – Ventura County
13. State or Industry – San Diego County

Figure 6. Location of case studies showing integration of aggregate data in land-use planning (source: Department of Conservation, Division of Mines and Geology).

Example 1: San Joaquin River – Fresno County
The SMARA mineral land classification study of the Fresno area (Wiedenheft-Cole & Fuller 1988) pointed out to local planners the importance of the aggregate resources in the San Joaquin River, and that they were the only local source of sand and gravel serving this rapidly growing area. The river was of interest to two other groups besides the mining industry, however. A major real estate developer was trying to get a permit to subdivide large portions of the bluffs and floodplain of the San Joaquin River and build expensive homes. Another community group in Fresno was opposed to this use of the river, and wished to create a development similar to the American River Parkway in Sacramento County, which is a recreational and wildlife/riparian habitat resource for the community. They formed a coalition called the Save the San Joaquin River Association (SSJRA). Upon the release of our SMARA classification report covering this area the SSJRA, realizing that mining was only an interim land use, joined forces with the mining industry to keep subdivisions out of the river. With SMARA classification data, Fresno and Fresno County decided against further subdivision in the river and opted to continue to allow in-stream mining with reclamation aimed at the subsequent use as a parkway. The coalition currently remains functional and is resurrected as needed. Mining is still taking place along the San Joaquin River.

Example 2: Ventura County
Aggregate classification in western Ventura County (Anderson et al. 1981) gave local planners an objective mineral resource data base for the first time. This allowed them to define long-term goals for in-stream mining and the management and protection of the Santa Clara River, which had been badly eroding farmlands, riparian habitat, and bridge abutments for the past decade. A regional plan has been developed and implemented to conserve and protect mineral and other important natural resources along the river. At the same time, the SMARA classification data identified alternative sources of aggregate in less environmentally sensitive areas so that a local source of aggregate, at low cost, is still available to the citizens of Ventura County. The SMARA classification work focused attention on resource issues in the area, provided unbiased resource and market data on construction aggregate, and allowed local government, under the leadership of the County, to resolve bottlenecks in the permitting process which had been years in the making. Now, 15 years after the initial study was completed, a re-mapping of the western Ventura County area has been finished, which shows the County is still maintaining a resource base capable of providing aggregate for approximately 100 years and a reserve base of 22 years. The forecasted consumption of aggregate in the combined western Ventura and Simi Production-Consumption Regions made in the original 1977 report was overestimated by about 18%, reflecting the decrease in construction in California resulting from the current economic recession.

Example 3: Sacramento/Fairfield area
Construction aggregate resources in Sacramento and Yolo counties (Dupras 1988) were in short supply in 1988. In Sacramento County, remaining mineral deposits were concentrated along the Highway 50 corridor and were being preempted from mining by subdivisions and light industry at an alarming rate (27 million ton-

nes/year). County planners were unaware of the existence of these resources. In Yolo County, agricultural, mining, real estate subdivision, recreational, and environmental interests were in a deadlock over the future of Cache Creek. Land-use decisions in both counties were being dealt with individually, with little apparent systematic or long range planning. As a result, little balance was achieved, critical natural resources were wasted, and every decision was a battlefield of controversy. The SMARA classification report covering these areas demonstrated their interconnection, quantified the mineral resource supply problem (defined the life span of the local mineral reserves), and identified alternative resources to serve local needs. It also got both jurisdictions to recognize the need for integrated long-term planning and to develop and implement those plans. In Sacramento County, the SMARA Classification project manager was invited by the Board of Supervisors to take part in an Aggregate Technical Advisory Committee (ATAC) for the County to develop those plans and advise in their implementation. Sacramento County hired an independent consultant to confirm the work of the State and accommodate specific County political realities. The findings and quantified data in the State SMARA aggregate report were confirmed and endorsed by the consultant, giving the County the added confidence to incorporate our work into their General Plan. The Yolo County portion of the resources is along Cache Creek. Political resolution of new mining in this area has not yet been achieved, and to date there remains a moratorium on the granting of new mining permits there, despite the confirmation of the State data and two subsequent independent consultant reports. Thus, the resource remains available to future generations, but will probably not be mined until the added price of imports from outside the county becomes economically unacceptable to local taxpayers.

Example 4: Mather Air Force Base
Large tonnages of aggregate resources underlie Mather Air Force Base in Sacramento (Dupras 1988). This was discovered during work on the Sacramento/Fairfield area discussed above, and mentioned in our final report as a potential future resource to the County. Surprisingly, five years later, announcement of the Base closure was made. The County Board of Supervisors requested participation by the SMARA Classification project manager to quantify and describe those resources to assist them in developing re-use plans for the property (Beeby 1989). The Governor's Office approved the request, and a plan for sequential land-use was developed and endorsed by the County (Hagan 1990). The plan calls for the avoidance and protection of vernal pools and other ecologically sensitive areas, and the mining of other selected portions of the base for aggregate and gold.

Reclamation of the mining pits will involve innovative land sculpting to allow for the affordable construction of a golf course, equestrian area, velodrome, shooting range, and an urban forest. These recreational second-uses of the land have been endorsed by Sacramento County, the City of Sacramento, and the City of Rancho Cordova Parks Departments. This alliance between the State and local government, and between environmental, mining, military, and recreational interest groups was begun and made possible by the SMARA classification project. Now, some four years after that endorsement of mining by the County and City decision makers, re-use of the Base still remains politically complex and unresolved.

Example 5: Yuba City – Marysville area

Between the 1880's and about 1950 a huge gold dredging field at Hammonton was developed along the Yuba River, 10-miles east of Yuba City/Marysville. This gold mining left behind more than two billion tons of a more modern mineral resource – construction aggregate and silica, covering several square miles of the river's flood-plain. While the old dredger field was well known to the local aggregate companies, the immediate market for dredging spoils appeared too small for the deposit to be of much interest, even though Yuba County had already zoned the area for mining, easing the way for future permits. Thus, it's future economic significance was unrec-ognized by all but a few. A mineral resource assessment of this area by the SMARA classification project was completed (Habel & Campion 1988), and the Yuba River aggregate deposits suddenly became the focus of considerable attention by both local government and the aggregate mining industry at the national level.

The deposits are within a few years of being within economic truck-haul distances to Yolo and Sacramento Counties, which should alleviate pressure upon the Cache Creek and the American River deposits. In addition, a special study of the Yuba River deposits has been conducted by Yolo County. A major Texas aggregate com-pany with experience hauling bulk construction materials by unit train, purchased a copy of our SMARA classification report, and conducted their own study after meeting with us several times. They leased property, are now the major aggregate producer from this deposit, and brought many jobs to this rural region. If this com-pany is successful in bringing economic unit-train aggregate transportation to Cali-fornia, local government planners in perhaps as many as two-dozen counties will have an economically and environmentally sensitive alternative solution to local ag-gregate shortages and can manage their own resources more deliberately. The depos-its are adjacent to Beale Air Force Base, and negotiations to cross the base by rail have been temporarily set back due to increased runway needs at Beale due to the closure of nearby bases in Sacramento County. The future of this deposit remains bright.

Example 6: Apperson Ridge – Alameda County

More than one billion tonnes of high-quality concrete grade aggregate were identi-fied on Apperson Ridge in Alameda County during work of the SMARA classifica-tion project on the South San Francisco Bay Area (Stinson et al. 1986, 1987a). All previous attempts to obtain a mining permit for this property for the past 30 years had been unsuccessful. Our study showed that the entire region was rapidly depleting its local supplies of aggregate and unless new properties were permitted, aggregate reserves would be exhausted within 17 years. Our study also identified many un-permitted but promising mineral deposits. With this regional perspective of both need and resource availability, local planners worked with the permit applicant on Apperson Ridge, and a mutually acceptable mining and reclamation plan was devel-oped and accepted. The presence of this large newly permitted deposit greatly in-creased the reserve base and increased the options of local planners in protecting sensitive environments. It simultaneously provided a source of vitally needed low-cost aggregate to maintain the urban infrastructure of county jurisdictions. Califor-nia's economic downturn has delayed quarry development, so the ultimate impact of this deposit remains to be demonstrated, but in this extremely urbanized and industrialized region of California this permitted resource will be an important future asset.

Example 7: Russian River area – Sonoma County
For the past 30 years or more, conflicts between in-stream aggregate mining and agriculture have been the norm along the Alexander Valley and middle reaches of the Russian River in Sonoma County. An aggregate resources management plan developed by the County in the 1970's was not working well and each new mining permit application was more controversial than the one before. The work of the SMARA classification project in the North San Francisco Bay Area (Stinson et al. 1987b) provided an objective non-political perspective on the locations, value and need for aggregate resources to Sonoma, Napa, Mendocino, and Marin counties. This regional perspective expanded the scope of the earlier Sonoma County plan. A major revision of that plan was conducted by the county to accommodate the new State data in the SMARA report. This is an example of the importance of basing land-use decisions on good data, and is applicable to almost all of the 133 SMARA lead-agencies now in California.

Prior to receipt of a SMARA mineral land classification report by a lead agency, each new mining permit application could be presented by the applicant as being critical – the last remaining minable deposit of quality aggregate in the county. Similarly, the anti-mining constituency could claim that one could mine similar or better material anywhere in the county, and that this particular site was of critical value for some other reason. Without their own in-house expertise, who should the Board of Supervisors believe? In this atmosphere, every decision had the potential to become a war. Objective and factual data from the SMARA classification project, with no special interest in any specific mine or in anything other than better planing through consideration of needed mineral resources, can and often has, calmed these troubled waters.

Example 8: Hospital Creek alluvial fan – San Joaquin and Stanislaus Counties
As part of the mineral land classification of San Joaquin and Stanislaus Counties (Jensen & Silva 1988), all existing and former producing construction aggregate deposits in the region were evaluated and plotted on compiled geologic maps. It was noticed that two older alluvial fans draining the Coast Ranges from Oristimba Creek and Garzas Creek were the sites of current or former sand and gravel mining. A previously unmined third fan of suspected similar age drained Hospital Creek farther to the north. It was evaluated in the field and appeared very close in lithology and degree of weathering to the Oristimba and Garzas creek fans. The field geologist felt that economic deposits of aggregate were likely to occur there by extrapolation from those known deposits farther south. The Hospital Creek fan was therefore mapped as MRZ-2b (containing inferred mineral resources), and the maps were publicly released in 1988. Three competing aggregate producers (Teichert Aggregates, RMC Lonestar, and Granite Construction Company) quickly leased portions of the fan and Teichert began a drilling program that substantiated our assumptions and proved the presence of the aggregate resources. Teichert petitioned the State to reclassify land adjacent to the distal end of the fan in response to the new drilling data, which we quickly did (Boylan & Loyd 1991), and released new maps showing the extent of the

larger deposit. Stanislaus County was very pleased to have the geologic inventory of their aggregate deposits and the identification of the new resource. Teichert Aggregates sent a letter to the Director of the Department of Conservation which states: 'Given the economic challenges facing the State of California it is a delight when government and business endeavors are resoundingly successful. Teichert is pleased to note that its aggregate reserves in the Vernalis area were permitted by San Joaquin County on December 3, 1992. The State's designation and classification process played an integral role in the permitting of these reserves.' This is a case not only where SMARA mineral land classification identified a previously unknown aggregate deposit, but which also attracted three new producers to the region and has resulted in one new permitted mine. A second mining permit has been granted conditionally by the County, pending resolution of transportation issues. In addition to the benefit of a broader resource base for future generations, other benefits included added tax revenues to local government, more local jobs, and a lower local aggregate price.

Example 9: Sisquoc River petition – Santa Barbara County
The State Mining and Geology Board maintains a five-year schedule for systematic SMARA classification projects. The schedule is approved at public meetings to give all lead agencies and mining companies a forum where they can request classification or elevate the priority of their jurisdictions. This schedule determines how DMG staff and fiscal resources are deployed. Deviation from this schedule (to handle situations where an imminent land-use permit decision affects availability of mineral resources) is possible in the law through a petition process. If a petition for mineral land classification of a specific property is brought by the property owner and accepted by the Board, it supersedes the routine systematic classification effort. Such cases occur a few times each year.

One of the most dramatic successes of the SMARA petition process occurred in 1986 in northern Santa Barbara County along the Sisquoc River. At the time the county had not been classified, but it was known to DMG that the County's sole source of PCC-grade concrete aggregate was along a seven-mile reach of the Sisquoc River. Coast Rock Products, Inc., one of the aggregate producers in this area, brought an Environmental Impact Report to our attention wherein a proposed major one meter diameter oil transmission pipeline had been approved by the County to cross under the gravel mining portion of the river. Had this gone unchallenged, the seven meter burial depth of the pipeline would have precluded mining about half of the permitted reserves and cut the mining properties in two, requiring new access, new mining plans, and a revised reclamation plan. Coast Rock's petition for classification was accepted by the Board and completed by SMARA staff shortly after. The final report (Cole & Jensen 1986) was transmitted to Santa Barbara County showing the valuable aggregate deposits. With the classification petition report in hand and DMG SMARA testimony supporting the value of the deposit, Coast Rock convinced the county to modify it's approval of the pipeline to require a reverse syphon crossing of the Sisquoc River. This action eliminated any loss of aggregate reserves. With the resolution of the conflict between the pipeline and mining, all parties were pleased and both projects went forward to completion.

Example 10: Incompatible adjacent land use – Los Angeles County
The Los Angeles County Regional Planning Commission held a hearing concerning a proposal to permit an expensive subdivision overlooking and adjacent to an existing aggregate mining operation (P.W. Gillibrand Company) in Soledad Canyon. The establishment of the subdivision, as proposed, would have resulted in the curtailment of the mining operation and the effective loss of mineral resources. DMG participation in the hearing was requested by the County, and SMARA staff testimony was given on the mineral land classification information from the DMG's preliminary Saugus-Newhall-Palmdale report (Joseph et al. 1987), even though it had not yet been formally released. The testimony confirmed that of the mining company, but because of DMG's credibility and lack of any vested interest in either the subdivision or the mine, our testimony was taken much more seriously by County officials. This clarified the land-use and resource conflict in the minds of the Planning Commission. The Commission's decision was to modify the subdivision plan to include an intervening ridge as a wide buffer.

Example 11: Incompatible adjacent land use – Contra Costa County
(While this is an example of the protection of a glass sand resource rather than construction aggregate, it is included to demonstrate SMARA effectiveness as well as the importance of open communication between the mining industry and State and local government.)

Unimin Corporation's only California mining operation to date is their glass sand mine near Byron in Contra Costa County. The geologic unit being mined (a member of the Domengine Formation) is a limited resource and is perfect for the production of silica sand to be used as the main ingredient in the manufacture of glass. Glass sand from Byron is used to manufacture more than 90% of all the wine bottles used in California. Many other products are produced, including brown, green, and clear containers, household glassware, flat window glass, fiberglass, and golf course sand.

In 1990 DMG received a request from Unimin asking for help with a problem. A decision had been made by the East Bay Municipal Utility District to increase the height of a local dam in the east bay. This decision required the re-alignment of a Contra Costa County road to prevent its being flooded, and the new road location was proposed to be across Unimin's Domengine sand reserves. The reserves were assumed by local supervisors to be unimportant because DMG had not classified them in its SMARA mineral land classification project. In fact, this part of Contra Costa county had not yet been classified due to higher priorities elsewhere in the State. Unimin was told that we had no production data on glass sand as a result of the company's policy on secrecy, and it would be impossible to make any objective statements about the importance of the Byron plant. Unimin was told that if they would authorize the USBM to provide DMG with the data, that we would look into it the problem and write to Contra Costa County if needed.

The data arrived shortly and was impressive in its quality and thoroughness. It was independently confirmed, and with the authorization of the DOC and the SMGB, DMG wrote a letter to Contra Costa County explaining SMARA and the importance of the sand deposit and Unimin's mine. DMG subsequently got a call of thanks from Unimin relating how responsive Contra Costa County had been to DMG's letter. The planned road re-alignment had been moved to another location, saving Unimin 20 years of glass-sand reserves.

As a result of this success, the Unimin Board of Directors decided to remove all restrictions on release of their proprietary annual production data to California, as well as to all other states. This openness provides a much clearer picture of the State's glass sand resources which are vital to so many ancillary industries and so many Californians. Unimin also changed its policy on allowing professional visits to its operations, inviting DMG for an in-depth tour of the mine and processing plant – the first plant visitors in 22 years!

Example 12: Mining use-permit extension – Ventura County
In 1979, the mining permit of the P.W. Gillibrand Company at their Simi Valley location was up for renewal. A preliminary version of DMG's Ventura County report (Anderson et al. 1981) had been sent to Ventura County and also to P.W. Gillibrand. Mr Gillibrand requested that Tom Anderson, the lead author of that report, testify on behalf of his aggregate pit at an upcoming Ventura County Planning Commission Meeting to assist him in getting his use permit renewed. Mr Gillibrand shared with DMG that he was requesting a 40-year extension, but was hoping to get, at the best, a compromise for a 20-year extension, while assuming, realistically, that he would be given only 10 more years. Our policy, at the time, was not to appear at use-permit hearings advocating on behalf of a specific operator, but we advised Mr Gillibrand that he was well armed with factual data on local aggregate availability and the life span of existing permitted reserves from our report and should make his own best case. (We subsequently modified that policy and now appear at use-permit hearings if we are invited to do so in writing from the lead agency responsible for the pending decision).

Mr Gillibrand followed our advice and spoke on his own behalf at the use-permit hearing, using our data as well as letters of support from local residents and pit neighbors. To the surprise of company officials, the Ventura County Planning Commission voted to grant the P.W. Gillibrand Company a 40-year extension to his mining permit. This early success demonstrated the potential influence of the State Geologist's SMARA reports to the mining industry of southern California, which became extremely cooperative when we subsequently requested access to visit their properties or to examine previously confidential geologic or engineering testing records.

Example 13: State or industry reports? – San Diego County
When the SMARA project first got underway in California, it was well understood that the initial $1.1 million provided by the legislature was not enough to provide resource data to everyone immediately. It became clear early on that patience would be needed by both local government and aggregate producers. With the concurrence of the DOC Director and the State Geologist, the State Mining and Geology Board directed DMG staff and fiscal resources into the two areas of the State where the land-use pressure on mining was felt to be most acute – the Los Angeles Basin and the San Francisco Bay region. The industry understood and supported this priority, but sand and gravel producer associations in other parts of the State, who had been following the legislation, were anxious to have DMG provide mineral-land-

classification data to their local jurisdictions. One of the more active aggregate associations in California at the time was the San Diego Rock Producers Association, and with the support of their membership, they decided they could not wait the scheduled three years for the State study to be completed. They therefore took a bold move in 1979 and hired a major geotechnical consulting firm to conduct a private 'SMARA-type' study to be paid for by members of the San Diego Rock Producers Association. The report was completed within six months and submitted to DMG for review. All parties had originally hoped that this arrangement could save the State time and money if the State could officially endorse the privately done 'SMARA-type' report, giving it the blessing of the State Geologist and officially starting the 12-month clock for incorporation into the County general plan. However, problems were discovered when the report was reviewed by DMG. First, while the report was generally quite sound, it was inconsistent with the two DMG completed SMARA reports released to date. Permitted reserves were not distinguished from total resources, which were all considered available for mining. As a result, tonnages of permitted reserves (resources controlled by aggregate companies and for which valid mining permits had been granted) were neither mapped nor quantified. Similarly, no attempt to quantify aggregate demand was made, so the life span of permitted reserves, which is an extremely important number to local government officials, was not possible to determine. Alternative aggregate resources were not considered, nor were current land-uses as they related to resources sectors.

As a result, the industry 'SMARA-type' report was not endorsed by the State Geologist, nor was the County required by law to use it. Most significantly, when the report and the consultant were used by the industry in mining-permit hearings before the County Planning Commission or local city jurisdictions, both were attacked by the anti-mining faction as lacking credibility and being somehow tainted because they had been paid for by the mining industry. On the positive side, within a year after our review of the report, our own independent SMARA study was completed in San Diego County (Kohler et al. 1982). Our report was made much faster to complete because of the data gathering of the San Diego Rock Producers Association's attempt to conduct their own study.

This incident provided a valuable lesson to the aggregate industry that public perception as to the quality of the finished SMARA report was directly dependent upon who had paid for its completion, and was a reminder to us in DMG that our policies of independence from the mining industry were both justified and necessary. The State could advocate the necessity of mining and of granting more mining permits but could not advocate a specific mining company permit or a specific deposit without stepping over the line with local government permit authority.

6 CONCLUSION: THE IMPORTANCE OF OPTIONS IN LAND-USE PLANNING

The case studies presented in this paper have purposely been selected to illustrate the range of uses of SMARA classification data. Because of local autonomy in land-use-decision making and the granting of use permits in California, generalizations from these case studies are risky. Each permit decision has its unique characteristics, and no parcel of land is without several advocates, often with highly different agendas. The most significant generalization, however, is easy to conclude. SMARA has elevated public awareness of the importance of mineral resources to a point where they can no longer be ignored, and their presence (or absence) must be dealt with openly in land-use decisions.

SMARA has also provided local government with the tools to make informed and rational decisions with respect to mineral resources. Almost all lead agencies have welcomed the SMARA data and most have used it well in their planning and permitting process. Some have undoubtedly disregarded the data and others have purposely misused it, claiming to an applicant, for example, that the state would not allow them to grant the use permit because it would preclude the eventual extraction of mineral deposits (the state lacks that authority). Most lead agencies have been empowered by the minerals data, though some have resented the State's intrusion. With 528 lead agencies, it would probably not have been difficult to predict these results back in 1975 when SMARA was first enacted. What would have been harder to envision has been the extensive use by national and international mining companies and those associated with the mining industry – investors, bankers, salesmen, educators, appraisers, and explorationists. SMARA reports have aided in the discovery of new mineral deposits and have attracted new mining companies to the State. New mines have attracted new manufacturing plants and have extended the life span of existing plants, providing a stronger tax base and needed jobs.

Use of mineral-land-classification data by the Federal Government has been a mixed success. Congress made only minor changes in proposed wilderness area boundaries based upon SMARA work, and only when active mines were involved. After initial resistance to state intrusion on Federal land, the US Forest Service (USFS) and the US Bureau of Land Management (BLM) land managers now recognize the value of the state data, and memoranda of understanding (MOUs) are in place between California and those two large Federal landholders. The MOUs acknowledge the SMARA data and reclamation regulations and agree to abide by the SMARA law, without actually relinquishing federal authority. The USFS has contracted with the US Geological Survey to conduct SMARA-like land classifications in at least one of their forests. The Department of Defense has used SMARA reports, as have sovereign Indian Nations on their lands. California's SMARA efforts have been successful enough to attract national and international attention. Governments of Australia, Canada, Chile, China, Japan, New Zealand, Poland, Taiwan, and the United States have studied the land-classification element of SMARA for application to their mineral resources, as have ten US states and three Canadian provinces.

With the data from a SMARA mineral land classification report, a local planner knows what the options are, and can weigh the advantages and disadvantages of opening a mine at any given site against any other, and against the needs of the overall jurisdiction or region. This information can then be weighed against local desire and the needs of the affected community. Without the SMARA classification report, the planner will usually see only what the project advocates or opponents wish to be seen. In this atmosphere, the conservation of natural resources may be left entirely to chance, a practice that has not been successful in the past. In the San Francisco Bay area, it has been estimated by the USGS, that 95% of all known mineral deposits are already built upon and are precluded from ever being mined. The presence of a mineral deposit may not be the most significant natural resource on any given piece of property, but to inadvertently preclude the possibility of its being mined because of ignorance of its mineral potential is a luxury we can no longer afford.

REFERENCES

Anderson, T.P. Loyd, R.C. Kiessling, E.W. Kohler, S.L & Miller, R.V. 1981. *Mineral land classification of Ventura County:* Division of Mines and Geology Special Report 145, 82 pp. 24 plates.
Beeby, D.J. 1988. Aggregate resources – California's effort under SMARA to ensure their continued availability: *Mining Engineering* 40(1): 42-45.
Beeby, D.J. 1989. Inferred mineral resources on Mather Air Force Base; Report to the Natural Resources Subcommittee of the Facilities Reuse Committee to the Sacramento Area Commission on Mather Conversion (SACOMC), Unpublished, 2 pp.
Boylan, R. & Loyd, R.C. 1991. *Mineral land classification of the south Tracy site, San Joaquin County, California*: Division of Mines and Geology Open File Report 91 03, 16 pp. 2 plates.
California Department of Conservation – Division of Mines and Geology, 1995. *Surface Mining and Reclamation Act of 1975*. Division of Mines and Geology Note 26, Revised 1/95. 27 pp.
California Department of Conservation – Division of Mines and Geology, 1996. Minefile database Division of Mines and Geology.
California Department of Conservation. 1983. *State Mining and Geology Board, 1983. California surface mining and reclamation policies and procedures:* Division of Mines and Geology Special Publication 51, 2nd Revision, 38 pp.
Cole, J.W. & Jensen, L.S. 1986. *Mineral land classification of a portion of the Sisquoc River, Santa Barbara County, California, for Portland-cement-concrete-grade-aggregate*: Division of Mines and Geology Open-File-Report 86-19, 16 pp. 1 plate.
Dupras, D.L. 1988. *Portland cement concrete grade aggregate in the Sacramento-Fairfield production-consumption area*: Division of Mines and Geology Special Report 156, 37 pp. 40 plates.
Evans, J.R. Anderson, T.P. Manson, M.W. Maud, R.L. Clark, W.B. & Fife, D.L. 1979. *Aggregates in the greater Los Angeles area, California:* Division of Mines and Geology Special Report 139, 96 pp.
Habel, R.S. and Campion, L.F. 1988. *Portland cement concrete grade aggregate in the Yuba City-Marysville production-consumption region*: Division of Mines and Geology Special Report 132, 19 pp. 12 plates.
Hagan, M.L. 'Dayo' 1990. Final Report to the Natural Resources Subcommittee of the Facilities Reuse Committee to the Sacramento Area Commission on Mather Conversion (SACOMC), Unpublished, 6 pp.
Jensen, L.S. & Silva, M.A. 1988. *Mineral land classification of portland cement concrete aggregate in the Stockton-Lodi production consumption region*: Division of Mines and Geology Special Report 160, 29 pp. 25 plates.
Joseph, S.E. Miller, R.V. Tan, S.S. & Goodman, R.W. 1987. Mineral land classification of the greater Los Angeles area: *Saugus-Newhall production-consumption region and Palmdale production-consumption region:* Division of Mines and Geology Special Report 143, Part V, 33 pp. 24 plates.
Kohler, S.L. Miller, R.V. Manson, M.W. & Lowry, P.A. 1982. Mineral land classification: *Aggregate materials in the western San Diego County production-consumption region, California*: Division of Mines and Geology Special Report 153, 28 pp. 39 plates.
Loyd, R.C. Hill, R.L. & Miller, R.V. 1994. *Mineral land classification in California*; Division of Mines and Geology: California Geology, 47(1):10-21.

O'Hare, M. with McKinney, J. Austin, D. & Kito, M.S. 1995. *The Effectiveness of the California Surface Mining and Reclamation Act; A report to the California legislature and the Governor pursuant to §2774.6 of the Surface Mining and Reclamation Act;* Graduate School of Public Policy, University of California, Berkeley..

Rapp, J.E. 1975 *Sand and gravel resources of the Sacramento area, California*: Division of Mines and Geology Special Report 121, 34 pp.

Stinson, M.C. Manson, M.W. & Plappert, J.J. 1986. Mineral land classification: *Aggregate materials in the San Francisco-Monterey Bay Area*: Division of Mines and Geology Special Report 146, Part I (project description) 20 pp. 3 plates.

Stinson, M.C., Manson, M.W. & Plappert, J.J. 1987a. Mineral land classification: *Aggregate materials in the San Francisco-Monterey Bay Area*: Division of Mines and Geology Special Report 146, Part II (South Bay area), 65 pp. 75 plates.

Stinson, M.C. Manson, M.W. & Plappert, J.J. 1987b, Mineral land classification: *Aggregate materials in the San Francisco-Monterey Bay Area*: Division of Mines and Geology Special Report 146, Part III (North Bay area), 49 pp. 65 plates.

Tooker, W.W. & Beeby, D.J. (eds) 1992. Industrial minerals in California: *Economic importance, present availability, and future development*: Division of Mines and Geology Special Publication 105, 127 pp.

US Bureau of Mines, 1995 and earlier years, California Annual Mineral Production Statistics, unpublished and proprietary except in summary.

US Bureau of Mines and US Geological Survey, 1980. *Principles of a resource/reserve classification for minerals*; US Geological Survey Circular 831, 5 pp.

Wiedenheft-Cole, J. & Fuller, D.R. 1988. Mineral land classification: *Aggregate materials in the Fresno production-consumption region*: Division of Mines and Geology Special Report 158, 21 pp. 18 plate 40.

The state of aggregate resource management in Ontario

DONALD A. STEWART
Planning Initiatives Ltd., Kitchener, Ontario, Canada

1 INTRODUCTION

This paper provides an overview of how the Government of Ontario has responded to the increasing pressures for additional regulation of aggregate extraction, and to increasing concerns about the physical, social, and economic environment related to aggregate resource management. During the 1960's and early 1970's there was relatively little regulation of the industry. The development of gravel pits proceeded, with comparatively little regulatory requirement and enforcement of their operations, and also to their ultimate rehabilitation after completion of extraction. During the 1970's and 1980's initial steps were taken by the government to regulate the industry through the Pits and Quarries Control Act (PQCA). At the same time, steps were taken to protect non-renewable aggregate resources through adoption of the Mineral Aggregate Resources Policy Statement (MARPS), a formal expression and statement of provincial interest. Following what were record breaking production levels in the late 1980's, more stringent regulations were put in place in 1990 with passing of the Aggregate Resources Act (ARA). The ARA now includes some of the most rigorous requirements in North America for the development, operation, and rehabilitation of both sand and gravel pits and quarries. While much remains to be done, there is evidence of increased environmental sensitivity and positive rehabilitation efforts on behalf of the aggregate industry.

The major issues of concern to both the aggregate industry and the provincial government in the mid 1990's are outlined, as are a number of the initiatives currently being undertaken to address those issues. Integration of resource management philosophies are resulting in major opportunities and challenges for the aggregate industry in the Province of Ontario.

The rapid economic growth in North America during the last quarter of the twentieth century has highlighted the need for wise management of this non-renewable resource. Equally important is the need for extracting the resource in a sensitive manner to minimize the impact on the surrounding natural and human environment. This is especially so because of the perception of aggregate resource extraction by many as being incompatible with sensitive uses such as residential development and various natural or environmental features.

In the case of Ontario, and especially southern Ontario, the high demand for aggregate related products in major urban centres has resulted in a need to extract the resource, often in close proximity to urban development and/or such areas such as floodplains along rivers and woodlot areas. With transportation costs currently accounting for about 50% of the ultimate cost to the consumer, sources of aggregate longer distances from the processing and end point will result in significantly higher construction costs.

These increasing pressures, and the need to minimize conflicts between aggregate resources and other resources or uses have resulted in a rapid evolution of the state of aggregate resource management. It appears that the decade of the 1990's will see the advancement of aggregate resource management to a very high level in the largest aggregate producing Province in a country with one of the highest per capita uses of aggregates in the world.

1.1 *High production and consumption*

Of all the provinces in Canada, Ontario typically has more production of aggregates than any other province. In the years 1988 and 1989, approximately 200 million tonnes of aggregates were extracted annually, coincidental with the peak of the rapid economic growth of the 1980's. In 1988, Statistics Canada figures indicate that Ontario aggregate shipments accounted for about 40% of all the aggregate movements in the country (PIL 1994). Sand and gravel produced and consumed within Ontario represented about 36% of all sand and gravel produced in Canada. Ontario production is followed by British Columbia, Alberta, and Quebec, all of which had less than one half the amount of the Ontario production. In terms of crushed stone, Ontario's production amounted to approximately 49% of all Canadian production. The next largest producers of crushed stone were Quebec, British Columbia, and Nova Scotia.

Prior to the economic boom in Ontario in the late 1980's, aggregate production and consumption had typically ranged from 100 to 130 million tonnes per year. During the recessionary period of the early 1990's following the economic boom, aggregate production dropped from a peak of 197 million tonnes to approximately 128 million tonnes in 1992 (MNR 1995). By 1994, production had risen again, slowly, to about 137 million tonnes.

The significant changes in levels of aggregate production clearly illustrate the strong direct correlation between aggregate production and economic growth. The rate, and fate, of aggregate extraction are closely tied to the construction of basic infrastructure, most notably, public roadways. In Ontario, almost 60% of the sand and gravel produced is used for roads, asphalt, and aggregate in asphalt. Another 20% is used for concrete aggregate (MNR 1995). The Aggregate Producers' Association of Ontario (APAO) indicate that about 24,000 tonnes of aggregate are used in the construction of every kilometer of two-lane roadway. Approximately 425 tonnes or about 20 tandem truckloads of aggregate are utilized in the construction of a typical single family home of approximately 2000 square feet (MNR 1995). About 17,000 tonnes of aggregate are used in constructing an average sized elementary school, and a high rise apartment building requires about 150,000 to 200,000 tonnes (MNR 1995).

In the last 25 years, aggregate consumption in Ontario on a per capita basis has ranged between 15 and 20 tonnes per capita (PIL 1995). This compares with an average figure of approximately 9 tonnes per capita on an annual basis in California.

2 EVOLUTION OF LEGISLATION TO 1980'S

Concerns regarding environmental and planning issues related to aggregate extraction date back to the 1960's in Ontario. The first major restrictions to extraction of aggregates occurred in the Niagara Escarpment Protection Act in 1970. Specific provisions were implemented to control extraction in the Escarpment, a major natural feature winding through some 700 kilometers of southern Ontario. In 1971, the PQCA was implemented to regulate aggregate extraction in designated areas, most of which occurred in the more densely populated areas of southern Ontario.

Over the next decade, a number of inadequacies and shortcomings of the PQCA became evident. A Mineral Aggregate Working Committee, appointed in 1975, produced a 10 point policy paper with appropriate policies to be included in municipal official plans. With implementation of the Planning Act in 1983, sand and gravel pits and stone quarries were recognized as a 'use of the land' but were not be considered as development of the land (GO 1988). The creation of a new pit or quarry, or an expansion of an existing operation, must conform to policies, procedures, and regulations as stipulated in official plans and zoning by-laws.

In 1986, the previously prepared mineral aggregate policies were refined and formally adopted by the Government of the Province as a Provincial Policy Statement entitled, 'Mineral Aggregate Resources Policy Statement' (MARPS). This statement of provincial policy provided a framework of basic principles for managing aggregate resources and stated that:

– Mineral aggregates are essential non-renewable natural resources,

– Mineral aggregates should be available to the consumers of Ontario at a reasonable cost,

– All parts of Ontario possessing mineral aggregate resources share a responsibility for meeting future provincial demand,

– Notwithstanding the need for mineral aggregates, it is essential to ensure that the extraction is carried out with minimal social and environmental cost.

The overall Policy Statement stipulates that:

1. All land use planning and resource management agencies within the Province shall have regard for the implications of their actions on the availability of mineral aggregate resources to meet future local, regional, and provincial needs,

2. Any planning jurisdiction, including municipalities and planning boards, identify and protect as much of its mineral aggregate resources as is practical, in the context of other land use planning objectives, to supply local, regional, and provincial need (Government of Ontario 1986).

Over several years, considerable discussion and legal proceedings ensued over the principle that the need for aggregate resources is a matter of provincial interest. Notwithstanding potential localized disturbance and inconvenience, all municipalities within the province must share in the responsibility for providing aggregate resources, where available at a reasonable cost. Secondly, other provisions included in the Planning Act, 1983, indicated that all municipalities and resource management agencies must 'have regard for' the implications of their actions on the availability of aggregates.

In spite of implementation of this Policy Statement, there were still several identified problems with the PQCA regarding the regulation of pits and quarries (PIL

1994). They included:
- – Enforcement difficulties due to vague wording and interpretation differences,
- – Lack of specific site plan requirements, and
- – An inadequate licensing system.

Work continued during the latter part of the 1980's to prepare a new Act to replace the PQCA and the various other legislation related to pits and quarries in the Province.

3 LEGISLATIVE/POLICY CHANGES IN THE 1990'S

Beginning with the implementation of the Aggregate Resources Act (ARA) the decade of the 1990's seems to be one destined for significant changes regarding aggregate resource management policies. The ARA was proclaimed January 1, 1990, replacing the PQCA, the Beach Protection Act, and Part 7 of the Mining Act. It is administered and enforced by the Ministry of Natural Resources (MNR). The purposes of the Act, as outlined in Section 2 are:

1. To provide for the management of aggregate resources of Ontario,
2. To control and regulate aggregate operations on Crown and private lands,
3. To require the rehabilitation of land from which aggregate has been excavated, and
4. To minimize adverse impacts on the environment with respect to operations (Government of Ontario 1990).

The ARA incorporates principles and methods to ensure that environmental factors are considered as part of the resource development process, and includes stronger enforcement powers to implement these objectives. The Act ensures that aggregate extraction is only an interim use of the land, and that mandatory progressive rehabilitation occurs throughout the life of the operation.

Under the new Act, aggregate licenses are divided into two categories – Class A licenses for operations extracting more than 20,000 tonnes per year, and Class B licenses for extraction of less than 20,000 tonnes. For new Class A licence applications, an Environmental Report is required to accompany the site plans, and is to include significant detail on a number of matters as outlined in Section 9 of the Act. They include consideration of potential environmental, social, and economic effects of the operations, as well as the quantity and quality of the resource, and the proposed rehabilitation of the property.

With the implementation of the new Act, annual licence fees were established for all sand and gravel and quarry operations, at the rate of $0.06 per tonne for the previous years production. This licence fee is divided between the local municipality, upper tier municipality, MNR, and a fund for rehabilitation of abandoned pits and quarries. Section 50(1) of the Act also provides for a rehabilitation security deposit for each pit to ensure the rehabilitation of the site. This deposit amounts to $0.08 per tonne of aggregate to a maximum of $6,000.00 per hectare for all areas that require rehabilitation.

For all pits and quarries that were licensed prior to the implementation of the ARA, the new act also contained a requirement that each of those operations must, within four years, have replacement site plans prepared. By the end of 1993 all licensed site plans were to have been brought up to the same standards.

3.1 Comprehensive set of policy statements

In the early 1990's, a number of other Provincial government initiatives, although not related to aggregate resources only, had a significant effect on policies and management of the resource. In 1991, a commission headed by John Sewell, former City of Toronto Mayor and outspoken planning critic, was created to 'recommend changes to the Planning Act and related policy that would restore the confidence and the integrity of the planning process, protect public interests, better define roles and responsibilities, focus more closely on protecting the natural environment, and make the planning process more timely and efficient' (Sewell et al. 1993). In two years, the Commission held broad ranging consultation activities, and made a total of 98 wide ranging recommendations relating to Provincial Policy Statements, including those related to aggregate resources, the natural environment, the provincial role in planning, and other planning related matters.

Many of the recommendations have been implemented through legislation and regulatory documents, and have already had a major impact on present and future aggregate resource management. With regard to Provincial interest, the report recommended preparation of a 'Comprehensive Set of Policy Statements' (CSPS) as expressions of Provincial interest. Several policy statements had been in effect previously, including Policy Statements on Mineral Aggregate Resources; Wetland and Floodplains; Land Use Planning and Housing; and Agriculture and Foodland Guidelines. Legislation in the Planning Act at that time indicated that municipalities and other agencies 'shall have regard for expressions of Provincial Policy interest'. However, the Sewell Report recommended that provision be strengthened to indicate that all agencies 'shall be consistent with Provincial Policy Statements'. These recommendations had the effect of helping to strengthen MARPS. At the same time however, this increased the dilemma about how to deal with aggregate applications or operations where being consistent with MARPS could also be in contravention to the policies within other policy statements.

Specifically, the Policy Statement relating to aggregates in the new CSPS maintained the earlier principle of MARPS. Other policies found within the CSPS of significance to aggregate extraction and aggregate resource management are in the Natural Heritage, Environmental Protection and Hazard Policies.

Relevant policy subsections indicate that: 'Development will not be permitted in significant ravine, valley, river, and stream corridors, and in significant portions of the habitat of endangered species and threatened species'.

'Development will not be permitted on adjacent lands if it negatively impacts the ecological functions of the features listed above' (OMMA 1996).

'All planning jurisdictions including municipalities, planning boards, and resource management bodies, shall protect provincially significant wetlands where they have been identified'.

'In prime agricultural areas, extraction of mineral aggregates on prime agricultural lands may be permitted as an interim use provided that agricultural rehabilitation of the site will be carried out whereby substantially the same areas and same average soil quality for agriculture are restored'.

On prime agricultural lands, extraction may occur below the water table, and complete agricultural rehabilitation is not required only if it is demonstrated that:

– There is a substantial quantity of mineral aggregate below the water table warranting extraction below the water table,

– Other alternatives have been considered by the application and found unsuitable. Other alternatives include resources in areas of classes 4 to 7 agricultural lands, resources on lands committed to future urban uses, and resources on prime agricultural lands where rehabilitation to agriculture is possible, and

– In those areas remaining above the water table following extraction, agricultural rehabilitation will be maximized' (OMMA 1996).

These excerpts from the different policy statements in the CSPS highlight many of the major natural/ecological areas of greatest concern or potential conflict for aggregate extraction in the 1990's. They also are the areas in which considerable efforts are being made to mitigate environmental impacts, and identify innovative designs to integrate the management of the different resources in a positive manner.

The CSPS were followed by Bill 163, proclaimed in March, 1995, to implement many of the changes recommended in the Planning Reform Study, and incorporate the CSPS into the Planning Act. Bill 163 also implemented the 'shall be consistent with Policy Statements' provisions as outlined in the Planning Reform Study, but the wording was changed back to have regard to in 1996 by subsequent amendments introduced by the Conservative government. This change in phrasing, although minor in actual word change, sparked major discussions among land use and resource planners and environmental groups about how rigid or flexible the policy statements were intended to be.

By the mid-1990's, the strong concern for the environment as expressed in various legislation, and the greater awareness and environmental education of the public, has resulted in a framework in which as many as 50% of the aggregate applications in some areas are referred to the Ontario Municipal Board (OMB). As a result, environmental issues have grown significantly in importance as they affect aggregate resource management, and will likely continue to be the key issues in the latter part of the 1990's.

Persons, agencies, corporations, and government bodies which feel aggrieved by land use decisions have the right of appeal to this quasi-judicial tribunal, the OMB. However, appeal to the Board in the past has been typified by extended delays of from 1 to 2 years before the appeals can be heard. It has also been very costly for individuals and corporations either making the appeals or defending the positions taken by them in their actions. Depending upon the complexity of the issues, the costs to retain the necessary experts and legal advice, prepare for and attend the hearing, can cost anywhere from $50,000 to $500,000 and in some cases, beyond one million dollars.

As a result of the large number of hearings, the lengthy delays in hearing the cases, and the costs to the various parties, a number of initiatives were taken during the early 1990's to reduce the backlog and the number of cases going to the Board. The Provincial Government established an 'Office of the Provincial Facilitator' to assist in resolving issues of dispute prior to them going to the OMB. By 1996, the previous backlog of some 3000 cases were considerably reduced, as was the time to wait for a hearing reduced to about 6 to 8 months.

4 ISSUES IN THE 1990'S

In 1990, following the implementation of the ARA, the MNR commissioned a two year study on aggregates entitled, *Aggregate Resources of Southern Ontario: A State of the Resource Study*. The objective of the study was to provide an up-to-date and comprehensive reference document for aggregate resource companies, government, ministry, and public use. It was to formulate statistical forecasts for aggregate demand in Ontario between 1990 and 2010, in the seven major market areas of southern Ontario. Finally, the study was to provide background information to serve as a benchmark for future strategic planning initiatives and policy formulation. The major findings of that study, completed in 1992 and published in 1994 were that:

1. Southern Ontario is moving towards a critical economic, social, and environmental situation in terms of protection of, and access to, aggregate resources required to meet the needs of Ontario residents,

2. Existing licensed aggregate reserves within some of the major market areas could be depleted within a few years in some areas if new reserves are not licensed, and,

3. A continuing and increased emphasis should be placed on a coordinated and balanced approach to aggregate resource management. Strategic long term land planning, management, and successful rehabilitation of pits and quarries should be encouraged to productive, appropriate land uses which achieve no net environmental loss, or a net environmental gain wherever possible (PIL 1994).

As can be seen from the major findings above, issues relating to the environment have been playing a key role in aggregate matters.

4.1 *Environmental issues*

The 1995 CSPS state the following as major goals in the Natural Heritage, Environmental Protection, and Hazard Policy sections:

'To protect the quality and integrity of ecosystems, including, air, water, land, and biota; and, where quality and integrity have been diminished, to encourage restoration or remediation to healthy conditions'.

Other legislation governing various aspects of the environment also impact aggregate resource extraction. Several of these are illustrated in Table 1.

4.2 *Water resource protection*

One of the most significant aggregate related issues in Ontario is the potential impact of aggregate extraction on both surface and groundwater resources. Given the geological origins and characteristics associated with aggregate resources, both sand and gravel and stone, it is not surprising that the frequency of occurrence of aggregate resources in close proximity to either surface or groundwater supplies is high. Notwithstanding the fact that the initial processing of unconsolidated sand and gravel re-

Table 1. Federal and Provincial Environmental Legislation relating to aggregate resources. Source: Ministry of Natural Resources, Aggregate Resources of southern Ontario: A state of the resource study, 1994.

Document	Relationship to aggregate resources
1. Environmental Assessment Act (EAA)	No direct impact unless designated under Section 3 by the Ontario Cabinet (no designations to date)
2. Environmental Protection Act (EPA)	Regulates discharge of contaminants, including noise, dust and water pollution
	Several guidelines regulate the operational aspects of aggregate extraction
	Specific sections/regulations on noise, dust/air emissions, discharges, spills and waste
	MISA program under the Act will regulate discharges
3. Ontario Water Resources Act (OWRA)	Specifically relates to water pollution and water taking permits
4. Fisheries Act (Federal)	Administered in Ontario by the Ministry of Natural Resources Relates to any potential impacts on fish or fish habitat
5. Conservation Authorities Act	Regulations within this mandate relate to aggregate extraction within the floodplain area of rivers
6. Lakes/Rivers Improvement Act	Any extraction near lakes, river or watercourses
7. Foodland Guidelines	Relates to aggregate extraction on Class 1-3 lands and specialty crops.
	Requires rehabilitation to agricultural after use
8. Trees Act	Regulates the removal of trees
	Does not currently apply to licensed properties
	Revisions are currently underway which may affect aggregate resource extraction

sources (crushing, screening, and washing), is one of the few industries that uses absolutely no chemicals in the processing itself, there has been a high instance of concern over aggregate extraction in areas of surface or groundwater sources. Tests by the provincial Ministry of Environment and Energy (MOEE) in the 1990's concluded that extraction operations themselves had little effect on adjacent surface water quality levels other than turbidity (MOEE 1990). Since aggregate extraction does not usually produce toxic chemicals in itself (such as those in base metal mining), potential water quality impacts are generally restricted to temporary increases in turbidity, and the potential for spills from petroleum related products used on the site.

From a hydrogeological viewpoint, the two areas of concern are groundwater quality and quantity, and groundwater temperature. With respect to aggregate operations, potential groundwater concerns are generally related to water quantity. Policy 1.1 of the CSPS indicates that 'development may be permitted only if the quantity and quality of the groundwater and surfacewater are protected. Development that will negatively impact on groundwater recharge areas, headwaters, and aquifers, which have been identified as sensitive areas, will not be permitted'. While groundwater and surfacewater concerns have always been important issues, the inclusion of these in the formal policy statements by the Provincial Government is reflected in the requirement for more detailed and stringent hydrological and hydrogeological studies required as part of extraction applications.

The impact of aggregate extraction on groundwater quantity is also of importance, especially in major areas for groundwater recharge and for the headwaters of significant water courses. With the implementation of the CSPS several regional municipalities moved quickly to include policies in their official plans to severely restrict or prohibit aggregate extraction in such areas. In other cases, municipalities have moved to restrict below water table extraction, and to limit extraction to 1.0 meter above the water table. The implementation of such local policies could result in sterilization of large areas of high quality aggregate resources. The approval of future below water extraction operations is becoming increasingly more difficult and costly because of the detailed studies now being required. Conversely, the additional attention being paid to the hydrogeological matters may result in more innovative operational and rehabilitation plans designed to attain improvements or net environmental gains as a pre-requisite to allowing below water extraction.

4.3 *Wetland enhancement*

Another environmental issue receiving great attention is the concern over the decreasing amount of wetlands in Ontario. The Wetlands Policy Statement approved by cabinet in 1992, and reaffirmed under the CSPS, states that there shall be no loss of Provincially Significant Wetlands. The conservation of other regionally and locally significant wetlands are encouraged.

One recent example of innovation is a case in which the operator of an existing licensed operation requested and obtained approval for extraction below water in one pit. In return, he amended the operation in another pit so that an adjacent Provincially Significant Class 1 Wetland could be expanded into the pit area, with improved fish, wild fowl, and wetland habitats being created (Fig. 1). In another innovative design, utilization of below water extraction, with a small water body and wetlands being created, will form a linkage between three adjacent woodlot areas on the same site.

Figure 1. Site rehabilitation to expand Class 1 Wetland and wild fowl habitat.

4.4 *Extraction in woodlands*

Aggregate extraction within or adjacent to woodlands is becoming more acute. In southern Ontario, where there is frequently no more than about 15% of the total area of any given municipality still covered by woodlands, steps are being taken to restrict or prohibit aggregate extraction in significant (+10 hectare) woodlots. While the woodlots are a renewable resource as compared to aggregates as a non-renewable resource, there is frequently the concern that replacement of the woodlots will not result in an equivalent environmental feature.

4.5 *Extraction in floodplains*

Extraction within floodplains is also facing increasing scrutiny and testing prior to approval because of related environmental issues. While there are constraints to extraction within the floodplain of major rivers or watercourses, such extraction, if properly designed, can be used to enhance not only the natural environment, but also the broader social, cultural, and economic environment. Two recent examples of innovative design are illustrated in the Kitchener area (Fig. 2). In the first case, a joint partnership arrangement between the local Conservation Authority and a private aggregate operator have resulted in the creation of a unique rivers edge environmental park area with cold and warm water ponds and other features created along the river. The rehabilitated site is now being considered for future designation as a man-made environmentally sensitive area. By the time the final rehabilitation has created the

Figure 2. Environmental Park Rehabilitation Project, Snyder Flats, Kitchener.

last of the environmental park area, not only will the natural environment in the area be improved, but the Conservation Authority will also have benefited by significant aggregate royalties from the operator, for aggregate extracted on the Conservation Authority lands.

In a second example, a similar joint venture between the City of Kitchener and a local aggregate operator will allow land owned by the City to be extracted with royalties being paid to the City. The operational plan and rehabilitation will direct the development of a small municipal park, canoe launch, fishing areas, and riverine/wetland habitat directly adjacent to one of the major entryways to the City.

4.6 *Major physiographic features*

Two physiographic regions within southern Ontario have been the subject of considerable discussion with respect to aggregate extraction in the early 1990s. The Niagara Escarpment, which stretches across much of southern Ontario is also the source of some of the best bedrock for the construction industry in all of Ontario and Eastern Canada. During a review of the Niagara Escarpment Plan in the early 1990s, lengthy hearings were held regarding the extent to which aggregate extraction should or should not be allowed within the Escarpment. As a result of the hearings and testimony between the different provincial ministries, environmental groups, and private associations, a further eighteen month study was completed to determine the extent of aggregate extraction to be allowed within the Escarpment. At the time of writing this paper, the final decision on the Niagara Escarpment Report has not yet been determined by the Minister.

Just as the Niagara Escarpment includes the best bedrock for construction material, the Oak Ridges Moraine, which winds through the northern part of the Greater Toronto Area, provides some of the best unconsolidated aggregate material immediately adjacent to and within the largest market area of all of southern Ontario. In a move similar to that regarding the Niagara Escarpment, the Province of Ontario, in 1993, instituted a temporary freeze on new 'green field' development in the Oak Ridges Moraine, including that of new aggregate extraction operations. The purpose of this freeze was to allow time for an intensive study into development within the Moraine, and the potential environmental implications of such development. As of early 1996, that report also was still before the Minister awaiting final resolution of the extent of future development to be allowed within the Oak Ridges Moraine.

The outcome of both of these studies have the very significant potential to greatly affect the supply of aggregates to the Greater Toronto Area (GTA) in the late 1990s and even more so in the early part of the 21st Century. The State of the Resource Study indicated that, in 1990 the GTA consumed some 65 million tonnes of aggregate, almost 1/3 of the total provincial aggregate production. However, the amount produced within the GTA (primarily within the Niagara Escarpment and Oak Ridges Moraine) was only about 18% of the total Ontario production. The 37 million tonnes produced in the GTA accounted for only 60% of the total requirement for the GTA. This fact clearly highlights another of the major issues in Ontario aggregate production in the 1990s, that being supply and demand.

5 SUPPLY/DEMAND ISSUES

5.1 *Major market area needs*

The State of the Resource Study prepared by Planning Initiatives Ltd. for MNR concluded that within all of the major market areas of southern Ontario, only one market area (Kitchener/Waterloo/Cambridge/Guelph) produced enough aggregate to meet the local demand. In that case, not only was the area self-sufficient but it was also able to 'export' a considerable amount of aggregate, most notably to the western portion of the GTA. Should the Niagara Escarpment and Oak Ridges Moraine studies result in significant restrictions to future aggregate operations in those areas, the GTA would be severely affected by a lack of accessible aggregates within a short distance of the major market area. In that case, the demand for aggregates would extend farther into areas such as the Kitchener/Waterloo/Cambridge/Guelph areas. Not unexpectedly, many local residents in those outlying areas tend to resent the aggregate operations and the related traffic in their area when the material is to supply markets farther away such as the GTA.

The Planning Initiatives Ltd. study further indicated that the demand for aggregates in the GTA, by the year 2010, will increase by approximately 33% from the 1986 to 1990 period. The recent recession during the early 1990's may however, lengthen the time before the full increase is reached. Similar, but lower, increases in demand in the other market areas will also result in numerous new areas being required to be licensed in order to meet the local demand. Overall, the State of the Resource Study indicated a projected increase in demand for aggregates for all of Ontario of about 25% over the period of 1990 to 2010.

5.2 *Lengthy lead times to licence properties*

The lead times often required to purchase new aggregate lands, obtain the necessary approvals, and initiate operations in the different market areas range from approximately 3.5 years to as many as 8 years (PIL 1994). Environmental studies now being required, such as watershed and hydrogeological studies which, in themselves may take two years to complete, also influence the processing times for aggregate applications. As part of these types of studies, the question or issue of potential cumulative effects is also being analyzed. However, to date, no definitive study has been done on what, if any, cumulative effects have resulted from several operations in a given area. Because of such lengthy lead times, it becomes more critical that advance planning and ongoing monitoring of existing licensed reserves is undertaken. Steps are being taken now through various avenues to arrive at a balance and workable means to require the necessary studies, while streamlining the approvals process.

5.3 *Accessibility to aggregate resources*

Another major issue surfacing in the mid 1990s is that of sterilization of commercial quality aggregate resources, or accessibility to those resources. Investigations done for the State of the Resource Study indicated that anywhere from 15% to 85% of the deposits in a primary aggregate resource area could be sterilized because of various constraints to the resources (PIL 1994). These constraints are illustrated graphically in Figure 3, the 'Accessibility Constraints to the Aggregate Resource Model' developed in that study.

In a few cases, urban development has encroached immediately into the area adja-

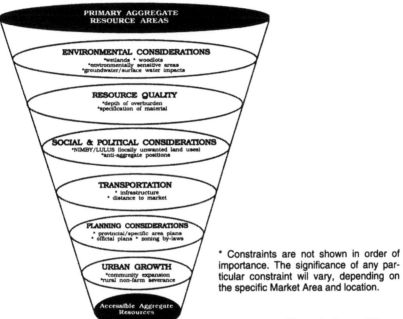

* Constraints are not shown in order of importance. The significance of any particular constraint wil vary, depending on the specific Market Area and location.

Figure 3. Accessibility constraints to the Aggregate Resource Model.

Figure 4. Residential development in London, Ontario, adjacent to gravel extraction.

cent to older existing pits as in Figure 4. In other cases, unlicensed properties in closest proximity to urban areas have become sterilized when urban development has spread over the resource prior to extraction occurring. In still other cases, the new CSPS will preclude extraction in Provincially Significant Wetlands and likely result in similar prohibitions in significant wooded areas, notwithstanding the fact the wetlands and wooded area are renewable resources as compared to the non-renewable aggregate resources.

6 MIDTERM REVIEW: THE STATE OF AFFAIRS IN 1996

The State of the Resource Study identified and analyzed the major issues facing aggregate resource operators and aggregate resource management, and concluded with a number of strategic initiatives for assisting in the future management of aggregate resources in the Province of Ontario. Those initiatives included the need to:
  – Identify provincially and/or regionally significant aggregate resource areas,
  – Prepare a comprehensive aggregate resource strategy for the Greater Toronto Area,
  – Prepare strategic plans for other market areas, geared to meet the specific market needs,
  – Provide for a minimum licensed supply to meet future demands for each market area, utilizing existing local resources where possible,
  – Prepare a joint Ministry of Natural Resources and Ministry of Municipal Affairs report to clarify and 'fine tune' the guidelines for the implementation of MARPS',
  – Formulate a 'model' process for official plan policies for aggregates,
  – Give priority consideration for new licence applications within critical market areas,
  – Create aggregate advisory committees (AAC) for aggregate rich areas,
  – Update and expand the coverage of aggregate resources mapping,
  – Monitor and update production, consumption, supply and reserve information.
  – Review long distance haulage opportunities from distant aggregate source areas, specifically from the Greater Toronto, including the identification of the potential for utilization of existing/abandoned railway lines, and increased use of waterways,
  – Implement initiatives with respect to conservation and wise management of aggregate resources,
  – Increase rehabilitation research and public education programs, including increasing awareness of the uses and need for the resource; encouraging aggressive rehabilitation programs for abandoned sites; formulation of technical guidelines and research to assist in achieving effective rehabilitation; promotion of rehabilitation projects; organization of aggregate workshops; and establishment of an environmental interpretation facility,
  – Encouraging aggregate extraction in urban areas prior to development to minimize loss of near market resources by:
  1. Identification and protection of the aggregate resources in major river corridor systems and floodplains, in or near urban areas,

2. Utilization of aggregate resources prior to development within residential plans of subdivision,

3. Priority utilization of aggregate resources in close proximity to the market areas rather than distant aggregate resource areas,

4. Continued efforts to refine the techniques of land rehabilitation,

5. Methods to facilitate accessibility to primary, secondary, and tertiary quality resources,

6. Increased attention to secondary resources within and near urban municipalities.

– Implement post-extractive land use opportunities which enhance environmental and/or social objectives including:

1. Creation or re-establishment of an 'ecologically diverse' environment,

2. Re-establishment of active, productive agricultural use,

3. Creation of recreational opportunities,

4. Rehabilitation to industrial/residential land uses in near urban or urban locations (PIL 1994).

The significance and importance of these and other issues to aggregate resource management in Ontario during the mid 1990s is reflected in the number of actions taken to address the various issues within two years of publication of the State of the Resource Study. Shortly after publication of the report, a workshop was convened by the MNR consisting of some 50 representatives from different stakeholders active in aggregate resource management. This included government and agency representatives, consultants, municipal officials and politicians, and others with expertise in specific fields. This workshop provided a forum to 'brainstorm' various issues, and to explore ways and means of addressing the issues facing those involved in aggregate resource management.

Over the next two years, a number of other initiatives by the MNR were also taken to advance the state of aggregate resource management in the Province. A Science and Technology Transfer Unit (STTU) was established to further the education and techniques on innovative and creative methods of rehabilitating aggregate lands.

During the same time, a number of aggregate advisory committees were established by representatives of the aggregate resource sections of different District Ministry offices. While the makeup of the different committees has varied, the overall objectives are to bring the different parties involved in aggregate resource management at the local level together in a forum other than the adversarial types of meetings and hearings frequently occurring as a result of site specific applications and issues. Generally these committees have been brought together to discuss aggregate resources in general, to educate each other regarding the different perspectives and interest groups, and to allow for a more open exchange of information between the parties. Early indications of the results of these committees are that these types of forums offer significant potential for gaining a better understanding of the concerns for each other. Communication between the parties on an open basis has been somewhat limited in the past and open communication and discussion may require some stimulation and time for it to be most beneficial.

In 1995, the Ministry also established an Aggregate Resources Working Group consisting of a broad range of stakeholders within aggregate resource management and other related environmental groups in the Province. It was brought together to evaluate the implementation of the Aggregate Resources Act in its five years since 1990, and to look at new strategic directions in light of some of the findings of the State of the Resource Study. The objectives and mandate of this working group are to undertake a comprehensive review of various stakeholder concerns, analyze and evaluate available options, and recommend new strategic directions to the Minister for necessary policy considerations (P. Umar personal communication).

With respect to the identification of major aggregate resource areas, many major aggregate resource areas in the Province had been previously identified and mapped by the Ministry of Northern Development and Mines (MNDM), Ontario Geological Surveys Branch during the 1980s. This mapping program had been undertaken on a township by township basis, with, among other things, areas of sand and gravel of primary and secondary significance being identified, and described in a series of documents entitled Aggregate Resource Inventory Papers. In 1995, a number of these initial inventory papers were updated and consolidated into larger scale documents covering the whole of individual counties and regions. This regional resource identification at the larger regional scale is in keeping with the broader ecosystem planning now taking place in the province.

7   FUTURE DIRECTIONS FOR THE NEXT CENTURY

7.1   *The need for a balanced approach to resource management*

In the fifteen year period from 1980 to 1995, the overall resource management 'pendulum' has swung from the relatively unrestrained and under-regulated days of the PQCA, to the highly environmentally conscious legislation of the 1995 CSPS passed by the Province of Ontario. Perhaps what is most needed for the next century, is a truly balanced approach to aggregate resource management. On the one hand, is a non-renewable finite resource which is in high demand for almost all facets of economic and infrastructure growth. At the same time, Ontario residents are among the highest per capita users of aggregates in North America and perhaps in the world. The utilization of and demand for the resource is very high. Recent studies in Ontario have indicated that, at best, current opportunities for recycling and substitution for aggregates can optimistically only hope to achieve savings of 3% to 5% of the total demand for aggregates (JEGEL 1992).

Although the total amount of aggregate resources within the Province is very large, accessibility to that resource is very restricted in many cases, often with restrictions on as much as 85% and 90% of the total resource. Those locations where the greatest amount of resource is frequently found include areas in and around glacial outwash channels now often occupied by major watercourses and their associated floodplains. Other major locations include those areas with hilly or rugged terrain created by numerous geological and geomorphological features which now lie under significant woodlot, wetland, or other environmentally sensitive areas.

This tendency of an overlapping location of the non-renewable aggregate resource with renewable, but sensitive, environmental and other natural or physiographic resources frequently creates conflict between management strategies for the two different resources. While such features as woodlands and wetlands are technically renewable resources, sensitivity to and of these resources is all too frequently seen as being a significant restriction to management and use of the non-renewable aggregate resources. Thus, a balanced approach is critical to the management of aggregate resources.

The preparation of integrated resource management strategies can utilize procedures and strategies in one resource area to benefit resource management in the other resource areas. In a recent study by this writer, an aggregate extraction program within a man-made recreational lake was formulated, with a rehabilitation program which would increase water storage capacity, create additional fish habitat and promote diversity for the recreational fisheries resource in the lake. Revenues received from the gravel extraction would be used to finance the fisheries and water resource improvements (PIL 1996).

7.2 *Accessibility to the resource is critical*

Accessibility to the resource must be maintained, especially in areas where demand is high, as transportation already accounts for at least 50% of the cost of aggregates. Significantly longer haulage distances for aggregates would undoubtedly have a considerable impact on the cost of construction of new infrastructure and new housing. At the same time, extraction of the aggregate resource must be designed to be certain that impacts on the environment are minimal.

7.3 *Comprehensive regional resource strategies required*

If a balanced approach to management of the various resources is to be achieved, a thorough understanding of the availability and location of various resources, aggregate and otherwise, is required. To this end, the preparation of comprehensive regional resource strategies is crucial to gaining an understanding of the importance of various types of resources. Identification of the location of aggregate resources within any given region is critical to assessing the importance of those resources compared to environmental resources in determining the proper balance for that particular region. Several steps have recently begun with the commencement of aggregate resource mapping by the relevant provincial ministry at the county and regional scale. Identification of the resources on a larger regional basis will be the first step toward assessing the regional importance of various deposits.

Recent legislation has begun to establish the importance of 'significant woodlots, wetlands, water courses, and natural heritage areas'. However, much more work must be done to quantify the relative importance and to establish criteria for determining the proper balance between the two types of resources.

Within Ontario, undoubtedly the area of greatest importance for establishing a regional resource management strategy is for the Greater Toronto Area, where as much as one third of the Provincial aggregates are consumed, but where only one sixth are actually extracted. This will also undoubtedly form one of the basic questions to be addressed by the recently appointed Aggregate Resources Working Group.

7.4 *Improve public awareness and understanding*

In order to achieve a balanced approach to aggregate resource management, there must be a greater public awareness and understanding of the importance of and need for aggregate resources, and the complexity of this industry. The State of the Resource Study recommended educational programs aimed at making groups aware of the uses and the need for the resource. While most people are generally aware of the numerous uses for aggregate products, most of the public are totally unaware of the full amount of aggregate required in houses, local construction, and building roadways. They are similarly unaware that at least half the cost of material is a result of transportation costs, and the potential financial implication of hauling that material for significantly longer distances.

Environmental awareness has increased considerably over the last decade, and has often been fueled by a lack of environmental sensitivity by some operators. Regulations with the Aggregate Resources Act have improved the degree of environmental sensitivity of aggregate operations, but much remains to be done. In particular, the industry as a whole must become more outspoken and more pro-active in demonstrating the improvements which have taken place in their industry. Individual operators can and will have to do more to demonstrate to their neighbours how their operations are environmentally conscious and sensitive.

Various initiatives are already being taken with the formation of Aggregate Advisory Committees and the Aggregate Resources Working Group as a first step in addressing this need for 'mutual understanding' between the parties. However, a lack of understanding, historical feelings of mistrust, and lack of communication between the parties will make this a slow process. Nonetheless, the early results of such initiatives do offer promising prospects.

After completion of the State of the Resource Study, a one day workshop brought together some 140 different individuals and agencies from across the Province with significantly different perspectives. Comments and observations from numerous people indicated this was the first time they had the opportunity to attend such a forum. Observations were that the coming together of the stakeholders with different perspectives without having site specific 'policies or positions' to adhere to, allowed for open discussion of many topics, and an open mind in considering other perspectives on aggregate resource management.

7.5 *More research required*

Looking back over the last six years since the Aggregate Resources Act was brought into effect, one can now see tangible and visible signs of refinements and improvements to management of the resource. New legislation has been introduced, and the requirement for progressive rehabilitation is beginning to show visible, positive results. However, because sand and gravel operations are normally formulated and planned in a minimum 15-25 year operational time frame, and quarries in a 25-75 year time frame, it is often difficult for the public to visualize and relate to end results in such a time frame. In the past, the public has unfortunately not seen the kinds of results that have now come to be expected. With the requirements that aggregate extraction is totally restricted or severely constrained in areas of wetlands, woodlands, surface and groundwater areas, and sensitive environments, significantly more field research is required. Such work will ensure that the resource can be managed to maintain access to the greatest amount of resource without detriment to the surrounding environment. This is best accomplished through innovative and creative pilot projects. Close monitoring of such projects and the widespread dissemination of the positive results, and failures will continue to expand the boundaries of first hand knowledge and projects that can achieve maximum utilization of aggregate resources.

A midterm review of the state of aggregate resource management in Ontario during the 1990s indicates that, with implementation of a new Aggregate Resources Act, and establishment and completion of a number of specific initiatives during the first half of the decade, considerable improvement in the overall resource management has occurred. The integrated resource management approach used in Ontario has focused on three main topic areas including:

1. Resource conservation,
2. Industry regulation, and
3. Planning for aggregates.

As we approach the turn of the century, the need and challenge will be to ensure that the integrated resource management can be refined to balance the need for aggregate resources, in a manner which will allow for the economical extraction of the resource, but which will still provide the maximum protection for the natural and social environment within which it takes place.

REFERENCES

Aggregate Producers Association of Ontario 1991. *Did You Know?* Aggregate Industry fact sheet.
Government of Ontario 1986. *Mineral Aggergate Resource Policy Statement.* Order in Council No. 1249-86, May. pp. 3.
Government of Ontario 1988. *Planning Act,1983.* Chapter 1, March, Queen's Printer, Section 34(1).
Government of Ontario 1990. *Aggergate Resources Act, 1989.* Chapter 23, Queen's Printer. April.
John Emery Geotechnical Engineering Limited 1992. *Mineral Aggregate Conservation Reuse and Recycling.* Ministry of Natural Resources. February, 38pp.
Ministry of the Environment and Energy 1990. *Stopping Water Pollution At its Source: MISA, The development Document for the Effluent Monitoring Regulation for the Industrial Minerals Sector*, Queen's Printer.
Ministry of Natural Resources 1995. *Statistical Overview, 1993,* Resource Stewardship and Development Branch, January.
Ontario Ministry of Municipal Affairs 1996. Comprehensive Set of Policy Statements. Queen's Printer, May, 1994. pp. 2, 14
Planning Initiatives Ltd 1994. Ministry of Natural Resources, *Aggregate Resources of Southern Ontario: A State of the Resource Study.* pages iv, 3-1, 3-3, 5-4, 5-13, 5-30, 10-6, 10-7.
Planning Initiatives Ltd 1996. Grand River Conservation Authority, unpublished report, April. 51 pp.
Sewell, Penfold & Vigod 1993. Commission on Planning and Development Reform in Ontario. *New Planning for Ontario: Final Report of the Commission on Planning and Development Reform in Ontario.* Queen's Printer, June.
Umar Pervez 1995. Ministry of Natural Resources. Personal Communication. October 27, 1995.

The geological conservation of glaciofluvial sand and gravel resources in Northern Ireland: An integrated approach using natural areas

JASPER KNIGHT

School of Environmental Studies, University of Ulster, Northern Ireland, UK

1 INTRODUCTION

Conservation is a technique of landscape management aimed at sustaining landscape resources and retaining their morphological, cultural and esthetic integrity (Kontturi & Lyytikainen 1985, O'Halloran et al. 1994). Geology and geomorphology form the framework or regional setting within which landscapes evolve (Duff 1994). Geology and geomorphology temper the nature of topography, the distribution of soils and vegetation, and influence the agriculture, water resources, cultural history, archaeology and prosperity of the landscape, and the pattern of human activity that takes place upon it (Cooke & Doornkamp 1974, Kontturi 1984, Kontturi & Lyytikainen 1985, Duff 1994) (Fig. 1). Conservation of geology/geomorphology in the UK is specified in the designation of local sites (few to 10 km^2-scale) of high scientific value, and is implied in the designation of regional areas (100-1000 km^2-scale) of high landscape or scenic value. However, neither of these scales of designation adequately reflect the breadth of landscape diversity, or landscape 'character'. This is because they are not seen as complementary, to involve more than one landscape feature, or to invoke humanistic landscape attributes such as beauty or 'sense-of-place' (Leopold 1969, Meinig 1979). In previously-glaciated regions, the link between geology/geomorphology and landscape is especially profound because glacial processes leave a strong erosional and depositional signature which influences the way in which the landscape evolves after deglaciation (Kontturi 1984, Kontturi & Lyytikainen 1985, Warren et al. 1985). Glaciofluvial (sand and gravel) landforms are important as an economic resource for sand and gravel extraction. Glaciofluvial landscape assemblages are also important as a landscape resource contributing to landscape diversity and 'character', and may be associated with ecological, cultural, esthetic and other landscape attributes (Cross 1992, Gordon & Campbell 1992, McCabe 1994).

1.1 *Scope of study and definition of terms*

This paper is concerned with the conservation of glaciofluvial landscape resources in Northern Ireland. This paper has three main aims. First, to discuss the attributes and role of landscape 'character'. Second, to assess the role of glaciofluvial sediments

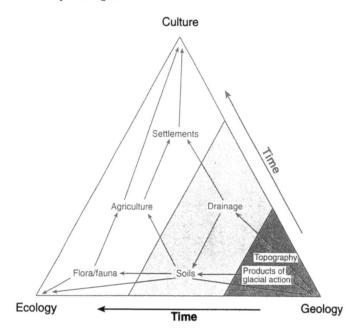

Figure 1. Ternary dia-
gram showing the de-
velopment over time of
landscape character.

and landforms as a resource and their impact in the landscape. Third, to construct 'natural areas' for the Northern Ireland landscape which describe: 1. Those areas distinct in morphology and other landscape attributes from adjacent areas, and 2. Landscape 'character'.

Landscapes are made up of a number of different *components*, such as landforms, each of which have particular *attributes*, such as their dimensions, their relationship to other components, and their perceived value as a resource. The presence or abundance of other landscape features, such as ecological, cultural, archaeological and esthetic features, is also a landscape attribute. Attributes can be split into two groups: 1. Physical or tangible, and 2. Non-physical or esthetic. Physical attributes include landform size and shape, landscape area and topographic characteristics. These attributes can be recorded objectively, i.e. their presence/absence in the landscape, and this record need not vary between observers. Non-physical attributes include landform grandeur, and landscape beauty, quality and openness. These attributes are observer-dependent, and their recording and evaluation may both vary considerably between individuals or landscape user groups. Physical and non-physical attributes both contribute to landscape 'character'.

2 LANDSCAPE CHARACTER

Human activities, carried out on the earth's surface, alter its attributes, such as its shape, vegetation, processes and the way it is perceived to function. These attributes in turn shape the nature of human activities, perceptions, cultures and identities (Meinig 1979, Drdos 1983, Zube 1987). Non-physical landscape attributes contributing to landscape character include 'sense-of-place', landscape greenness, grandeur and beauty (Leopold 1969, Calvin et al. 1972, Meinig 1979, Zube 1987). Physical landscape attributes include features of landscape cultural history such as archaeological sites, churches, farming and settlement patterns, and landscape design features such as field boundaries and open spaces (Warnock & Evans 1992, Duff 1994). Regional landscapes are important in promoting a sense of shared culture and identity. Local landscapes are important in promoting 'sense-of-place' and personal belonging. Therefore landscapes mean different things to different individuals or user groups depending on the scale on which they are observed (Fig. 2). Preferences for particular landscapes or features can be quantified, evaluated or compared on the basis of their attributes and characteristics (Leopold 1969, Garrod & Willis 1994). Calvin et al. (1972) found that 'natural scenic beauty' accounted for 62% of the variation between preferences for different landscapes. Landscape 'character' describes the sum of attributes that distinguish landscapes from one another. Some attributes, such as landscape morphology and 'greenness' contribute more to landscape 'character' than other attributes, such as agricultural patterns (cf. Calvin et al. 1972). Therefore conserving the principle attributes underlying the landscape can help conserve or enhance landscape character (Warnock & Evans 1992, Duff 1994, Countryside Commission 1995, Somper 1995).

In Britain, regions assigned a particular 'character' include the Lake District, the Cotswolds, and the Breckland area of East Anglia (cf. Warnock & Evans 1992, Duff 1994). In order to conserve or enhance regional landscape character in Britain, the Countryside Commission (1995, Somper 1995) is aiming to encourage landscape diversity, multipurpose land usage, sustainable rural development and the coordination of landscape policies with volunteer bodies and the public. The Countryside Commission use 14 variables to assess landscape 'character' (Somper 1995) including archaeology, cultural history, agriculture, ecology, landforms and geology/soils. Use of these variables means that decisions on landscape conservation and planning are founded on a wide range of landscape attributes. The decision-making process of landscape management (Hill & Tzamir 1972, Idle 1995) is therefore likely to be better informed, applicable to landscapes on a region-by-region basis, and to be pragmatic (Warnock & Evans 1992). This also means that particular landscape attributes, and particular landscapes as a whole, can be prioritized over others (Hill & Tzamir 1972, Zube 1987).

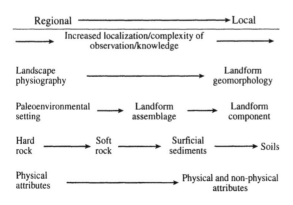

Figure 2. Changing continua of landscape characteristics with changing scale of observation.

3 PROBLEMS OF LANDSCAPE CONSERVATION

Two main problems to be addressed in the conservation of glaciofluvial landscapes are: 1. How information on landscape character should be collated and evaluated, and 2. How this information can be used to subdivide landscapes on the basis of their character in order to maximize conservation benefit.

3.1   *Landscape evaluation*

Landscape character is observed by different user groups and on different spatial scales. In order for landscapes in different areas or of different scales to be compared to one another, the same criteria of observation have to be used (Table 1). Observations listing the presence/absence or dimensions of particular landscape components and attributes may be objective and value-free, and act as a database for a range of applications including economic and environmental planning, conservation and agriculture. Observations are evaluated according to the purpose or user group for which the evaluation is required, but should be based on both landscape components and their attributes, i.e. holistic in nature (Leopold 1969, Zube 1987) (Table 1). Leopold (1969) used 46 criteria to assess the character of fluvial landscapes. Calvin et al. (1972) quantified responses from keyword descriptors to examine landscape perception. 'Willingness to pay' for the maintenance of landscape components (Garrod & Willis 1994) also means that they can be assigned relative or absolute values.

Table 1. Landscape attributes considered in the assessment of landscape 'character'.

Environment type	Landscape attribute/characteristic	Example
Physical: Natural environment	Existing natural boundaries	Rivers, mountains
	Landscape configuration	Relief, topography, slope, aspect, elevation
	Geological structure	Faults, igneous intrusions
	Landform distribution	Eskers, drumlin fields
	Resource distribution	Hard rock and drift lithologies
	Agriculture	Landuse, intensity of production
	Ecology	Vegetation cover type, (rare) plant distribution
	Climate	
Built environment	Existing landscape boundaries	Urban areas, administrative areas
	Existing landscape protection/ conservation schemes	Sites of Special Scientific Interest (SSSIs), Areas of Outstanding Natural Beauty (AONBs)
	Archaeological features	Standing stones, burial mounds
	Historical/cultural features	Estates
	Architectural features	Bridges churches, other buildings
Non-physical: Humanistic environment	Social	Ethnic groups, traditions
	Esthetic/cultural features	Landscape/landform intactness, vistas, visual dominance, physical diversity, beauty

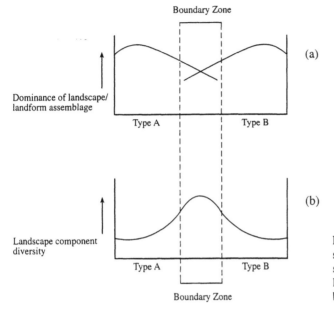

Figure 3. Sketches of the schematic change in (a) landscape dominance, and (b) landscape diversity across a boundary zone.

3.2 *Landscape subdivision*

Observation establishes that landscape components and attributes are non-uniformly distributed. Evaluation establishes that these elements also mean different things to different user groups. Subdivision is the process of breaking a landscape down into smaller units so as to bring order or rationality to the distribution of their components and attributes. Subdivision constructs boundaries between adjacent areas of different landform distribution or landscape character (Holland et al. 1991). Because 'character' is related to the strength of the landscape geomorphic 'signature', a stronger character is recognized in the center of distribution of any landscape type than in the boundary zone separating types (Fig. 3). The boundary zone is therefore the most sensitive part of the landscape system to morphological change (Holland et al. 1991) because it determines whether areas are included in one landscape subdivision or another.

4 BACKGROUND TO LANDSCAPE AND GEOLOGICAL CONSERVATION IN THE BRITISH ISLES

Landscape and geological conservation in the British Isles became important in the late 1970s (Gordon 1987, 1994) due to the perceived fear of a world-wide resource shortage particularly in non-renewable resources (Hodges 1995). In Britain, the Geological Conservation Review (GCR) used a standardized methodology of site assessment for geological conservation purposes. Twenty geological-based variables were used (Gordon 1994), but other landscape attributes at geological sites, such as ecology and archaeology, were not considered. The National Parks and Access to the

Countryside Act (1949) and the Wildlife and Countryside Act (1981) gave powers to the Nature Conservancy Council (NCC) to establish National Nature Reserves (NNRs) and Sites of Special Scientific Interest (SSSIs) covering specific high-quality geological and ecological sites (NCC 1990). Larger regions of high overall landscape or scenic value are also protected as Areas of Outstanding Natural Beauty (AONBs) (Gordon & Campbell 1992). By September 1994 NNRs totaled 157 and SSSIs 3811 in Britain (Idle 1995). This site network is supplemented by non-statutory Regionally Important Geological/Geomorphological Sites (RIGS) (Harley 1994). These are evaluated, selected and monitored by regional or county geological societies.

Landscape conservation in Northern Ireland is carried out by central government through the Department of the Environment (Northern Ireland) which administers the designation and management of AONBs and Areas of Special Scientific Interest (ASSIs) (Wilcock 1995). AONB designations totaled 11 and ASSIs 45 in 1995 (ibid.). Wilcock (1995) argued that statutory conservation schemes directed by central government should be integrated with local schemes organized by volunteer groups and district councils. There are no existing RIGS schemes in Northern Ireland. Furthermore, Wilcock (1995) identified finance as the main factor controlling the development of conservation policy in Northern Ireland.

4.1 *Problems of landscape conservation in the British Isles*

Conservation practices in the British Isles are split between small areas of perceived high scientific value (i.e. SSSIs/ASSIs), and larger areas of perceived high scenic or landscape value (i.e. AONBs) (Gordon & Campbell 1992, Gordon 1994, Wilcock 1995). These differences in spatial scale and conservation intent (Fig. 2) have three main problems. Problems and their possible solutions are: 1. Policy co-ordination is needed to ensure that large and small sites, and statutory and non-statutory sites, can manage landscape resources effectively (cf. Wilcock 1995), 2. Conservation schemes covering different landscape attributes, such as geological, ecological and archaeological features, need to be compatible so that maximum landscape benefit can be derived more efficiently (Gordon & Campbell 1992, Wilcock 1995). The association of different landscape attributes with one another, such as raths (prehistoric Irish settlement/religious sites) and native broadleaf woodland on Irish eskers (Warren et al. 1985, Cross 1992, McCabe 1994) shows that high-value sites often cover several landscape attributes, and 3. Description of landscape character needs to be implicit in landscape conservation schemes. 'Character', the product of a range of landscape attributes, can describe the meaning and significance of landscapes in ways that other designations cannot (Warnock & Evans 1992) because character specifies the importance of non-physical (humanistic) landscape attributes (Table 1). This means that the spatial units describing landscape character have to be: 1. Different to those used in other landscape designations, and 2. Capable of describing landscapes on different scales. For these reasons, a hierarchy of 'natural areas' is proposed as the best means to: 1. Describe landscape character, 2. Describe the relationship of one landscape to another, and 3. Divide up a landscape for conservation purposes.

5 NATURAL AREAS

The concept of 'natural areas' grew from the perceived need to conserve wilderness landscapes, such as National Parks in the USA, 'as bench marks for assessing the extent of man's impact' (Moir 1972, p.397). Minimum disturbance was an important characteristic of natural areas (Moir 1972). Natural areas are also being used at the present time by English Nature as the basis for landscape and geological conservation (Duff 1994). English Nature define natural areas as 'tracts of land unified by their underlying geology, landforms and soils, displaying characteristic vegetation types and wildlife species, and supporting broadly similar land uses and settlement patterns' (Duff 1994, p.9). English Nature divide England into 76 areas on the basis of these criteria. The definition of natural areas therefore appears to have changed in meaning over time from 'natural' (cf. Moir 1972) to 'the rural landscape within which we live' (cf. Duff 1994). This implies that the conservation of landscape character is also related to the nature of human activity in the landscape. Conserving landscape character promotes and is promoted by sustainable or traditional rural industries, and planning controls on unsustainable or character-changing processes or activities (Countryside Commission 1995, Somper 1995). Natural areas can therefore be used as monitors of environmental change and the efficacy of landscape management (Moir 1972, Tjallingii 1974, Gehlbach 1975, Duff 1994). Landscape change on different spatial scales can be evaluated through landscape hierarchies. Hierarchies require that the components and attributes of regional-scale landscapes are held in common by the local-scale landscapes of which they are comprised, but that each local landscape has a unique overall 'signature' that differentiates it from adjacent landscapes of the same hierarchical level (cf. Cooke & Doornkamp 1974, Tjallingii 1974, Urban et al. 1987). This means that within the context of the hierarchical level of observation (i.e. whether landscapes are regional or local), all natural areas may be considered as 'homogeneous'. Landscapes with a shared geomorphic signature, such as previously-glaciated areas, can be connected through landscape hierarchies because they were once linked to regional erosional and depositional systems (cf. Demek 1978). Local landscapes can be described by natural area hierarchies as these relict regional landscapes are broken down by processes of present-day landscape change. The concept of 'geotopes' in the European literature (Wiedenbein 1994, Erikstad et al. 1995) represents the smallest spatial scale at which landscape homogeneity is maintained (Sturm 1994, Wiedenbein 1994). Therefore the smallest landscape area recognized by natural area criteria (the *subunit*, discussed later) is equivalent to the geotope.

6 THE NATURE OF GLACIOFLUVIAL LANDSCAPES

Previously-glaciated landscapes are particularly sensitive to environmental change (Kontturi 1984) because their components are no longer accumulating. They also represent a valuable non-renewable aggregate and landscape resource to be exploited by human activity (Kontturi 1984, Kontturi & Lyytikainen 1985). Glaciofluvial landscapes have distinctive sedimentary, morphological and esthetic/cultural attributes that differentiate them from other landscape types (Kontturi & Lyytikainen 1985,

McCabe 1994). Also, soils developed on glaciofluvial sediments give rise to ecological assemblages different to those formed on other materials (Kontturi & Lyytikainen 1985, Warren et al. 1985, Cross 1992). The character of glaciofluvial landscapes in Ireland is intimately associated with human activity and cultural development since the last deglaciation (Warren et al. 1985, Cross 1992, McCabe 1994) (Fig. 1). Therefore the conservation of glaciofluvial landscapes has special requirements (Kontturi & Lyytikainen 1985) that might not be met by the broad brush of existing landscape conservation schemes.

7 GLACIOFLUVIAL LANDFORMS AND LANDSCAPES IN NORTHERN IRELAND

Almost all the north of Ireland was ice-covered during the late Weichselian (Midlandian) glacial event of 25,000-15,000 yrs BP (McCabe 1985, 1987, 1993). Ice flowed from linked domes or divides situated over Donegal, the western Sperrin

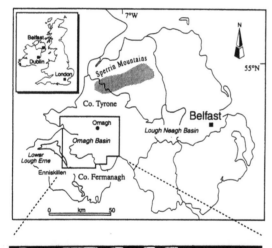

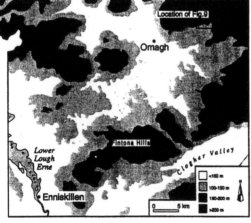

Figure 4. Map of Northern Ireland showing the location and topography of the Omagh Basin and the location of Figure 9.

Mountains and the Lough Neagh Basin, dominantly in a north or northeast, and south or southwest direction (McCabe 1985) (see Fig. 4 for locations). Areas of ice-marginal glaciofluvial sediments deposited in the subsequent deglacial cycle of 17,000-15,000 yrs BP (McCabe 1985, 1993) can be split into: 1. Coastal, and 2. Inland systems.

1. *Coastal systems* were influenced by downdraw of the ice margin into the marine environment, and changes in relative sea level, giving glaciomarine deltas and subaqueous morainal banks (McCabe 1987, 1993). These features have discrete morphologies which correspond to discrete depositional systems and events, and are difficult to correlate with inland depositional events or each other. However, these features often have a high scientific and esthetic value due to sedimentary exposure, often by coastal erosion, and occupy topographically-dominant positions due to glacioisostatic rebound (McCabe 1994).

2. *Inland systems*, characterized by glaciolacustrine deltas, eskers and moraines (Dardis 1982, 1986, McCabe 1993), were formed during ice marginal stagnation and retreat into lowland ice centers. These features, overlying a drumlinized landscape, can be broadly correlated to give relative-age ice margin positions during deglaciation (Dardis 1982, 1986).

The glacial geomorphology of the Omagh Basin, southwestern Northern Ireland (Figs 5 and 6) reflects the regional setting of inland depositional systems. Glaciofluvial sediments in the Omagh Basin (Fig. 6) were deposited in a variety of subaqueous and ice-contact depositional settings (Dardis 1980, 1982). There are three main depositional systems associated with these different settings (Fig. 6, see Fig. 4 for locations): 1. In the Clogher Valley, topographically-controlled esker-delta assemblages record a north and northeastward direction of ice retreat, 2. Recessional kame-deltas

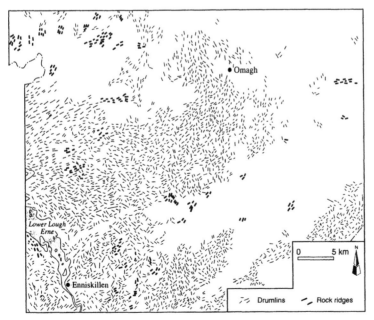

Figure 5. Distribution of streamlined subglacial bedforms in the Omagh Basin.

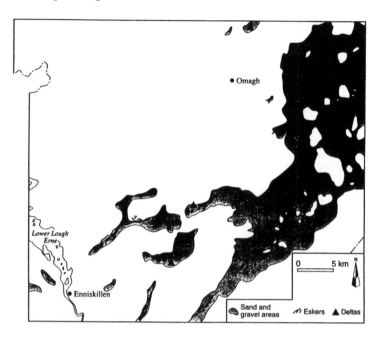

Figure 6. Distribution of glaciofluvial sand and gravel in the Omagh Basin.

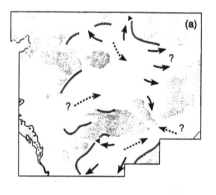

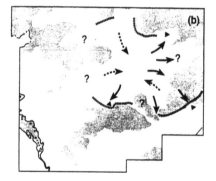

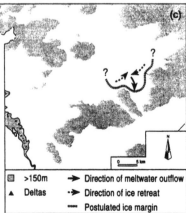

Figure 7. Schematic chronology [(a) is the earliest and (c) the latest] of the last (late Weichselian) deglaciation of the Omagh Basin.

were deposited onto the north side of the Fintona Hills as ice retreated into the central Omagh Basin, and 3. Cross-valley moraines and outwash complexes to the east of the Omagh Basin record stages of ice marginal retreat and interlobate activity of the Omagh Basin and Lough Neagh Basin ice lobes towards the west and east respectively (Dardis 1980, 1986).

Interpretations of landform development can be used to construct a deglacial chronology of the Omagh Basin (Fig. 7). After ice mass extension associated with drumlinization, ice downwasted and the Fintona Hills emerged as nunataks (Fig. 7a). This change in configuration altered the ice mass balance, steepened the long profile and caused active backwasting to dominate. High level deltas were built up against the bedrock uplands. Marginal retreat continued (Fig. 7b) and ice was confined to the north side of the Fintona Hills by withdrawal between interfluves. The lack of ice-marginal evidence to the west of the Omagh Basin, and active moraine emplacement to the east between the Omagh and Lough Neagh Basin ice lobes (Dardis 1980, 1982) indicate an asymmetric ice sheet profile. Final ice mass disintegration (Fig. 7c) took place in the area south of Omagh, with ice activity maintained until a late stage.

8 NATURAL AREAS APPLIED TO NORTHERN IRELAND

The distribution of glacigenic landforms in the Omagh Basin (Figs 5 and 6), reconstruction of their former relationships (Fig. 7), and evaluation of landscape attributes (Table 1) can be used to assemble a hierarchy of natural areas reflecting landscape character. Also because Northern Ireland shows a strong geomorphic signature of the last glacial event (McCabe 1987), it is appropriate to construct natural areas for the wider landscape (Fig. 8) of which the Omagh Basin is part.

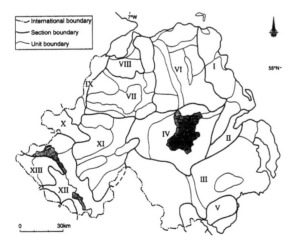

Figure 8. Map showing the sbudivision of the Northern Ireland landscape into a hierarchy of glacigenic natural areas (Sections I-XIII). I. Antrim Plateau, II. Lagan Valley, III. South Ulster Lowlands, IV. Lough Neagh Basin, V. Mourne Mountains, VI. Antrim Lowlands, VII. Sperrin Highland Complex, VIII. Limavady Basin, IX. Foyle Valley, X. East Donegal Uplands, XI. Omagh Basin, XII, Lough Erne Lowlands, XIII. Leitrim Hills.

Northern Ireland is divided into three spatial scales of natural areas on the basis of landform distribution, landscape attributes and other factors (Table 1). *Sections* (total 13) broadly correspond to physiographic regions based mainly on solid-rock geology and topography. Sections are split into *units* (total 55) which represent particular sediment systems or landform assemblages (Fig. 8). Units are split into *subunits*, such as individual eskers or deltas. Subunits are similar to the 13 types of sand and gravel landscape morphologies recognized in Northern Ireland by McCabe (1994). Subunits are the smallest spatial scale recognized by natural areas criteria (cf. 'geotopes'). As such, subunits can be used to monitor or evaluate the rate of landscape change (Gehlbach 1975). A case study of the Murrinmaguiggan area, 12 km northeast of Omagh (Fig. 4 for location) demonstrates how these concepts can be applied on the ground.

9   THE MURRINMAGUIGGAN MORAINE-OUTWASH COMPLEX

9.1 *Description*

The area is composed of an assemblage of ice-marginal moraines, discontinuous esker fragments and kettled outwash sediments (Dardis 1980, 1982) (Fig. 9). The main outwash surface, at 180-190 m a.s.l. was deposited during retreat of the Omagh Basin ice mass towards the south and southwest, causing a fall in proglacial lake levels (Dardis 1982). Ice retreat was associated with moraine emplacement between bedrock uplands. Sedimentological data are available at a number of locations (Fig. 9). These data show facies characteristic of highly variable braided proglacial deposi-

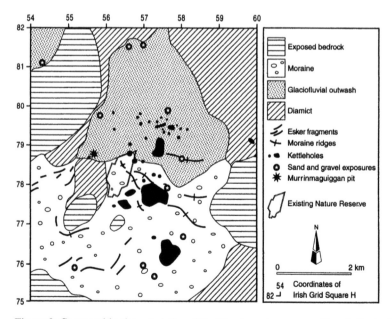

Figure 9. Geomorphic characteristics of the Murrigaguiggan area, Omagh Basin, Northern Ireland.

tional environments and include planar, cross-bedded and channelized sands and gravels, rippled sands, and massive sands, gravels and silts, typically in fining-up sequences. These facies are often overlain by gravely mass flow diamictons. Ice melt-out has resulted in large kettleholes up to 550 m in diameter. Localized meltwater erosion also dissected moraine ridges. Postglacial peat then developed on both out-wash and moraine surfaces to a depth of 2-3 m.

9.2 *Landscape conservation issues*

The Murrins Forest Nature Reserve, designated in 1975 on the basis of broad ecological and geomorphic diversity, covers 54 ha of the Murrinmaguiggan area (Fig. 9). This area includes two water-filled kettleholes, and has a variety of topographic and geological attributes. Overall, the area is of high scenic value with intact peat supporting a variety of ecosystems and habitat types. An increase in physical diversity at the frontal margin of the morainic system gives visual focus and landscape dominance (cf. Leopold 1969). Morainic segments are up to 30 m higher than the surrounding landscape and can be viewed from up to 8 km away.

Landscape pressures at the present time are mainly those generated by human activity. Sand and gravel extraction is concentrated in the Murrinmaguiggan pit (Fig. 9). This pit is up to 800 m long, covers c. 25 ha, and has encroached almost to the Nature Reserve boundary. Furthermore, a number of trial excavations have recently (1994-5) been made in the intact frontal moraines to the east of this pit. Despoilment and landscape degradation may result from future extraction at these sites. Other landscape pressures include intensification of agriculture, resulting from field enlargement, hedgerow removal and slope gradient reduction. Also, kettleholes are used as reservoirs for domestic water usage. Water extraction reduces watertable levels in the surrounding peatlands. Therefore there may be future conflicts of interest between the needs and requirements of different landscape user groups, including conservationists and aggregate producers.

A natural areas approach identifies the outwash and morainic systems (Fig. 9) as two separate subunits within the Fintona-Sixmilecross sands and gravels unit of natural area section XI (Omagh Basin) (Fig. 8). This subdivision is supported by the relative uniformity of topography, sedimentology and landscape characteristics within each subunit, i.e. the integrity of the flat, kettled outwash surface, and the morainic area with larger kettleholes, ridges, and isolated dissected hills. A natural areas approach could be used to summarize the essential characteristics of individual subunits, and to promote conservation at the subunit level. Also, in recognizing that kettlehole drainage has an adverse effect on peatland ecosystems, natural areas can help co-ordinate the policies or needs of individual landscape users so as to manage landscape resources more effectively.

10 DISCUSSION

'Landscape resources' is an umbrella term for a wide variety of components within the landscape system that are subject to human usage. These include esthetic, cultural, archaeological, hydrological and geological landscape attributes that contribute to the holistic nature of landscape 'character'.

Landscape resource protection and management/planning are end members of the continuum of human usage of the landscape (cf. Hill & Tzamir 1972). Resource protection results from a policy to 'minimize resource scarcity'. This usually involves setting up strong landscape boundaries (cf. Fig. 3) to safeguard a discrete area, yet these boundaries also isolate this area from its regional context. This approach is often taken when particular landscape attributes are restricted to relict niches, such as in the distribution of rare plant species or habitat types. Management/planning or conservation approaches result from a policy to 'maximize resource abundance'. This approach is often taken when boundaries between landscape areas are transitional, and is important in landscapes of high overall value where individual landforms would otherwise be difficult to single out. The optimal method of conservation is difficult to identify in landscapes with diverse attributes and different user groups. Therefore landscape or geological conservation is most an issue at the interface between 'natural' and 'human' systems, i.e. where natural landscapes come under human pressure.

Natural area concepts can prioritize particular landscape over others by identifying which components and attributes contribute most to landscape character. The frontal moraines at Murrinmaguiggan are important to local and regional character. A suitable conservation strategy in this area would be to extend the existing Nature Reserve across the moraine ridges. Alternatively, planning controls could be tightened about those landforms which are particularly sensitive to landscape change and the degradation of environmental quality. Planning controls could also be used to encourage sand and gravel extraction elsewhere in the outwash system, such as farther to the north (Fig. 9) where a number of pits have already been excavated and successfully restored by backfilling. This procedure of natural area assessment emphasizes that integrated landscape resource planning requires pragmatic and adaptable policies to be applied within and between landscape subdivisions on different scales so that the landscape and its 'character' can be conserved.

11 CONCLUSION

Landscape conservation practices, as a management process, are needed to resolve conflicts between landscape user groups and to sustain landscape resources including its character. Natural areas describing the components and attributes that make up landscape character can be used to monitor local environment change (cf. Moir 1972). Policies and techniques required to manage local natural areas can be applied to more intricate regional landscapes. The example of the Murrinmaguiggan moraine-outwash complex shows that local natural areas can describe both physical and non-physical landscape attributes, and can be related in a landscape hierarchy to regional landscapes.

Conservation practices in Britain at the present time appear to be centered on 'unique' areas of high perceived scientific value such as SSSIs (Gordon & Campbell 1992, Gordon 1994). Conservation of these areas should aim to preserve landscape components and attributes. The wider landscape should also be conserved if it is perceived to act as a functioning, integrated system. Conservation of regional landscapes of a lower scientific value but of high scenic or landscape value is therefore also necessary, i.e. AONBs. In Britain, the Countryside Commission (1995, p.9) is preparing 'a new map of England to reflect... landscape character'. This may indicate a progression towards more integrated landscape conservation policies, practices and planning methods as have been initiated elsewhere in Europe (Kontturi 1984, Kontturi & Lyytikainen 1985, Sturm 1994, Erikstad et al. 1995).

ACKNOWLEDGEMENTS

Marshall McCabe is thanked for commenting on a draft of this paper. An anonymous reviewer and Martin Sharp are thanked for their constructive criticism which improved clarity. Mark Millar and Kilian McDaid drew the diagrams, and information on the Murrins was provided by Ian Enlander (DoE). This work is supported by a DENI CAST award with the Department of the Environment (N.I.).

REFERENCES

Calvin, J.S. Dearing, J.A. & Curtin, M.E. 1972. An attempt at assessing preferences for natural landscapes. *Environ. Behavior* 4: 447-470.
Cooke, R.U. & Doornkamp, J.C. 1974. *Geomorphology in Environmental Management.* Clarendon: Oxford.
Countryside Commission. 1995. *Quality of countryside: Quality of life.* CCP 470: Cheltenham.
Cross, J.R. 1992. The distribution, character and conservation of woodlands on esker ridges in Ireland. *Proc. Royal Irish Acad.* 92B: 1-19.
Dardis, G.F. 1980. The Quaternary sediments of central Ulster. In K.J. Edwards (ed.), *County Tyrone. INQUA Field Guide no.3*: 19-31. INQUA: Dublin.
Dardis, G.F. 1982. Sedimentological aspects of the Quaternary geology of south-central Ulster, Northern Ireland. Unpublished PhD Thesis, Ulster Polytechnic, Jordanstown.
Dardis, G.F. 1986. Late Pleistocene glacial lakes in south-central Ulster, Northern Ireland. *Irish J. Earth Sci.* 7: 133-144.
Demek, J. 1978. The landscape as a geosystem. *Geoforum* 9: 29-34.
Drdos, J. 1983. Landscape research and its anthropocentric orientation. *GeoJ.* 7: 155-160.
Duff, K. 1994. Natural Areas. *Earth Heritage* 1: 8-12.
Erikstad, L. Andersen, S. Ingolfsson, O. Lundqvist, J. Pedersen, S.S. Salonen, V.-P., Selonen, O. & Vilborg, L. 1995. New strategies in Quaternary geotop conservation – presentation of the Weichselian ice margin project, the Nordic countries. In *Terra Nostra: International Union for Quaternary Research XIV International Congress, Abstracts*: 76. INQUA: Berlin.
Garrod, G.D. & Willis, K.G. 1994. Valuing biodiversity and nature conservation at a local level. *Biodiv. Conserv.* 3: 555-565.
Gehlbach, F.R. 1975. Investigation, evaluation, and priority ranking of natural areas. *Biolog. Conserv.* 8: 79-88.
Gordon, J.E. 1987. Conservation of geomorphological sites in Britain. In V. Gardiner (ed.), *International Geomorphology 1986, Part II*: 583-591. Wiley: Chichester.
Gordon, J.E. 1994. Conservation of geomorphology and Quaternary sites in Great Britain: An overview of site assessment. In C. Stevens, J.E. Gordon, C.P. Green & M.G. Macklin (eds), *Conserving our Landscape*: 11-21. JNCC: Peterborough.
Gordon, J.E. & Campbell, S. 1992. Conservation of glacial deposits in Great Britain: A framework for assessment and protection of sites of special scientific interest. *Geomorphol.* 6: 89-97.

Harley, M. 1994. The RIGS (Regionally Important Geological/geomorphological Sites) challenge – involving local volunteers in conserving England's geological heritage. In D. O'Halloran, C. Green, M. Harley, M. Stanley & J. Knill (eds), *Geological and Landscape Conservation*: 313-317. Geological Society: London.

Hill, M. & Tzamir, Y. 1972. Multidimensional evaluation of regional plans serving multiple objectives. *Papers Reg. Sci. Assoc.* 29: 139-165.

Hodges, C.A. 1995. Mineral resources, environmental issues, and land use. *Science* 268: 1305-1312.

Holland, M.M. Risser, P. G. & Naiman, R.J. (eds) 1991. *Ecotones: The role of landscape boundaries in the management and restoration of changing environments.* Chapman and Hall: New York.

Idle, E.T. 1995. Conflicting priorities in site management in England. *Biodiv. Conserv.* 4: 929-937.

Kontturi, O. 1984. The development of the state of glaciofluvial landscape in Finland. *Fennia* 162: 63-80.

Kontturi, O. & Lyytikainen, A. 1985. Assessment of glaciofluvial landscapes in Finland for nature conservation and other multiple use purposes. *Striae* 22: 41-59.

Leopold, L.B. 1969. Landscape esthetics. *Nat. History* 78: 36-45.

McCabe, A.M. 1985. Glacial geomorphology. In K.J. Edwards & W.P. Warren (eds) *The Quaternary History of Ireland*: 67-93. Academic Press: London.

McCabe, A.M. 1987. Quaternary deposits and glacial stratigraphy in Ireland. *Quat. Sci. Rev.* 6: 259-299.

McCabe, A.M. 1993. The 1992 Farrington Lecture: Drumlin bedforms and related ice-marginal depositional systems in Ireland. *Irish Geogr.* 26: 22-44.

McCabe, A.M. 1994. Sand and gravel landforms in Ireland – a scientific basis for conservation? *Earth Heritage* 2: 18-22.

Meinig, D.W. (ed.) 1979. *The Interpretation of Ordinary Landscapes.* Oxford University Press: New York.

Moir, W.H. 1972. Natural areas. *Science* 177: 396-400.

Nature Conservancy Council. 1990. *Earth science conservation in Great Britain – a strategy.* NCC: Peterborough.

O'Halloran, D. Green, C. Harley, M. Stanley, M. & Knill, J. (eds) 1994. *Geological and Landscape Conservation.* Geological Society: London.

Somper, C. 1995. Character building. *Landsc. Design* no.242:48-50.

Sturm, B. 1994. The geotope concept: Geological nature conservation by town and country planning. In D. O'Halloran, C. Green, M. Harley, M. Stanley & J. Knill (eds), *Geological and Landscape Conservation*: 29-31. Geological Society: London.

Tjallingii, S.P. 1974. Unity and diversity in landscape. *Landsc. Plann.* 1: 7-34.

Urban, D.L. O'Neill, R.V. & Shugart, H.H., Jr 1987. Landscape Ecology – A hierarchical perspective can help scientists understand spatial patterns. *BioSci.* 37: 119-127.

Warnock, S. & Evans, S. 1992. Assessment and conservation of landscape character: A practical approach developed by the Warwickshire Landscapes Project. In A. Cooper & P. Wilson (eds), *Managing Land Use Change. Geographic Society of Ireland Special Publication* 7: 1-33. Geographic Society of Ireland: Coleraine.

Warren, W.P. O'Meara, M. Daly, E.P. Gardiner, M.J. & Culleton, E.B. 1985. Economic aspects of the Quaternary. In K.J. Edwards & W.P. Warren (eds), *The Quaternary History of Ireland*: 309-352. Academic Press: London.

Wiedenbein, F.W. 1994. Origin and use of the term 'geotope' in German-speaking countries. In D. O'Halloran, C. Green, M. Harley, M. Stanley & J. Knill (eds) *Geological and Landscape Conservation*: 117-120. Geological Society: London.

Wilcock, D. 1995. Top-down and bottom-up approaches to nature conservation and coutryside management in Northern Ireland. *Area* 27: 252-260.

Zube, E.H. 1987. Perceived land use patterns and landscape values. *Landsc. Ecol.* 1: 37-45.

Geomorphic effects of gravel extraction in the Russian River, California

JOAN FLORSHEIM & PETER GOODWIN
Philip Williams & Associates Ltd., San Francisco, California USA

LAUREL MARCUS
California State Coastal Conservancy, Oakland, California USA

1 INTRODUCTION

Alluvial rivers, such as the Russian River in northern California, adjust their morphology to both sediment supply and the magnitude and frequency of floods. Morphologic adjustments maintain the dynamic balance between sediment and hydrologic inputs as the fluvial system evolves in dynamic equilibrium over the long-term. Dynamic equilibrium is defined as the balance maintained by changes in flow and sediment supply around average conditions of the system as it evolves (Hack 1960, Knighton 1984). A reduction in sediment supply, due for example to gravel extraction, can cause both rapid channel incision and channel instability. Numerous deleterious effects arise from rapid channel incision, including increased bank heights, bank erosion, and a lowered groundwater table. Channel incision has negative effects on riparian habitat as the groundwater table falls below the level of the plant roots and as steep streambanks, which cannot support vegetation, are eroded. Incision has economic significance because it results in land loss, scour beneath bridge piers, and increased cost of pumping groundwater from the alluvial aquifer. A reduction of sediment supply can reduce the relative elevation differences between riffles and pools and thereby decrease habitat diversity.

A century of land use activities in the Russian River, including massive in-channel gravel extraction, agriculture, grazing, road construction, urbanization, and construction of dams in the upper basin, have resulted in geomorphic and hydrologic changes. This paper examines the evolution of channel morphology over this period and uses geomorphic techniques to illustrate that the channel is not currently in dynamic equilibrium. The study area is the 13 km long Middle Reach of the Russian River (Fig. 1), which flows through a broad alluvial valley and has experienced significant incision over the past century. Our data include photographs, longitudinal profiles, cross-sections, and water flow and sediment discharge records. These data provide a basis for comparing of the dominant discharge, calculated for the present sediment and hydrologic regime, with the morphologic bankfull discharge, and for analyzing empirical geomorphic equations that predict equilibrium width, depth, and pattern of the channel. Although gravel extraction is only one of the land use activities that has affected channel morphology in the Russian River in the past century, it is examined because mining could be most effectively managed to minimize the

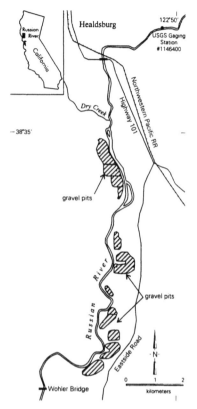

Figure 1. Location map of the Middle Reach of the Russian River.

negative effects to the river. The data and the accompanying analysis illustrate a history of channel instability that is characteristic of numerous other rivers in the western United States caused by gravel extraction (Collins & Dunne 1990). Documentation of the past evolution of channel form is essential in predicting future trends and in aiding management decisions about disturbed systems.

2 PHYSICAL CHARACTERISTICS OF THE RUSSIAN RIVER BASIN

The Russian River, California drains an area of 3,846 km^2 and flows through a series of broad northwest-trending alluvial valleys separated by narrow bedrock canyons. About 20,000 years ago, during the Wisconsin glaciation, sea level was about 100 m lower than at present, and the Russian River flowed in a deeply incised valley. As sea level rose following the glacial period, the Russian River filled its valley with gravel, sand, silt, and clay (California Department of Water Resources 1983). These alluvial deposits are an important groundwater source and have been extensively mined for aggregate.

Historically, the alluvial valley bottom of the Russian River contained numerous 'side channels and sloughs', and the main channel had riffles and deep pools (California Coastal Conservancy 1992). Historical photographs show an active channel that was close in elevation to the floodplain and had relatively low banks. Riparian habitat once extended across the broad valley. Logging, which began before 1900, may have increased the supply of sediment to the channel from hillslopes. However, this effect has been greatly overshadowed by the recent reduction in sediment supply from gravel extraction. Dam and stock pond construction also reduce the sediment supply to the Russian River. The volume of gravel extracted exceeds the volume supplied from upstream reaches. Today, the Russian River flows in an incised channel and is relatively narrow and straight. The active channel is isolated from the floodplain, and there are fewer overbank flows than a century ago.

3 CHANGES IN THE RUSSIAN RIVER: LATE 1800 TO THE PRESENT

3.1 *Aerial photographs*

Changes in channel morphology resulting from gravel extraction and related road construction are illustrated in a series of aerial photographs (Fig. 2). In 1958, the channel was sinuous and wide and former traces of the channel indicate frequent switching. In 1962, the channel was dredged for gravel, creating pits that lowered the local base level. Subsequent flow filled the pits. By 1991, the low flow channel was relatively straight with riparian vegetation established in the former low flow channel. The 1991 photo further indicates that a portion of the active channel, cut off by construction of the gravel haul road in 1962, was now a floodplain gravel extraction pit. After the environmental effects of channel dredging were identified in the 1960's, in-stream gravel extraction was mainly limited to bar skimming (excavation of bars to the low flow water surface), while off-channel excavation occurred as floodplain pits. There are currently 12 floodplain gravel pits adjacent to the river in the Middle Reach (see Fig. 1), many of which are deeper than the thalweg of the river (Sonoma County Planning Department 1994).

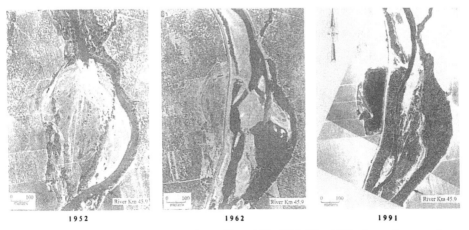

1952   1962   1991

Figure 2. Sequence of aerial photographs at River Mile 28.5. The 1952 photo shows a sinuous channel with remnants of former channel traces. The impacts of in-stream dredging for gravel and haul road construction are evident in the 1962 photo. The 1991 photo shows channel adjustments that have occurred since the cessation of large scale in-stream dredging.

3.2 *Topographic maps*

Successive topographic maps show the straightening and narrowing of the Middle Reach caused by historic land use activities (Fig. 3). The channel mapped in 1864 is relatively wide, consistent with the description of the channel provided in recent interviews with long-time residents of the area. Before watershed disturbances, the river was probably in dynamic equilibrium and had side channels, sloughs, and abundant riparian vegetation. A narrowing of the riparian zone is evident between 1864 (average width about 408 m) and 1990 (average width about 215 m). The decrease in meander amplitude evident in Figure 3 can be quantified as a decrease in sinuosity (channel length divided by straight-line valley length) from 1.3 in 1864 to 1.1 at present.

3.3 *Longitudinal profiles*

A longitudinal profile of the Middle Reach of the Russian River, surveyed by the US Army Corps of Engineers in 1940, provides a baseline for assessing recent channel changes. Thalweg elevations from cross-section surveys conducted by the Sonoma County Water Agency and the Sonoma County Planning Department in 1968 and 1991 are plotted with the 1940 data to illustrate the magnitude of channel incision in the Middle Reach (Fig. 4). Up to 6.1 m of incision occurred between 1940 and 1991.

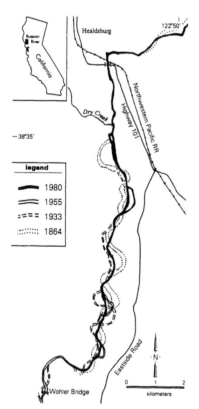

Figure 3. Historic changes in channel planform in the Middle Reach. Channel planform illustrated includes 1864 (general land office map), 1933, 1955, and 1980 (USGS topographic map).

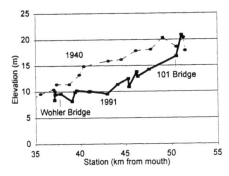

Figure 4. Longitudinal profile of the Middle Reach of the Russian River in 1940 and 1991.

About 60% of this incision occurred between 1940 and 1968 (Florsheim & Goodwin 1995). Average bank height (distance between thalweg and bank top) increased by about 1.4 m between 1982 and 1991 (or 0.1 m/y) as a result of the incision. The steep high banks, are composed of unconsolidated floodplain deposits, and are easily eroded during floods by shear stress on the banks, undercutting, or seepage failure.

Changes in channel gradient in the past half century indicate that incision may have ceased in the relatively low gradient downstream portion of the Middle Reach, whereas incision is likely to continue in the steeper upstream portion. In 1940, the average gradient in the Middle Reach was about 0.0005. By 1991, the gradient of the downstream portion (River Mile 23 to 27) had dropped to 0.0003, but the gradient of the upstream portion (River Mile 27 to 32), had increased to 0.0009. Future incision in the lower Middle Reach will be limited by the downstream bedrock constriction at Wholer Bridge (see Fig. 1).

Local base level changes occurred following the massive in-stream dredging in the 1950's and 1960's. Upstream of each in-stream pit, the slope was steeper and the velocity higher. Bed and bank erosion increased as the channel adjusted to the lower local base level. The erosion progressed upstream similar to the process of gully headcutting. Downstream of each pit sediment supply was reduced, due to deposition of sediment in the pit, also causing incision and channel widening.

As the main channel along the Middle Reach incised over the past half century, tributaries also incised in response to a lowered base level. Figure 5 shows incision on Dry Creek upstream of its confluence with the Russian River between 1964 and 1978. Completion of Warm Springs Dam on Dry Creek in 1982 will continue to limit sediment supply and may cause further incision. Accounts of long-time residents indicate that other smaller tributaries along the Middle Reach have also incised.

3.4 *Cross-sections*

Cross section data show that channel bar height has increased or remained constant while the thalweg has incised over the past decades in the Middle Reach. There was little change in the volume of stored sediment in the Middle Reach as a result of the 1986 flood in the Russian River as shown on Figure 6 (channel bar height increased and thalweg elevation decreased). These changes in bar topography are consistent with observations in laboratory flume experiments which document the effects of decreased sediment supply (Dietrich et al. 1989, Lisle et al. 1993). A decrease in sedi-

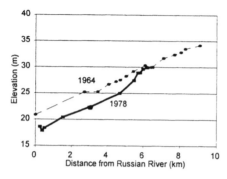

Figure 5. Longitudinal profile of Dry Creek in 1964 and 1978 (surveyed by the US Army Corps of Engineers).

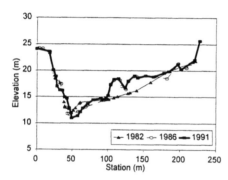

Figure 6. Cross-sections of the Middle Reach 1982-1986.

ment supply, constricts the active zone of bedload transport along the thalweg of the channel, and there is an inactive zone of transport on bars along the margins of the channel. The coarsening is caused by accretion of coarse particles onto emerging bar heads, and winnowing of fine material from inactive areas of the bed. Lisle et al. (1993) found that bed incision is concentrated at riffles (the cross-overs between pools). This may account for the relative flattening of the bed in the Middle Reach manifested in the loss of deep pools in recent decades. Eventually, coarsening of the bed could reduce the rate of incision.

4 DYNAMIC EQUILIBRIUM VS. INSTABILITY

4.1 *Dominant discharge*

Channel geometry is a product of the interactions among flow, the quantity and character of sediment in transport, the character of bed and bank material, and vegetation. In gravel-bed alluvial channels in dynamic equilibrium, the channel-forming flow, or 'dominant discharge', is the flow that over time, transports the majority of the sediment and is responsible for the characteristic size and shape of the channel (Wolman & Miller 1960, Leopold et al. 1964, Knighton 1984). 'Bankfull discharge' is the flow that fills the channel from bank to bank before spreading over the floodplain. In channels in dynamic equilibrium, the bankfull flow is similar to the channel-forming flow; these flows commonly have recurrence intervals of about 1.5 years in relatively humid environments (Leopold et al. 1964, Leopold 1994).

Geomorphic field studies of fluvial systems often assume that the channel-forming flow can be inferred from channel morphology. However, in channels in which reduced sediment supply causes incision, bankfull discharge increases as the channel deepens and widens. In such cases, bankfull flow is an inaccurate measure of the channel-forming flow and the dominant discharge will have no morphologic expression. Therefore, a method of quantifying the dominant discharge is needed that does not rely on channel morphology.

Wolman & Miller (1960) introduced a method of determining the channel-forming flow based on the magnitude and frequency of floods and sediment transport. The frequency of occurrence of a particular flow is determined from the stream flow record at a gauging station. The overall work performed by, or the effectiveness of a flow of a particular magnitude is represented as the product of the frequency of an event and the rate of sediment transport. We tested a range of sediment transport concentration predictors (Florsheim & Goodwin 1995) and evaluated rates of sediment transport measured at United States Geological Survey gauging stations. There are few bed load measurements for Russian River gauging stations; therefore, bed-load was estimated as a percentage of suspended load. During the period of record, available measured bed load ranged from 3 to 15% of the suspended load. In this study, we assumed that the bedload transport was the maximum 15% of the suspended sediment load. Estimates of suspended load and bedload were summed to provide a rating curve for total load. Total sediment discharge was determined from the transport capacity equation:

$$Q_s = cQ^n$$

where Q_s is the total sediment discharge in kg/day; Q is the associated water discharge in m³/s; and c and n are constants derived by fitting a power function to the plotted data. Details of this analysis are provided in Florsheim & Goodwin (1995).

The post-dam dominant discharge for the Middle Reach is about 450 m³/s and has a recurrence interval of 1.4 years (Fig. 7). It is likely that a range of flows near the dominant discharge is responsible for the characteristic channel shape, but in this study, the average value was used to describe the relationship between channel morphology, sediment transport, and flow frequency.

We simulated the dominant discharge at surveyed cross sections in the Middle Reach using a simple one-dimensional flow model (Hydrologic Engineering Center 1990) to compare the water surface elevation of the dominant discharge with the morphologic bankfull elevation. On average, the dominant discharge depth is about 60% of the bankfull depth. This percentage may be somewhat low since tributary flow is not included in the analysis. The result illustrates that the Middle Reach of the Russian River is not in dynamic equilibrium, as the existing channel morphology does not reflect the character of the dominant discharge. Over the long term, the river will attempt to reestablish a balance between sediment transport, the discharge regime, and channel morphology through a series of adjustments which are likely to include bank erosion and widening.

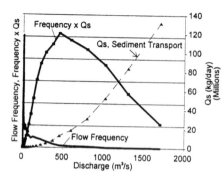

Figure 7. Dominant discharge for the Middle Reach, calculated from US Geological Survey Gaging Station flow data at Healdsburg (Gage #1146400).

4.2 *Comparison of the existing and equilibrium channels*

Rivers are dynamic and do not generally stay static or fixed in position over time. A river in dynamic equilibrium may adjust its width, depth, slope, and other characteristics. The tendency toward adjustment of river variables toward a balanced state is well accepted (Knighton 1984). This study evaluates whether historical changes in channel morphology along the Russian River indicate instability related to land-use activities or are the expected adjustments of a river in dynamic equilibrium. Geomorphic techniques define the cross-section geometry that conveys the dominant discharge, and this geometry was compared to the present channel form. An evaluation of whether or not the river is in dynamic equilibrium must acknowledge that the floodplain is an integral component of the fluvial system, and that the river will migrate over time. The geomorphic parameters used in this study to evaluate stability in the Middle Reach include:

– The equilibrium channel geometry, or the width and depth of the channel that conveys the dominant discharge, and

– The equilibrium channel slope which predicts the likelihood of future aggradation or incision and meandering.

4.2.1 *Equilibrium channel geometry*

There are consistent relations among the width, depth, velocity, and discharge of a river (Leopold et al. 1964, Dunne & Leopold 1978). The mutual adjustment of fluvial variables may be evaluated using a series of 'hydraulic geometry' relationships (Leopold & Maddock 1953, Wolman 1955):

$$w = aQ^b$$
$$d = cQ^f$$
$$v = kQ^m$$

where w, d, v, are the top width, hydraulic mean depth (cross-sectional area divided by top width), and mean velocity, respectively. From the continuity equation:

$$Q = wdv = (aQ^b)(cQ^f)(kQ^m)$$

it follows that $ack = 1$ and $b + f + m = 1$.

Table 1. Hydraulic geometry relationships for gravel-bed rivers.

Source of hydraulic geometry equation	Equation for mean width	Equation for mean depth
Emmett (1975) Upper Salmon River, Idaho.	$w = 2.86$ Qbf 0.54	$d = 0.26$ Qbf 0.34
Bray (1982) gravel-bed rivers in Alberta, Canada.	$w = 4.80$ Q2 0.53	$d = 0.26$ Q2 0.33
Dury (1976) 'stereotype' gravel-bed stream in humid area.	$w = 2.99$ Q1.58 0.55	$d = 0.44$ Q1.58 0.36

Fluvial geomorphologists have developed empirical hydraulic geometry relationships for stream channels in various regions of the United States and other countries. Langbein & Leopold (1964) inferred that, in the equilibrium case, $b = 0.55$ and $f = 0.36$. The coefficients a, c, and k depend on the sediment characteristics of the river. Table 1 illustrates a range of applicable relations for gravel bed rivers (Emmett 1975, Dury 1976, Bray 1982). Equations such as those in Table 1 imply that the width and depth of channels are controlled by the magnitude of the dominant discharge, using bankfull ($Q_{1.5}$), the mean annual flood ($Q_{2.33}$), or the most probable flood ($Q_{1.58}$) as the channel forming flow. For the Middle Reach of the Russian River, the equations predict that the channel width that conveys the dominant discharge in dynamic equilibrium is, on average, 86 m. This predicted width is much smaller than the 134 m average width of the current dominant discharge. The predicted dynamic equilibrium depth of the channel that conveys the dominant discharge is, on average, 2.6 m, shallower than the current depth of the dominant discharge, 3.1 m. The existing channel is thus wider and deeper than it would be if it were in dynamic equilibrium.

4.2.2 *Equilibrium channel slope*
Yalin (1992) suggests a method to predict how a river will evolve during the time between disturbance and equilibrium, as the system adjusts to the new hydrologic and sediment transport regime. Slope is used to predict channel change because it adjusts more slowly than width, and therefore the temporal evolution of the slope can be assumed to occur at constant width (Yalin 1992). The historic equilibrium condition, represented by the subscript 1, can be characterized by the dominant discharge $(Q_{dom})_1$, channel width $(w_e)_1$, sediment load $(Q_s)_1$, and channel slope $(S_o)_1$. Over the long term, these variables will adjust to changes to the hydrologic system by creating a new condition of dynamic equilibrium represented by the subscript 2, e.g. $(w_e)_2$ and $(S_o)_2$. Slope equilibrium in the new state $(S_o)_2$ can be achieved by meandering alone, incision alone, or by a combination of the two:

If $S_o < (S_o)_2$ or the existing slope is less than the equilibrium slope, the channel will aggrade, unless base level changes,

If $S_o > (S_o)_2$ or the existing slope is greater than the equilibrium slope, the channel will incise, the river will meander, or both. Meandering lowers the slope by increasing the length of channel between two points.

Two methods of estimating the equilibrium slope were used, namely the theoretical method of Yalin (1992) and the empirical method of Leopold et al. (1964, Fig. 7-23, p. 246). Yalin (1992) derives the equilibrium slope (S_R) using the following equation:

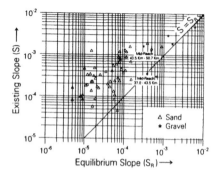

Figure 8. Relationship between present measured channel slope of a river and calculated equilibrium slope for a range of rivers (after Yalin 1992).

$$S_R = 0.42D^{1.07} Q^{-0.43}$$

where D is the median particle size (D_{50}, m) and Q is the dominant discharge (m²/s). The relationship between the present channel slope and the dynamic equilibrium slope for a range of natural rivers is shown in Figure 8. Most of these rivers including the upper part of the Middle Reach of the Russian River should reduce their slopes by incision or meandering to achieve equilibrium. The present slope of the lower part of the Middle Reach, which is upstream of the Wohler constriction, is similar to the estimated equilibrium slope, and assuming an adequate supply of sediment, future degradation is not as likely. Because upstream in-channel gravel mining continues and upstream dams trap sediment, future changes in the Middle Reach are likely to include further bank erosion and widening as existing meanders migrate downstream.

5 DISCUSSION

Historic accounts and photographs suggest that the Russian River was probably close to dynamic equilibrium prior to development of the basin in the 1800's. The undisturbed river system would respond to episodic events such as large floods or sediment input from landslides, but would tend back toward an equilibrium form.

   In future years, the Russian River will adjust to the changes in the basin including continuing gravel extraction in tributaries and in the main channel, trapping of sediment behind dams and variation in flow regime due to reservoir operation, and increased urbanization. The river channel will continue to incise due to the reduced sediment supply, but the rate of incision is likely to be less than in the past. This is because gravel extraction activities no longer include large-scale dredging operations, a gravel management plan was recently completed for a portion of the river, because some of the tributaries and upper reaches of the river have already incised to bedrock, and because the bed material coarsens as sediment supply is reduced. Bank erosion will continue and the channel will widen as the steep high banks are undercut by moderate flood flows.

   As the channel deepens and widens, it will adjust its slope and pattern in an attempt to find a semi-stable configuration.

If the channel meanders, and sediment supply is sufficient, the river may eventually create a new floodplain at the elevation of the dominant discharge, below the existing terraces. Over time, if the sediment load and flow regime remain constant, the river would approach a new dynamic equilibrium. This process of adjustment toward a stable form could take many decades or even centuries.

If current gravel extraction levels continue in tributaries and in the main channel of the Russian River (in the case that management guidelines are not implemented), the channel will not attain equilibrium because sediment supply is insufficient. The channel will continue to incise until it reaches bedrock or is limited by downstream base level controls, and then it will widen, by bank erosion. The river will migrate across its floodplain in episodes of widening, migration, and deepening as long as the sediment supply is less than the capacity of the river to transport sediment. Continued channel incision will lower the groundwater table and decrease the potential water supply in the alluvial aquifer. Finally, as the river migrates and widens, it will remove the remaining narrow strip of riparian habitat on a time scale which is too rapid for a natural recovery of the ecosystem.

The geomorphic effects of gravel extraction in the Russian River, are similar to well documented effects in other rivers in California affected by gravel extraction. Examples include Cache Creek (Sandecki 1989, Collins & Dunne 1990), Mad River (Klein 1993, Lehre 1993), and Stony Creek (Kondolf & Swanson 1993). Collins & Dunne (1990) also cite examples from rivers in Washington and New Zealand. Detailed monitoring and measurement of actual gravel replenishment combined with active management to ensure that the extraction rate does not exceed the replenishment rate, could reduce the incision, bank erosion, channel instability, and the ecological effects of gravel extraction.

6 CONCLUSIONS

Modification of the Russian River basin and changing land-use practices in the past century have altered dominant discharge, slope, width, and sediment supply. Historical data, as well as empirical and analytical relationships, show that the Russian River is not currently in dynamic equilibrium. Channel width, depth and slope have adjusted in response to a reduced sediment load.

– Historical data illustrate a narrowing and reduction in sinuosity in the riparian corridor, incision in the main channel and tributaries, and a reduction in bed topography.

– The dominant discharge, or the current channel-forming flow, is not in equilibrium with the present channel morphology.

– The present channel is wider and deeper than the predicted dynamic equilibrium channel.

– The present channel has become more isolated from the floodplain due to incision.

The geomorphic analysis and predictive methods outlined in this paper provide insight which is essential in developing an effective management and restoration plan for the Russian River. These methods can be applied to other disturbed river systems to guide management strategies.

ACKNOWLEDGMENTS

The authors acknowledge useful discussions with Karen Gaffney. Data was provided by: Robert Gaiser, Sonoma County Planning Department; Cathy Crosset, Caltrans; Bill Cox, California Department of Fish and Game; Mike Sandecki, California Department of Conservation; and Robert Morrison and Doris Anderson, Sonoma County Water Agency. Assistance with data analysis and production was provided by Tamara Rose, Aaron Mead, Johnny Lin, Ande Bennett, Marcia Greenblatt, and the Philip Williams & Associates, Ltd. support staff. Reviews by John Clague, Ian Spooner, and Robert Coats significantly improved the manuscript.

REFERENCES

Bray, D.I. 1982. Regime Equations for Gravel Bed Rivers. In R.D. Hey, J.C. Bathurst & C.R. Thorne (eds), *Gravel Bed Rivers*, p. 517-580.

California Coastal Conservancy 1992. Interviews with Long-time Residents of the Russian River. Unpublished data.

California Department of Water Resources (DWR) 1983. *Upper Russian River Gravel and Erosion Study*, 93 pp.

Collins, B. & T. Dunne 1990. Fluvial Geomorphology and River Gravel Mining: a Guide for Planners, Case Studies Included. CA Dept. of Conservation Division of Mines and Geology, 29 pp.

Dietrich W., J. Kirchner, H. Ikeda & F. Iseya 1989. Sediment Supply and the Development of the Coarse Surface Layer in Gravel-bedded Rivers. *Nature* 340:215-217.

Dunne, T. & L. Leopold 1978. *Water in Environmental Planning*. Freeman and Co., San Francisco, CA, 818 pp.

Dury, G. 1976. Discharge Prediction, Present and Former, from Channel Dimensions. *Journal of Hydrology* 30: 219-46.

Emmett, W. 1975. *The channels and waters of the Upper Salmon River Area, Idaho*. US Geological Survey Professional Paper 870-A, 116 pp.

Florsheim, J. & P. Goodwin 1995. *Geomorphic and Hydrologic Conditions in the Russian River, CA: Historic Trends and Existing Conditions*. Prepared for the California State Coastal Conservancy, Mendocino County Water Agency and Circuit Rider Productions, Inc., 41 pp.

Hack, J.T. 1960. Interpretation of erosional topography in humid temperate regions. *Amer. Jour. Sci.* 258A: 80-97.

Hydrologic Engineering Center 1990. HEC-2 *Water Surface Profiles User's Manual*. US Army Corps of Engineers. CPD-2A.

Knighton, D. 1984. *Fluvial Forms and Processes*. Arnold, Baltimore, MD, 218 pp.

Klein, R. 1993. Channel bed and bank erosion. Mad River Technical Supplement. Program EIR on Gravel Removal from the Lower Mad River. Vol. II. Appendices. Humboldt County Planning and Building Department.

Kondolf, G.M. & M.L. Swanson 1993. Channel adjustments to reservoir construction and gravel extraction along Stony Creek California. *Environmental Geology* 21: 256-629.

Langbein, W. & L. Leopold 1964. Quasi-equilibrium States in Channel Morphology. *Amer. Jour. Sci.* 262: 782-794.

Lehre, A. 1993. Estimation of Mad River Gravel recruitment and analysis of channel degradation. Mad River Technical Supplement. *Program EIR on Gravel Removal from the Lower Mad River*. Vol. II. Appendices. Humboldt County Planning and Building Department.

Leopold, L.B. 1994. *A View of the River*. Harvard University Press, MA, 298 pp.

Leopold, L.B., M.G. Wolman & J.P. Miller 1964. *Fluvial Processes in Geomorphology*. Freeman Co, San Francisco, CA, 511 pp.

Leopold, L.B. & T. Maddock 1953. *The Hydraulic Geometry of Stream Channels and Some Physiographic Implications.* USGS Professional Paper, 252. 57 pp.

Lisle, T., F. Iseya & H. Ikeda 1993. Response of a Channel with Alternate Bars to a Decrease in Supply of Mixed-size Bed Sediment Load: a Flume Experiment. *Water Resources Research* 29(11): 3623-3629.

Sandecki, M. 1989. Aggregate Mining in River Systems. *California Geology* 42(4): 88-94.

Sonoma County Planing Department 1994. *Aggregate Resources Management Plan and Environmental Impact Report, November 1994.*

Wolman, M.G. 1955. *The Natural Channel of Brandywine Creek, PA.* USGS Professional Paper 271, 56 pp.

Wolman, M.G. & J. Miller 1960. Magnitude and Frequency of Forces in Geomorphic Processes. *Journal of Geology* 68: 54-74.

Yalin, M.S. 1992. *River Mechanics.* Pergamon Press, Oxford, 219 pp.

The geology, exploration characteristics, and resource potential of sand and gravel deposits in Alberta, Canada

W.A. DIXON EDWARDS
Alberta Energy and Utilities Board, Edmonton, Alberta, Canada

1 INTRODUCTION

The mineral aggregate industry in Alberta is comprised of about 300 public and private sector producers operating several thousand pits. Total provincial mineral aggregate production in 1991 was 45,484,836 tonnes worth $153,226,689. This places Alberta fourth in total mineral aggregate production in Canada. The annual per capita consumption of mineral aggregate in Alberta for 1991 was 18.7 tonnes, considerably higher than the Canadian average of 10.4 tonnes (Edwards 1995).

Ninety-nine percent of Alberta's mineral aggregate is produced from sand and gravel deposits. Most production comes from Recent alluvial deposits (floodplain and valley terraces), glaciofluvial deposits (outwash), and preglacial fluvial deposits (Table 1). Alberta is divided into two primary geographic regions, the Rocky Mountain Front Ranges/Foothills and the Alberta Plains. Almost all production is from the Plains. The Plains are characterized by scattered hills and ridges, broad flat plains, and incised river valleys usually containing several levels of terraces. The preglacial gravels are found as caps on the hills and ridges and the glaciofluvial deposits are scattered about the plains.

Sand and gravel in Alberta is widespread and less expensive to mine and beneficiate than crushed stone. The average cost of one tonne of sand and gravel at the pit in Alberta was $3.37 in 1991. This is slightly higher than the national average sand and gravel cost of $3.16 per tonne but lower than the average cost of crushed stone at $5.78 per tonne (Vagt 1994).

The known and exploitable amount (reserves) of sand and gravel in Alberta are dwindling (Edwards 1995). The president of the Alberta Sand and Gravel Association (1995) noted that sand and gravel is being consumed at twice the rate at which it is being found. Mapping and discovery of new deposits is essential in the interests of resource cost and public resource management. In the interest of maintaining low-cost supplies it is important for industry to have available as many useable sand and gravel deposits as possible. Mineral aggregate from privately owned land in Alberta is a private resource but in the future more aggregate is expected to come from Crown lands. It is necessary for provincial resource managers to have a better understanding of the aggregate resource in order to plan and manage the Crown re-

Table 1. Estimated occurrence and sources of production of sand and gravel in Alberta in 1988 (Edwards 1991).

Type of deposit	Occurrence (%)	Production (%)
Preglacial	20	25
Glacially derived	70	30
Recent, alluvial	10	45
Total	45×10^9 tons	42×10^6 tons

source for greatest public benefit. Effective exploration for sand and gravel deposits requires a detailed understanding of the distribution and origin of deposits.

Detailed sand and gravel deposit maps (1:50,000 scale, or level 3 detail) are available for about 18% of Alberta. Reconnaissance sand and gravel deposit maps (1:250,000 scale, or level 4) are available for another 20% of Alberta (Edwards & Chao 1989). Approximately one hundred and fifty publications describing sand and gravel deposits can be purchased from Alberta Geological Survey Information Sales and a useful index to these publications is also available (Edwards & Chao 1989).

2   GEOLOGICAL CHARACTERISTICS

Sand and gravel deposit types producing mineral aggregate in Alberta include preglacial, glaciofluvial, alluvial, glaciolacustrine, eolian, colluvial, and alluvial fan deposits. The geological characteristics, setting, and age of each of these types of deposits are shown in Table 2.

2.1   *Preglacial deposits*

The preglacial deposits record the erosional history of the Plains over a period of about 50 million years from the Early Tertiary to Late Quaternary prior to the last continental glaciation (hence the name 'preglacial sands and gravels' or 'preglacial deposits'). River systems ran eastward or northeastward from the mountains across the bedrock surface and eroded into it. The granular materials deposited by these rivers are today's sand and gravel deposits. During this long period of fluvial action, continental uplift gradually elevated the entire plains surface. The oldest deposits were uplifted for the longest period and became isolated remnants occupying topographic highs (Table 2, Unit 4). Progressively younger deposits occur at decreasing elevations. The youngest preglacial deposits are found in channels carved through the bedrock at elevations below surrounding plains level (Table 2, Unit 1).

Edwards et al. (1994) separate the preglacial deposits into four age classes. The oldest category includes the Cypress Hills Formation and deposits on the western Del Bonita Uplands and the Swan Hills. Deposits in this category (Table 2, Unit 4) are considered to be Early to Middle Tertiary in age and cap the highest hills in Alberta east of the Foothills and Rocky Mountains. Deposits capping the Wintering Hills, Whitecourt Mountain, Halverson Ridge, and the Pelican Mountains probably were deposited during the Late Tertiary and are listed as Unit 3 in Table 2. Deposits such as those near Grimshaw, Entwistle, Lacombe, and Cluny occur at or near plains level, were formed during the Quaternary, and are identified as unit 2 in Table 2. The youngest preglacial deposits were formed during the late Quaternary (Table 2, Unit 1). Dates on fossil materials recovered from deposits near Villeneuve, Simonette, and Watino are in the range of 22,000 to 40,000 years before present (Edwards et al. 1994).

The preglacial deposits generally are gravelly or even cobbly. The coarse nature of these sediments is the result of their fluvial deposition in high energy systems and results in attractive deposits for the production of coarse aggregate. The coarse sediments are derived primarily from sandstone and quartzite formations presently found in Omenica Terrane in British Columbia or the Rocky Mountains of British Columbia and Alberta. Plains bedrock such as soft sandstones, shale, and ironstone forms a minor, deleterious component of most preglacial deposits.

2.2 *Glaciofluvial deposits*

Glaciofluvial deposits were formed from meltwater flowing in contact with glacial ice (ice contact forms) or meltwater flowing away from the glacier (outwash). Mineral aggregate in Alberta is produced from outwash plain, valley train, and meltwater channel deposits. Ice contact deposits that produce mineral aggregate include kames, kame terraces, eskers, and crevasse fillings.

Outwash plain deposits are very large sheets formed from unconfined meltwater flow and in Alberta they are composed primarily of sandy material. The deposits with greatest aggregate value occur on the Alberta Plains and have a Laurentide glacial origin. Glacial movement was from the northeast so that a significant rock component in the outwash is derived from the Canadian Shield in northern Alberta or the North West Territories.

A valley train deposit is similar in genesis to an outwash plain but the deposit is confined by valley sides. Valley train deposits in Alberta formed in front of valley or alpine glaciers in Rocky Mountain Front Ranges valleys or Foothill corridors. Valley train deposits in the Canmore Corridor are excellent sources of mineral aggregate (Edwards 1979).

As the continental glacier covering the Alberta Plains melted streams formed and carried away sand and gravel. After the glacier receded these streams became much smaller or dried up. The channels which these streams left behind are called meltwater channels and can contain terraces, point bars, or channel bars. Meltwater channel deposits are widespread across Alberta.

Eskers are long, linear ridges of sand and gravel formed by meltwater streams flowing in contact with and confined by glacial ice. Kames are irregular mounds or hills of mixed sands, gravel, till, or stratified drift formed by meltwater in contact with the glacier. Kames terraces are deposits which are built against a linear wall of ice. Crevasse fillings are deposits of sand and fine materials which filled glacier crevasses. The deposits provide a relict pattern of the crevasses.

2.3 *Alluvial deposits*

Alluvial (post glacial fluvial) deposits are formed by the deposition of sand and gravel in Recent rivers. Alluvial deposits in Alberta include point bars, channel bars

Table 2. Setting and geological characteristics of sand and gravel deposits in Alberta and characteristics useful in exploration.

Geological feature	Setting	Origin	Material[1]	Age	Occurrence[2]	Characteristics
Preglacial						
Unit 4	Highest hills on plains	Fluvial	G, S & G	Mid Tertiary	Very limited	Exposed, extensive
Unit 3	Hills, ridges on plains	Fluvial	G, S & G	Late Tertiary	Limited	Buried, extensive
Unit 2	Plains level	Fluvial	G, S & G	Quaternary?	Limited	Buried, extensive
Unit 1	Buried channels	Fluvial	S & G, S	Quaternary	Common	Buried, extensive, elongate form
Glaciofluvial						
Outwash plain	Plains	Outwash	S, S & G	Quaternary	Very common	Exposed, sheet form
Valley train	Mountain valleys	Outwash	G, S & G	Quaternary	Very limited	Buried, elongate form
Meltwater channel	Plains dry valleys	Outwash	S, S & G	Quaternary	Very common	Exposed, point bar
Esker	Plains	Ice contact	S, S & G	Quaternary	Limited	Exposed, linear ridge
Kame	Plains	Ice contact	S & G	Quaternary	Very common	Buried, conical hill
Kame terrace	Mountain valley sides	Ice contact	S & G, G	Quaternary	Very limited	Buried, elongate form
Kame delta	Plains	Ice contact	S, S & G	Quaternary	Very limited	Exposed, fan shaped
Crevase filling	Plains	Ice contact	S	Quaternary	Limited	Very low, curvilinear ridges
Alluvial						
River	Plains and mountains	Fluvial	S & G	Recent	Very common[3]	Exposed, point-channel bars
Terrace	Major valleys	Fluvial	S, S & G, G	Recent	Very common	Buried, elongate form
Glaciolacustrine	Plains, mountain valleys	Lacustrine	C, S[4]	Quaternary	Common	Exposed, extensive
Eolian	Plains, mountain valleys	Eolian	S	Recent	Very common	Exposed, sheet form
Colluvial/Fan	Mountain valley sides	Gravity/fluvial	G	Recent	Common	Exposed, fan shaped

1. S = sand, S & G = sand and gravel, G = gravel, C = clay and silt, 2. Very limited = < 5 areas mined, limited = 5-15 areas mined, common = 15-100 sites mined, very common = > 100 sites mined, 3. Formerly very common, sites in or adjacent to rivers are being limited by environmental regulations, 4. Silt and clay is raw material for manufactured aggregate.

and terraces. Point bars occur on the inside of bends on meandering rivers at, or up to several meters above, the average water surface elevation of the river. Channel bars are linear or tear shaped in form and occur in the river or at its edge on more or less straight river stretches. These bars occur slightly above or below river level depending on the time of year.

During the last several thousand years, many rivers have incised into the glacial drift and Plains bedrock and during this process have left linear deposits of sand and gravel to mark this otherwise erosional event. These deposits are referred to as alluvial terraces. Terrace deposits in Alberta range from several meters to over 100 m above present day river level. The same section of a river valley may contain terraces at several different elevations. The different levels often are grouped by elevation and referred to as 'low terraces', 'intermediate terraces', or 'high terraces' depending on the number and relative elevation of the terrace levels. The identification of different levels of terraces is important in exploration.

The rock component of alluvial deposits generally is derived from three sources: preglacial materials (mountain origin), glacial materials (Laurentide glacier materials derived from the north or northeast, or Cordilleran glacier materials from the west), and Plains bedrock.

2.4 *Glaciolacustrine deposits*

Melting of both the Laurentide and Cordilleran glaciers formed glacial lakes in Alberta. Fine sand, silt, and clay were the primary materials deposited in these glacial lakes. Meltwater streams flowing into these bodies of water occasionally deposited gravel in deltas. Glaciolacustrine deposits in Alberta containing coarse sand and gravel are rare and generally thin.

2.5 *Eolian deposits*

Eolian deposits were formed by the deposition of fine sand and silt by wind as dunes or sheets. The mean grain size in a dune field east of Edmonton is 0.19 mm (Edwards et al. 1985). Eolian sand may cover very large areas. In the Edmonton-Lloydminster region eolian sand accounts for about 10% of the sand mapped in the region.

2.6 *Colluvial and alluvial fan deposits*

Colluvial deposits were formed by the deposition of rock fragments by gravity as a sheet or triangle at the base of a cliff or steep slope in the mountains or foothills. Alluvial fan deposits were formed by the deposition of rock fragments by intermittent streams in the form of fans. These deposits may occupy very large areas and contain extensive volumes of material (Edwards 1979). They occur along the valley sides of all main mountain corridors.

3   EXPLORATION CHARACTERISTICS

Conceptual models were developed by Alberta Geological Survey (AGS) staff to explain the occurrence of sand and gravel deposits, to find other deposits of the same origin, and to estimate material quality and deposit size. Deposit characteristics important in sand and gravel exploration and development in Alberta are shown in Table 2.

3.1   *Preglacial deposits*

Deposits were formed by laterally active fluvial systems during a period of continental uplift. The oldest preglacial deposits occupy topographic highs and progressively younger deposits occur at decreasing elevations. The youngest preglacial deposits, emplaced just prior to continental glaciation, are incised into the bedrock and occur below plains level. An older set of preglacial deposits cannot overlie a younger sand and gravel deposit. Preglacial deposits typically contain hard, coarse aggregate (Table 3).

Deposits can be extensive and exceed 10 m in thickness. They usually are buried by till, clay, or sand. Common rock components in preglacial deposits are sandstone and quartzite (Table 4). Constituents generally are hard and tough. The coarse aggregate from the preglacial deposits yields some of the highest quality mineral aggregate produced in Alberta.

3.2   *Alluvial deposits*

A basic change in most fluvial systems is a fining of sediment size downstream. This change is critical in exploration for granular resources. The change can be observed in alluvial bars and terraces. For example, the grain size of material in alluvial ter-

Table 3. General grain size distribution for samples collected at ten preglacial deposits (Edwards et al. 1985, 1994).

Size fractions		Preglacial deposits* (values as %)									
		1	2	3	4	5	6	7	8	9	10
Gravel		58	66	75	78	70	63	81	80	77	70
	Boulders	–	–	–	3	–	–	2	–	–	–
	Cobbles	1	–	5	20	10	3	24	10	7	3
	Pebbles	57	66	70	55	60	60	55	70	70	67
Sand		41	33	23	19	28	36	17	19	21	29
	Coarse	17	9	2	4	3	8	1	3	5	10
	Medium	15	8	5	5	7	7	2	4	4	8
	Fine	9	16	16	10	18	21	14	12	12	11
Fines		1	1	2	3	2	1	2	1	2	1

*deposits located on or near: 1. Halverson Ridge, 2. Grimshaw, 3. Watino, 4. Swan Hills, 5. Pelican Mountains, 6. Villeneuve, 7. Lacombe, 8. Wintering Hills, 9. Cluny, 10. Del Bonita Uplands. Data for all sites except deposit 6 from (Edwards et al. 1994), data for deposit 6 from (Edwards et al. 1985).

Table 4. Rock types in the gravel (19-38 mm) fraction of preglacial deposits.

Rock type	Preglacial deposits* percentage of rock type									
	1	2	3	4	5	6	7	8	9	10
Quartzite	33	23	24	54	31	10	35	26	57	29
Sandstone	37	28	68	35	59	67	49	42	9	32
Conglomerate	–	6	4	11	1	6	1	<1	6	14
Shale**	–	–	<1	<1	7	3	3	2	–	1
Chert	4	5	2	<1	2	12	4	1	6	<1
Carbonate	–	–	–	–	–	–	8	29	20	–
Argillite	1	1	–	–	–	–	<1	–	–	23
Igneous***	–	11	<1	–	–	–	–	–	–	1
Quartz	25	14	2	–	–	2	–	<1	2	–
Other metamorphic***	–	13	–	–	–	<1	–	–	–	–

* Deposits located on or near: 1. Halverson Ridge, 2. Grimshaw, 3. Watino, 4. Swan Hills, 5. Pelican Mountains, 6. Villeneuve, 7. Lacombe, 8. Wintering Hills, 9. Cluny, 10. Del Bonita Uplands, ** Includes mudstone and ironstone, *** Igneous and metamorphic rocks of mountain origin, not Canadian Shield origin.

races along the Peace River becomes progressively finer in a downstream direction. Terraces in the Peace River valley for 90 km downstream from Peace River Town are primarily gravel, from 90 km to 230 km downstream from Peace River Town are mixed sand and gravel, and over 230 km downstream from Peace River Town the terraces are sand (Scafe et al. 1989, Fox et al. 1987).

Many rivers incised into the plains bedrock and left linear alluvial terrace deposits of sand and gravel along the valley sides. A series of terraces belonging to a former river level often can be traced along the river valley. Series or sets of terraces along the same valley can be composed of materials with quite different grain sizes, for example sand or coarse gravel, and the identification of different levels of terraces and their grain size character is important in exploration. The terrace level will decrease in elevation downstream and increase in elevation upstream and often will change in relation to the height above the present river level. The Beaver River is an excellent example of grain size differences between terrace sets and illustrates how a geological model showing these differences can be an aid in the exploration for specific material such as gravel (Edwards & Fox 1980). Along the Beaver River Valley the intermediate level terraces are coarser in composition than either the high or the low level terraces. Terraces which formed in major river valleys commonly contain gravel. Alluvial terraces often contain a lower percentage of fines and a higher percentage of gravel than point bars currently forming in the river.

Gravel is often found along rivers in Alberta in point bars and channel bars. Point bars occur on the inside of meander bends. They can contain gravel but the material commonly is interbedded with fine sand, silt and clay and often is overlain by a bed of fine materials. Channel bars are linear or tear shaped and occur in the river or at its edge on more or less straight stretches of the river. Channel bars commonly are composed of gravel if the river system contains coarse materials. These bars occur slightly above or below water level, depending on the time of year.

Although the grain size distribution of alluvial deposits within Alberta is highly variable and can range from silty deposits to boulder deposits, trends can be established within a single river system and exploration models can be developed. Variability is due to differences in the sediment supply and the fluvial conditions. Coarse material from alluvial deposits can be high quality aggregate but these deposits are more variable in quality across the province than the preglacial deposits.

3.3 *Glaciofluvial deposits*

Outwash deposits can be excellent sources of aggregate. They can be very large in area (> 100 ha), contain huge volumes of material (> 10 million m^3), but are often sandy in grain size (Edwards & Fox 1980). Canadian Shield derived granite, gneiss, and schist commonly account for more than 50% of the coarse fraction in outwash deposits in central and northern Alberta and 35 to 55% in southern Alberta (Edwards & Fox 1980, Shetsen 1980). Near the foothills or near preglacial channels, rocks of mountain origin (particularly quartzite) are predominant and stones derived from the Canadian Shield account for less than 25% (Edwards et al. 1985, Shetsen 1980). Rocks of local origin, often deleterious, usually account for < 10% of the pebble fraction of most deposits (Edwards & Fox 1980, Edwards et al. 1985, Shetsen 1980).

Meltwater channel deposits represent small, but highly useful, sources of sand and gravel. In the Cold Lake area, nine meltwater channel deposits with an average volume of about 1 million m^3 and area of about 30 ha were mapped (Edwards & Fox 1980). All the deposits in the Cold Lake area are described as sand or gravelly sand but several are worked for pockets of gravel. Meltwater channel deposits can occur as terraces or point bars.

Valley train deposits are similar to outwash plain deposits in volume but may contain a higher percentage of gravel. The deposit below Grotto Mountain in the Canmore Corridor has about 14 million m^3 of material with 69% gravel (Edwards 1979). The gravel is high in carbonates (> 80%) and sandstones-quartzites (5-20%).

Ice contact deposits have a high to very high potential for the occurrence of granular materials. Some types (kames, crevasse fillings) may be very poorly sorted and difficult to exploit. Ice contact deposits in the Cold Lake area have an average size of ~12 ha, an average volume of ~300 000 m^3, and an average composition of gravelly sand (Edwards & Fox 1980). These deposits all have distinctive shapes which can be useful in exploration. Kames are conical hills, eskers are narrow ridges and crevasse fillings are low and curvilinear.

3.4 *Glaciolacustrine deposits*

Glaciolacustrine and lacustrine deposits are primarily fine materials which have little application as conventional mineral aggregate. The margins of these deposits may have been reworked by waves or currents to winnow out the finer materials to leave coarser beds of sand and gravel (bars or beaches). Areas where rivers or meltwater streams may have entered the lake should be checked for deltas. Deltas are triangular or fan shaped and can contain sand and gravel. A glaciolacustrine deposit in the Edmonton area has supplied silt and clay for use in the manufacture of synthetic (expanded) aggregate.

3.5 *Eolian deposits*

Eolian sand occurs as large, thin sheets or as dunes. Sand areas often are covered by open pine forest. The fine sand from sheet or dune deposits has limited use as a mineral aggregate but it is important to recognize eolian sand deposits, if only to eliminate them during a gravel search.

3.6 *Colluvial and alluvial fan deposits*

These deposits can occupy very large areas and contain extensive volumes of material. The material is generally too coarse and poorly sorted to be used. Development of these deposits can result in flash flooding and disturbance of the groundwater movement (Edwards 1979).

4 RESOURCE POTENTIAL

Very little information has been published on the sand and gravel reserves available in Alberta. This type of information is contained in the files of Alberta Transportation and Utilities and private operators for confidential use. The only public sand and gravel reserve data are the blue-line maps and open file reports sold by Alberta Geological Survey Information Sales. These publications delineate potential sources of sand and gravel for mineral aggregate and provide some volumetric estimates. These volumetric data identify prospective resources, not reserves. The AGS data were gathered consistently, are displayed on maps in a standard manner, and cover a large portion of the province. The amount of mineral aggregate produced in Alberta and its major uses are described elsewhere in this volume (see 'A review of mineral aggregate production and operating conditions in Alberta, Canada').

4.1 *Preglacial deposits*

The preglacial deposits, deposited at various times over the last 50 million years by rivers flowing from the mountains, generally are gravelly or even cobbly, extensive, thick (may exceed 10 m thickness) and can produce aggregate of high quality. Preglacial deposits often require significant overburden removal (Units 1 and 2 in Table 2). Extraction may require dredging or pumping water from the pit as preglacial deposits also may be aquifers (Unit 1 in Table 2). They are ideal for hosting large operations and supplying major markets. Preglacial deposits supply both the Calgary and Edmonton markets. These deposits are critical resources (Table 5) and one such deposit northwest of Calgary has become a land use battlefield. Edwards (1995) provides a detailed description of events surrounding this case.

4.2 *Glaciofluvial deposits*

Glacially derived deposits (outwash, ice contact, meltwater channel) are the largest potential source of aggregate in Alberta (Table 1). In the Edmonton-Lloydminster region 289 of 585 known deposits are of glaciofluvial origin (Edwards et al. 1985).

Table 5. The estimated distribution and setting of sand and gravel deposits in Alberta.

Feature	Aggregate source[1]	Suitability	Volume	Development potential
Preglacial				
Unit 4	Very rare	Excellent	Very large	Moderate/poor
Unit 3	Rare	Excellent	Very large	Good/excellent
Unit 2	Rare	Excellent	Very large	Excellent
Unit 1	Rare	Excellent	Very large	Moderate/good
Glaciofluvial				
Outwash plain	Very common	Very good	Very large	Excellent
Valley train	Very rare	Very good	Large	Poor
Meltwater channel	Very common	Good	Small	Excellent
Esker	Rare	Good	Small/medium	Good
Kame	Very common	Fair	Small	Excellent
Kame terrace	Very rare	Very good	Medium	Moderate/poor
Kame delta	Very rare	Good	Medium	Good
Crevasse	Very rare	Poor	Very small	Excellent
Alluvial				
River	Common[2]	Very good	Medium/small	Poor
Terrace	Very common	Very good	Medium/large	Moderate/poor
Glaciolacustrine	Very rare	Very poor[3]	Large	Good
Eolian	Common	Poor	Very large	Good/moderate
Colluvial/fan	Common	Poor	Medium/small	Poor

1. Very rare = <5 areas mined, rare = 5-15 areas mined, common = 15-100 sites mined, very common = >100 sites mined, 2. Formerly very common, sites in or adjacent to rivers are being limited by environmental regulations, 3. Silt and clay is raw material for manufactured aggregate.

Unfortunately, most glaciofluvial deposits are sand. In the Edmonton-Lloydminster region only about 19% of the glaciofluvial deposits are gravel, the remainder are sand or gravelly sand. Processing is required to separate the sand from the gravel and extra costs ensue for discarding the excess. Ice contact deposits have a high to very high potential for the occurrence of granular materials, but some types (kames, crevasse fillings) may be very poorly sorted and difficult to mine. Kame deposits commonly are poorly sorted and require selective recovery. Resulting operations generally are small in scale and create a 'moonscape' appearance to the pit.

4.3   *Alluvial deposits*

Although the grain size distribution of alluvial deposits within Alberta is highly variable, coarse material from alluvial deposits can produce high quality aggregate. Alluvial deposits will decline in relative importance, regardless of the amount of potential aggregate in them, as a result of operating restrictions due to proximity to watercourses. Settling ponds are incorporated in operations on many types of deposits but they are especially important in alluvial deposits near river courses. It is essential that sediment does not escape into natural water bodies.

4.4 *Eolian and fan deposits*

Eolian deposits have limited use as mineral aggregate because of the fine grain size. One deposit is being used as a source of silica sand. If colluvial fans with intermittent streams are excavated, changes to dry channels can alter the water flow during spring run-off. Such diversions can endanger other land uses on the fan.

4.5 *Associated minerals*

Minerals with economic value may be present in sand and gravel deposits. These may be recovered as by-products during aggregate processing and sold for a higher unit value than the aggregate. Heavy minerals such as gold, platinum, garnet, and magnetite have been recovered from sands and gravels in Alberta. Silica also has been separated and sold. Preglacial, glaciofluvial, and alluvial deposits can contain viable by-product heavy minerals or silica. Preglacial deposits may even contain diamonds. The bedrock under the sand and gravel deposit also should be evaluated for possible ceramic, crushed stone, ammolite, coal, bentonite or other potential. The overburden should be examined for ceramic or silica sand potential.

5 CONCLUSION

Although the amount of sand and gravel present apparently is vast (Table 1), available reserves actually are much smaller and mineral aggregate resources are being consumed at twice the rate at which they are being discovered (Alberta Sand and Gravel Association 1995). No public mapping has taken place in Alberta for the last five years and existing resources are being removed from access through restrictions near watercourses and by other land uses. Mineral aggregate producers report that most supplies of sand and gravel available to them now will be consumed in the next 30 years.

Perhaps the greatest deficiency in understanding the amount of reserves available is simply that there is no public accounting of mineral aggregate reserves, no broad initiative to generate data, and little co-ordination of existing information. These deficiencies are now being identified and industry is starting to work closely with the Alberta government to address the concerns of both.

REFERENCES

Alberta Sand and Gravel Association. 1995. Annual General Meeting, January 13, 1995. Presidential address by Mr J. Moquin.
Edwards, W.A.D. 1979. *Sand and gravel deposits in the Canmore Corridor area, Alberta*. Alberta Research Council, Earth Science Report 79-2, 30 pp.
Edwards, W.A.D. 1991. The Geology of sands and gravels in the Bow Valley Corridor, Alberta. *27th Forum on the Geology of Industrial Minerals*. Alberta Research Council Open File Report 1991-23: 103-107.
Edwards, W.A.D. & Chao, D.K. 1989. *Bibliographic index and overview of aggregate resource publications*. Alberta Research Council, Open File Report 1989-13, 90 pp.

Edwards, W.A.D. 1995. *Mineral aggregate commodity analysis.* Alberta Geological Survey, Open File Report 1995-08. 54 pp.

Edwards, W.A.D. & Fox, J.C. 1980. *Sand and gravel resources of the Cold Lake area, Alberta.* Alberta Research Council Open File Report, 1980-8, 45 pp.

Edwards, W.A.D., Scafe, D.W. & Hudson, R.B. 1985. *Aggregate resources of the Edmonton/Lloydminster region.* Alberta Research Council, Bulletin 47, 64 pp.

Edwards, W.A.D., Scafe, D.W., Eccles, R., Miller, S., Berezniuk, T. & Boisvert, D. 1994. *Mapping and resource exploration of the Tertiary and preglacial formations of Alberta.* Alberta Research Council Open File Report 1994-06, 123 pp.

Fox, J.C., Richardson, R.J.H. & Sham, P.C. 1987 *Aggregate resource potential by geological ranking and reserve estimates; Peace River-High Level area, Alberta.* Alberta Research Council map.

Scafe, D.W., Edwards, W.A.D. & Boisvert, D.R. 1989. *Sand and gravel resources of the Peace River area.* Alberta Research Council Open File Report 1991-21, 47 pp.

Shetsen, I. 1980. *Sand and gravel resources of the Lethbridge area.* Alberta Research Council Earth Sciences Report 81-4, 41 pp.

Vagt, O. 1994. Mineral aggregates. In *Canadian Minerals Yearbook.* Mining Sector, Natural Resources Canada: 32.1-32.14.

Environmental effects of aggregate extraction from river channels and floodplains

G. MATHIAS KONDOLF
Department of Landscape Architecture and Environmental Planning,
University of California, Berkeley, USA

1 INTRODUCTION

River channels and floodplains are important sources of aggregate in many settings by virtue of the durability of river-worked gravels and their sorting by fluvial processes. The relative importance of alluvial aggregates is a function of the quality, location, and processing requirements of alluvial aggregates and alternative sources present in a given region. In Washington State, excellent upland sources of aggregates occur in extensive glacial outwash deposits convenient to many markets (Leighton 1919), and riverine sources account for less than 17% of the state's production (Collins 1995). By contrast, of the 120 million tonnes of construction aggregate produced annually in California (Carillo et al. 1990, Tepordei 1992) virtually all is derived from alluvial deposits, as indicated by the coincidence of mines with river courses (Fig. 1). Annual aggregate production from alluvial deposits in California exceeds estimated annual average bedload sediment production from the entire state by an order of magnitude (Kondolf 1995). The active tectonics of the region lead to high sediment yields, while the Mediterranean climate results in high inter-annual variability in flow and sediment transport. For example, Knudsen et al. (1992) calculated annual sediment loads for the Sisquoc River near Santa Maria, California for the period 1930-1990. Over this 60-year period, 85% of the total sediment load was transported in only two years, 1967 and 1969, high runoff years following an extensive fire in the watershed in 1966.

Aggregate is extracted from the active channel itself (*instream mining*), from pits excavated on the floodplain or terraces, termed *wet pits* if they intersect the water table, *dry pits* if they lie above it (Fig. 2).

While aggregate mining along rivers involves many of the same transient impacts as upland quarries (noise, dust, traffic, and contaminant spills), of more fundamental concern are the environmental effects that are unique to the dynamic riverine environment and that have no counterpart in upland quarries. By removing sediment from the active channel bed, instream mines interrupt the continuity of sediment transport through the river system, disrupting the sediment mass balance in the river downstream and inducing channel adjustments (usually incision) extending considerable distances (commonly one km or more) beyond the mine site itself. Concurrent with incision may be coarsening of bed material and loss of gravels used for spawning by

113

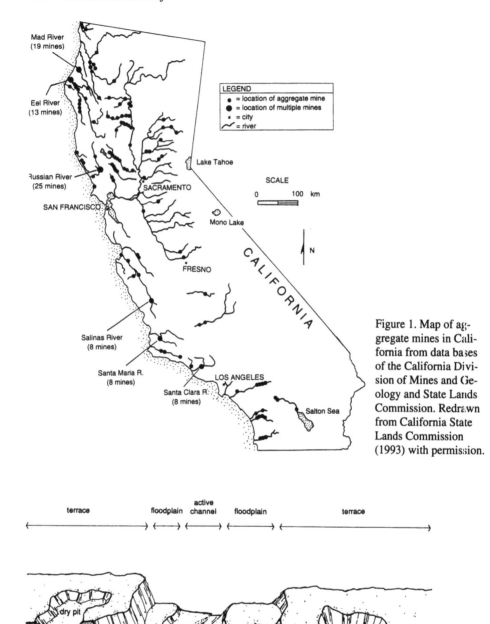

Figure 1. Map of aggregate mines in California from data bases of the California Division of Mines and Geology and State Lands Commission. Redrawn from California State Lands Commission (1993) with permission.

Figure 2. Diagram depicting alluvial deposits exploited for aggregate in relation to river channel morphology and alluvial water table (from Kondolf 1993 and 1994a, used with permission of Carfax Publishing Co. and Elsevier Science Publishers).

salmonids, salmon and trout (Kondolf & Wolman 1993). These effects may be considered as *cumulative effects* because they may become obvious only over time and extend beyond the limits of the mine site itself. Moreover, the effects of one mine may interact with those of other mines, yielding a net cumulative effect not apparent from a single mine.

Floodplain and terrace pits convert large areas of riparian habitat or farmland to open-water ponds, effecting a massive transformation of the landscape. In many cases, abandoned gravel pits have captured the active channel, transforming a formerly lotic environment into a lentic environment.

Until recently, these effects unique to riverine extraction were largely unrecognized. For example, in reporting on the quality and quantity of aggregates available in Arizona (all described as alluvial deposits), Keith (1969) made no mention of possible environmental impacts of their extraction. In a recent comprehensive review volume on aggregates published by the Geological Society of London, the section on 'Environmental Considerations' discussed only noise, dust, blasting, nuisance, visual impact, and restoration (Smith & Collis 1993:95-97). Elsewhere in the volume, the section on 'Fluvial deposits' included a paragraph noting that instream mining 'may change the dynamic equilibrium of a river: it may improve land drainage but increase scouring and erosion of the channel, as well as cause damage to bridge abutments' (Smith & Collis 1993, p.16-17). However, there was no treatment of the topic beyond this brief statement.

2 EFFECTS OF INSTREAM MINING

In upland environments, quarries can be isolated from the (essentially static) environment during extraction and the site reclaimed to another use upon completion of mining. By contrast, the riverine environment is dynamic. Instream mines cannot be isolated from the surrounding environment, and geomorphic effects typically propagate upstream and downstream of the mine site. Floodplain mines may be geomorphically isolated when constructed, but they may become connected if the pits capture the active channel.

The principal geomorphic effect is channel incision (commonly accompanied by bank undercutting and bed coarsening), which, in turn, may undermine structures, reduce the frequency of overbank flooding, and lower groundwater levels. The effects of instream mines can be understood only in context of effects of other activities along the river, including other mines, upstream dams, canalization works, and changes in runoff or sediment yield resulting from land use changes.

2.1 *Transient effects*

Aggregate mining operations in rivers have a number of transient impacts in common with upland quarries, such as noise, traffic, dust and other emissions, and potential spills of diesel fuel or other contaminants. Along large rivers, the mines may be located some distance from settlement, so noise and dust may generate less public opposition than if they occurred closer to upland settlements. However, some of these transient impacts can be considered more serious when they occur on a river, because of sensitive aquatic species present, and because of the role of water in transporting contaminants to sensitive receptors. For example, spills of hazardous materials may be more serious because of the exposure of aquatic organisms and potential contamination of water supplies. Similarly, the noise of gravel extraction and processing operations may affect holding, feeding, or migratory behavior of fish, although this topic has not been directly addressed in the literature (Klein & Kondolf 1994).

Extraction from the channel within the water suspends fine sediment, usually at times of year when high concentrations do not normally occur and when the river is unable to disperse the suspended sediments. The resulting turbidity and siltation of the downstream channel can reduce (and change composition of) macroinvertebrate populations and induce a change in fish populations to those tolerant of high suspended sediment concentrations (Forshage & Carter 1973).

During the period of mine operation, noise, truck traffic, and clearing of riparian vegetation can be expected to effect wildlife utilization of the riparian corridor. Similarly, for the period of mine operation and over a subsequent period of recovery, the processing facilities (usually located on the floodplain) displace former uses, typically riparian habitat or agriculture.

2.2 *Channel incision*

The form and dimensions of alluvial river channels are largely functions of the discharge (amount and distribution on a seasonal and inter-annual basis) and sediment load (amount, caliber, and temporal distribution) supplied from the basin (Leopold et al. 1964). By directly altering the channel geometry and elevation, instream mining induces channel adjustments. Moreover, by harvesting the river's *bedload* (the sand and gravel transported along the riverbed by rolling, sliding, and bouncing), mining disrupts the sediment mass balance of the river. From geomorphic principles, we would predict that this change in independent variables should induce a channel response, and along many rivers the channel has been observed to erode its bed and banks.

In most rivers experiencing instream mining, there are other human influences that could conceivably induce similar channel responses, such as upstream dam construction, bank protection and flood control works, or increased peak runoff from land use changes in the catchment. However, attribution (at least partial) to instream gravel mining is often justified because of the scale of extraction relative to bedload sediment supply: extraction commonly exceeds supply by an order of magnitude or more (e.g. Collins & Dunne 1989, Kondolf & Swanson 1993, Kondolf 1995).

2.2.1 *Sediment starvation and kickpoint migration*
By interrupting the continuity of sediment transport through the river system, mining can induce channel incision downstream and upstream. The pit created in the river bed traps bedload sediment transported from upstream, allowing sediment-starved water to pass downstream where it is likely to erode the channel bed and banks to regain some part of its former sediment load. This effect is analogous to the incision documented downstream of dams, especially small dams that trap bedload sediment but do not significantly reduce flood magnitude (Williams & Wolman 1984).

At the upstream end of the pit, the over-steepened bed is an unstable *knickpoint*, which will migrate upstream, extending incision in the upstream direction as well. In many cases, upstream knickpoint migration can be controlled by installation of check-dams and energy dissipaters, but installation of such structures does not solve problem of sediment-starved water downstream.

Incision of over one meter extending over more than one km (downstream and/or upstream of the extraction site) has been documented in many rivers, with incision of up to nine metres and extending as much as 20 km (from multiple mines in Cache Creek) reported. By analogy to incision from sediment starvation downstream of dams, incision from gravel mines can be expected to propagate farther in finer gravels and sands than in coarse gravels, because of armoring effects in the latter.

2.2.2 *Undermining of structures*

Perhaps the most visible manifestation of channel incision is the undermining of bridges, pipelines, and river control works, and the exposure of buried intakes for water supply systems. Undermining of bridges from incision produced by instream mining has been documented in many settings, including England (Sear & Archer 1995), Arizona (Bull & Scott 1974), Colorado (Stevens et al. 1990), and California (Kondolf & Swanson 1993, Fig. 3).

The California Department of Transportation has rated all 1200 bridges over water in the state for vulnerability to scour. About 1% are considered 'critically' threatened by scour, and most of these are associated with instream gravel mines upstream or downstream (Cathy Crossett, California Department of Transportation, personal communication 1991). Incision from instream mining has already resulted in serious damage or failure of numerous bridges in the state, including Stony Creek (Kondolf & Swanson 1993), Cache Creek and the Russian River (Collins & Dunne 1990), the Mad River (Lehre et al. 1993), Tujunga Wash (Scott 1973), the San Luis Rey River (Parsons et al. 1994), the San Diego, Santa Clara, and Kaweah rivers, and Kelsey,

Figure 3. Undermining of the Highway 395 bridge over the San Luis Rey River during the floods of January-February 1993. In-channel gravel pits (20 m deep) existed about 1 km upstream and 1 km downstream of the bridge prior to 1993 (photograph by G. Kondolf, March 1993).

Putah, San Juan, and Temecula creeks (unpublished data in files of the California Department of Transportation, Structure Hydraulics, Sacramento). Ironically, the Department of Transportation has itself been a principal customer for the gravel produced from many of the mines to which the incision has been attributed. Since 1992, the Department has been required to purchase aggregate (and aggregate products) only from contractors meeting permit requirements under the state's Surface Mine and Reclamation Act (SMARA) of 1975, although some of these mines have historically-based rights to extract aggregate near bridges. An amendment to SMARA requires that the Department of Transportation be notified and provided opportunity to comment if a new gravel mine is proposed within one mile of a highway bridge (Kondolf 1994b).

Intakes within alluvial gravels for water supply systems have been exposed or the filtering effects of gravel have been reduced by incision and loss of overlying bed material on the Mad (Lehre et al. 1993) and Russian (Collins & Dunne 1990) rivers. Water and utility pipeline crossings were exposed and damaged along the San Luis Rey River (Parsons et al. 1994). Because of potential incision-related damage to the siphon crossing of the Tehama-Colusa Canal under Stony Creek, a nearby gravel extractor was required by Glenn County to post a bond to cover costs of siphon repair in the event of damage.

Mining-related incision was initially thought to improve channel capacity, but ultimately undermined flood protection works along the Otaki River, New Zealand (Soil and Water 1985).

The costs to repair or replace structures damaged by incision from instream mining have not been assessed to the mines responsible for the damage, and thus the costs have not been incorporated into the price of the aggregate produced from active channel deposits. Because these deposits typically require less processing than upland or even floodplain/terrace sources, and because the environmental costs of their extraction are externalized, these instream sources are commonly cheaper to produce than alternative sources. However, the damage from incision has led to prohibition of instream mining in many countries, including Switzerland, Germany, France, and England.

2.2.3 Groundwater effects

During periods of baseflow, alluvial banks commonly drain to the river level. Incision of the channel and a consequent drop in river level may result in a corresponding drop in the water table in the alluvial banks, especially given the high hydraulic conductivities characteristic of alluvial gravels. Aquifer storage may be reduced, with losses in small river valleys with instream mines in Lake County, California estimated at between 1-16% (Lake County 1992). Lowered water tables can result in increased pumping costs and increased drought stress on riparian vegetation.

2.3 Channel instability

Instream mining can result in channel instability through the direct disruption of pre-existing channel geometry (Collins & Dunne 1990) or through the effects of incision and related undercutting of banks. In the Lake Tahoe basin in California, the channel of Blackwood Creek was relocated and straightened in 1960 so gravel could be extracted from pits in the former meander bends. The pits subsequently captured the channel, incision propagated upstream and downstream, resulting in massive bank erosion and increasing the stream's sediment load to Lake Tahoe fourfold (Todd 1989). Similarly, on San Juan Creek, California, incision upstream of a gravel pit induced bank erosion and channel widening (Chang 1987).

As knickpoints migrate upstream, they may supply sufficient sediment to downstream reaches to induce aggradation and instability (Sear & Archer 1995). Even if extraction rates are small, disruption of the surface pavement layer (Parker & Klingeman 1982) may result in increased bed mobility at lower flows than formerly.

2.4 *Use of gravel mining for flood control*

The change in sediment mass balance affected by instream gravel mining can be utilized as a tool for river control on aggrading rivers. The braided Waimakariri River, draining the Southern Alps of New Zealand, is aggrading in its lower reaches, threatening to fill the present channel and avulse (adopt a new channel course), most likely to occupy a former channel course running through the city of Christchurch. From 1929 to 1973, the Lower Waimakariri River aggraded an average of 2.9 m, despite aggregate extraction averaging 5.9 m (Griffiths 1979). Government geologists have recommended that instream gravel mining be maintained or increased to reduce the threat of aggradation and avulsion (Basher et al. 1988).

Flood control has been cited as a justification or benefit of instream mining projects (e.g. Bissell & Karn 1992), although in some cases the need for flood control was identified subsequently to need for aggregate. The State of Washington Department of Natural Resources (DNR) charges a royalty on gravel removed from rivers except when the removal is for purposes of flood control (DNR 1989). Although flood capacity can be increased by instream mining, the resulting incision can also undermine flood control works (Soil and Water 1985).

2.5 *Bed coarsening*

Incision (from any cause) may be accompanied by coarsening of the bed material (provided a range of sizes was present initially) because smaller, more mobile fractions are transported first, leaving behind a relatively immobile lag deposit of cobbles, boulders, or exposed bedrock. Gravel mining often selectively removes gravels of approximately the same size as needed for spawning by salmonids, with median diameters typically between 15 and 45 mm (Kondolf & Wolman 1993), thus reducing the availability of this resource.

Since construction of Shasta Dam in 1942, the Upper Sacramento River has had its supply of gravel from upstream cut off. Moreover, entire gravel bars were excavated to provide the 5.5 million m^3 needed for dam construction. Since then, intense aggregate mining in tributaries has caused many of these channels to scour to bedrock and has further reduced delivery of gravel to the mainstream Sacramento (Fig. 4). As a result, gravel bars and riffles have disappeared from the Upper Sacramento River, eliminating much formerly important salmon spawning habitat (Parfitt & Buer 1980).

Figure 4. Tributary to the Upper Sacramento River, scoured to bedrock as result of intense instream gravel mining, and now undergoing floodplain extraction (Photograph by G. Kondolf, January 1989).

2.6 *Bar skimming effects*

Bar *skimming* (or *scalping*) is extraction of gravel from the surface of gravel bars, usually down to a specified level above the river's water surface at the time of extraction (typically 0.3-0.6 m). Fish and wildlife agencies in California and Washington typically require that the bar, which would commonly have a steep margin and relatively flat top, be left after skimming with a smooth slope upwards from the edge of the low water channel at a 2% gradient (Collins 1995) to avoid stranding fish in shallow holes after high flows that inundate the bar. The rates of extraction from skimming operations are generally lower than other instream extractions (because the extraction depths are limited) and thus their impacts can be expected to be less, but impacts do occur from skimming.

By removing most of the gravel bar above the water level, the confinement of the low water channel is reduced or eliminated, changing the patterns of flow and sediment transport through the reach. One potential effect is reduced efficiency of sediment transport through the newly-unconfined reach, inducing deposition and thereby triggering instability. Just as the upstream end of submerged riffles serve as hydraulic controls for upstream pools at low flow, the upstream end of bars may serve as controls for upstream reaches at bankfull flow (see Leopold et al. 1964 for discussion of bankfull flow). Thus, removal of the bar may alter channel hydraulics upstream as well as at the mined site itself. Pauley et al. (1989) documented scour in riffles upstream of a skimmed bar on the Puyallup River, Washington, apparently because of the lowering of the downstream hydraulic control. This potential effect has raised concerns in Washington because of the potential loss of incubating salmon embryos in the scoured gravels (Bates 1987).

The current channel may be abandoned and a former channel adopted instead following bar removal (Dunne et al. 1981). Skimming may eliminate side channels, which are important habitats for juvenile salmonids (Pauley et al. 1989). Removal (or simple disruption) of the surface layer on the bar may render the remaining gravel mobile at lower discharges because the effects of the coarse pavement layer in regulating bedload transport (Parker & Klingeman 1982) are lost.

A little-recognized effect of gravel bar skimming is the potential for increased establishment of willows on skimmed gravel bars in western North America and other semiarid regions in which moisture availability during the growing season limits plant survival. Where undisturbed gravel bars are more than 1 m above the low water elevation of the river, the bars may remain largely unvegetated because seedlings that establish there are likely to die from desiccation by virtue of the depth to water table during the dry summer and fall. By lowering the top of the bar, bar skimming may create shallow water table conditions in which the willows can establish (Fig. 5). Establishment of vegetation on the bar may create habitat but it also may reduce flood capacity of the channel.

2.7 *Lag in channel response*

Bedload sediment transport occurs as a power function of discharge, so variations in discharge produce even greater variations in sediment transport. These variations are extreme in Mediterranean climate rivers, with their large seasonal and inter-annual variations in discharge. In central and southern California, sediment transport can be considered *episodic*, in that most sediment transport occurs during a very small percentage of the time (e.g. Knudsen et al. 1992).

The effects of instream gravel mining may not be obvious immediately because active sediment transport is required for the effects (e.g. incision, instability) to

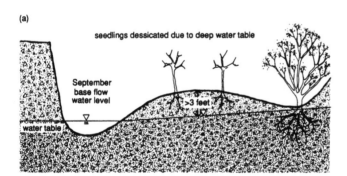

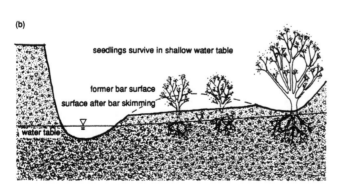

Figure 5. Diagram showing potential effect of gravel-bar skimming on establishment of willow seedlings. a) Top of the un-skimmed gravel bar is too high for willow establishment because the depth to groundwater is too great. b) After skimming, the surface of the gravel bar is closer to the water table, permitting survival of seedlings.

propagate upstream and downstream. Given that geomorphically-effective sediment transporting events are infrequent on many rivers, there may be a lag of several or many years before the effects of instream mining are evident and propagate along the channel. Thus, gravel mines may operate for years without apparent effects upstream or downstream, only to have the geomorphic effects manifest years later during high flows. Similarly, rivers are often said to have 'long memories', meaning that the channel adjustments to instream extraction or comparable perturbations may persist long after the activity itself has ceased.

3   EFFECTS OF FLOODPLAIN PITS

The principal effect of wet pit mining on floodplains (and terraces) is the conversion of land uses during and after the mining operation. Where an aggregate pit lies entirely above the water table (a dry pit), it is possible to reclaim the pit to agriculture or housing, similar to other quarries or open-pit mines. However, floodplain pits typically intersect the water table for at least part of the year (wet pits), resulting in land-use conversion from farmland or riparian habitat to open-water pond. The scale of the landscape transformation effected by this pit mining can be appreciated by flying over river floodplains in light aircraft.

Even if the levees separating the pit from the channel remain intact, there is typically a strong hydrologic connection among the river, the pit, and the alluvial water table such that conditions in (or contamination of) the pit waters can affect water quality in the alluvial aquifer. Along the Russian River in California, the potential for contamination of groundwater by contamination of exposed waters in floodplain pits has been raised as an issue.

Another potential effect of floodplain pits arises if the pits capture the active river channel, effectively transforming the floodplain pits into instream pits. Pit capture has been documented on Tujunga Wash, California (Scott 1973), along the Yakima River, Washington (Dunne & Leopold 1978), and at 12 of 25 floodplain mining sites studied in northern Alaska (Woodward-Clyde Consultants 1980). In general, pit capture is probably most likely in sites where the gravel pit lies in a shortcut for the flooding river, such as the inside of a meander bend.

Pit capture can result in incision propagating upstream and downstream, and a large 'lake' may be created within the river, transforming a lotic environment into a lentic environment, thereby inducing changes in the ecology of the reach. Captured gravel pits in the Naugatuck River, Connecticut, are now virtual lakes with seasonally stagnant water and depressed dissolved oxygen levels; based on estimated bed-load transport rates, the pits are expected to persist for hundreds of years (MacDonald 1988).

Along the Merced and Tuolumne rivers in California, captured gravel pits constitute habitat for exotic warm-water fish species that prey extensively on salmon smolts migrating oceanward. Over 70% of juvenile chinook salmon migrating oceanward were estimated to be lost to predation along an 8-km reach of the Tuolumne River (EA 1992). Predation in captured gravel pits is recognized as a serious obstacle to restoring native salmon runs in the San Joaquin River system, but refilling or re-isolating captured pits is expensive. Thus far, funding has been allocated to repair failed levees to re-isolate one gravel pit on the Merced River at a cost of US$ 361,000 (Kondolf et al. 1996) and refilling of two pits on the Tuolumne River has been proposed at a cost of $5.3 million (McBain & Trush 1996).

4 RECLAMATION OF FLOODPLAIN AND TERRACE PITS

Dry pits can be reclaimed to agriculture, as is done at the Aspen Mine, which exploits older terrace gravels of the American River southeast of Sacramento. The gravel is removed, the topsoil replaced, and the resulting ground surface (presently used for agriculture) is about 6 m lower than the original surface (Sacramento County 1987). On wet pits, reclamation to agriculture is not possible unless the pit is refilled so the resulting land surface is above the water table. In the Aggregate Resource Management Plan adopted in 1980, Sonoma County, California, intended to direct floodwaters into floodplain gravel pits along the Russian River, so the pits would refill from deposition of sediment. The process of deposition has begun to occur in one pit, which is separated from the river by only a low berm and which therefore floods frequently. However, the Department of Fish and Game prohibited flood waters from being directed into the pits because of the potential for fish to be carried or swim into the pits, only to become trapped as floodwaters receded (Marcus 1992). Moreover, the time required for such refilling could be quite long, depending upon the river's sediment load and caliber, and the hydraulic conditions at the approach and entrance to the pit. The resulting core of sediment deposited in the pit would probably be composed mostly of suspended sediment (sand and silt), considerably finer than the gravel and sand removed. Thus, groundwater flow patterns through the aquifer may be locally altered by creating lenses of reduced hydraulic conductivity within the wider floodplain aquifer of high conductivity gravels.

The pit could be refilled with other materials. However, to preserve the conductivity of the aquifer medium would require filling the pit with something like gravel and sand, which would presumably need to be derived from a mine elsewhere. Abandoned gravel pits have been utilized as landfills in some areas, but it is difficult to imagine a less favorable site environmentally than a floodplain gravel pit, with its high water table, the high hydraulic conductivity of the floodplain gravels, and the resultant threat posed to water supply and aquatic ecological resources.

If the pits are not refilled, they can be used for swimming, as is done in Helena, Montana, and Santa Clara Valley, California, or for boating and water skiing, as in the Hedeland district of Denmark (Schultz 1990). Pits can also be used to recharge groundwater, especially in proximal alluvial fan settings (Fig. 6).

Increasingly, reclaimed gravel pits are being used provide riparian wetland habitat, thereby partly mitigating historical losses of wetland habitat as high as 91% in western North America since the mid-19th century (NRC 1992). The potential of former pits as wildlife habitat was emphasized in a recent publication of the aggregate industry in California (CVRSGA 1995): 'There is a satisfying symmetry between sand and gravel mining and wetlands reclamation, a balance between the development of one resource (construction aggregates) and the creation of a new resource (wetlands)'.

Figure 6. Gravel pits flanking Alameda Creek, Fremont, California. The pits are used to recharge groundwater through the proximal fan gravels (Photograph by G. Kondolf, July 1989; from Kondolf 1993, used with permission of Carfax Publishing Company).

4.1 *Pit geometry*

Research on habitat values of abandoned gravel pits in the UK has identified shallow waters (< 1 m) and gently sloping banks as providing the most productive habitat because sunlight can penetrate to the bottom in shallow waters, supporting growth of aquatic macrophytes, and emergent banks with shallow water tables can support wetland plants (Andrews & Kinsman 1990, Giles 1992). The plants provide habitat and food for aquatic and riparian species. While waterfowl require some open water (consistent with deeper waters in pits), there is evidence that waterfowl avoid swimming near steeply sloping banks because of the threat posed by terrestrial predators that may lurk directly above the waters along a steep bank (Tom Griggs, The Nature Conservancy, Hamilton City, California, personal communication 1995).

Along the floodplain of the Great River Ouse near Milton Keynes, England, the Great Linford gravel pits provide excellent waterfowl habitat, and they are visited by large numbers of bird watchers (Fig. 7). Because the gravel deposits here were only about 2 m thick, the original pits were quite shallow, and shallow water marginal habitats were expanded further through regrading during restoration. Based on experience at Great Linford and other sites in the UK, Andrews & Kinsman (1990) recommended that pit margins be sloped at 7% or less over at least 20 m (measured normal to the shoreline) to provide a minimum of 15 m of water < 1 m deep even with seasonal water table fluctuations of 0.3 m. As an alternative to sloping banks, benches can be cut in the pit margins to provide both shallow aquatic habitat and exposed surfaces for establishment of riparian vegetation (Baseline Environmental Consulting 1992).

Despite the importance of these shallow water marginal habitats, most gravel pits presently abandoned on the landscape have steeply sloping banks, providing only a narrow band along which riparian vegetation can establish in between deep waters and steep, xeric uplands (Fig. 8). A deep, steep-sided pit maximizes the aggregate production from a given area. To create gently-sloping or stepped banks requires ei-

Figure 7. Restored gravel pits at Great Linford, along the Great River Ouse, near Milton Keynes, England (Photograph by G. Kondolf, July 1993).

Figure 8. A steep-sided shore along the Russian River (the Benoit Pit), support-ing little marginal ri-parian vegetation (Photograph by G. Kondolf, April 1993).

ther enlarging the area of disturbance (to maintain the same yield of aggregate) or re-ducing the aggregate yield (to maintain the same area of ground disturbance).

4.2 *Water level fluctuations*

Water table fluctuations pose another constraint upon creation of shallow water habitat. The Andrews & Kinsman (1990) recommendation of a 20 m wide sloping bank is based on an assumed water level fluctuation of 0.3 m, a value that may be typical of humid climates with relatively uniform seasonal distribution of precipita-tion and perennial streamflow. However, in more hydrologically variable climates such as the Mediterranean climate of California, river stage and alluvial water level

Figure 9. Gravel pit on Middle Creek, Lake County, California. Water levels fluctuate 7 m seasonally in this pit, posing obstacles to establishment of riparian vegetation and creation of aquatic or riparian habitat (Photograph by G. Kondolf, October 1994).

fluctuations are typically greater. The most extreme fluctuations occur along intermittent streams, such as Middle Creek, a tributary to Clear Lake, California, where water table fluctuations of seven meters occur (Fig. 9). Maintaining shallow water habitat in pits with such fluctuations is clearly infeasible, and establishing riparian vegetation may be impossible without irrigation.

4.3 *Spawning and rearing habitat for salmon*

Floodplain gravel pits have been successfully developed as off-channel spawning and rearing habitat for salmon and trout in Idaho (Richard et al. 1992) and Washington (Partee & Samuelson 1993). Such habitats are more likely to be successful to the extent that their geometries resemble natural side channels (typically groundwater-fed), which are used for spawning by salmonids in western Canada and Alaska (e.g. Vining et al. 1985). For spawning habitat, extractions should be linear, relatively narrow and shallow to create flowing water conditions. For rearing habitat, deeper pools may be appropriate. The Weyco-Brisco ponds along the Wynoochee River, Washington, were created by extractions that maximized habitat quality upon reclamation rather than extraction of aggregate from the site (Partee & Samuelson 1993). Off-channel habitat such as this is unlikely to be beneficial for salmon in areas with warmer summers, such as California, because the off-channel ponds are likely to provide habitat for warm-water species that predate upon salmon smolts.

5 CONCLUSIONS

Extraction of aggregate from rivers and floodplains involves a host of environmental effects that are distinct from, and in addition to, the environmental effects of upland quarries. The effects of upland quarries are mostly transient or easily mitigated upon reclamation. In contrast, instream gravel mining may induce effects that propagate upstream and downstream, including incision, loss of gravels and coarsening of the bed, undermining of structures, lowering of alluvial water tables, and channel instability. These effects may be cumulative from multiple mines and over a period of time, inducing channel adjustments that may persist for decades after cessation of mining.

Because they commonly intersect the water table, floodplain mines commonly transform the land from agriculture or riparian habitat to water-filled pits. These pits can have a variety of uses, but increasingly, creation of wetlands habitat upon reclamation is encouraged by resource agencies and emphasized by the aggregate industry itself. To be most effective as habitat, pits should have gently sloping (or benched) sides to maximize shallow water habitat. In Mediterranean climates, extreme seasonal variability in water levels may severely constrain the opportunities for habitat creation in reclaimed pits.

REFERENCES

Andrews, J. & Kinsman, D. 1990. *Gravel pit restoration for wildlife: A practical manual.* The Royal Society for Protection of Birds, Sandy, Bedfordshire.

Baseline Environmental Consulting 1992. Reclamation plan for the Kaiser Sand and Gravel Company Piombo Pit. Draft environmental impact report submitted to Sonoma County Planning Department, Santa Rosa, California.

Basher, L.R., Hicks, D.M., McSaveney, M.J. & Whitehouse, I.E. 1988. The Lower Waimakariri River floodplain: a geomorphic perspective. A report for North Canterbury Catchment Board, Department of Scientific and Industrial Research, Division of Land and Soil Sciences, Christchurch.

Bates, K. 1987. Fisheries perspectives on gravel removal from river channels. In Realistic approaches to better floodplain management. Proceedings of the Eleventh Annual conference of the Association of State Floodplain Managers, Seattle, June 1987. Natural Hazards Research and Applications Information Center, Special Publication No. 18: 292-298.

Bissell & Karn 1992. Project description, Coast Rock Products, Inc., Master plan for mining and reclamation along the Sisquoc and Santa Maria Rivers in Santa Barbara and San Luis Obispo Counties. Bissell & Karn, San Ramon, California.

Bull, W.B. & Scott, K.M. 1974. Impact of mining gravel from urban stream beds in the southwestern United States. *Geology* (2): 171-174.

Carillo, F.V., Davis, J.F. & Burnett J.L. 1990. California: Annual Report. US Department of Interior, Bureau of Mines. Washington, D.C.

Chang, H.H. 1987. Modeling fluvial processes in streams with gravel mining. In C.R. Thorne, J.C. Bathurst & R.D. Hey (eds), *Sediment transport in gravel-bed rivers*: 977-988. John Wiley.

Collins, B. 1995. Riverine gravel mining in Washington state, overview of effects on salmonid habitat, and a summary of government regulations. Unpublished report to the US EPA, Seattle.

Collins, B. & Dunne, T. 1989. Gravel transport, gravel harvesting, and channel-bed degradation in rivers draining the Southern Olympic Mountains, Washington, USA. *Environmental Geology and Water Science* (13): 213-224.

Collins, B. & Dunne, T. 1990. Fluvial geomorphology and river gravel mining: a guide for planners, case studies included. California Division of Mines and Geology Special Publication 98, Sacramento, California.

CVRSGA (Central Valley Rock, Sand & Gravel Association) 1995. Wetlands reclamation in California's Central Valley. Brochure prepared by CVRSGA, Los Banos, California.

Dunne, T. & Leopold, L.B. 1978. Water in environmental planning. W.H. Freeman and Co., San Francisco.

Dunne, T., Dietrich, W.E. Humphrey, N.F. & Tubbs, D.W. 1981. Geologic and geomorphic implications for gravel supply. Proceedings for the Conference on Salmon-Spawning Gravel: A Renewable Resource in the Pacific Northwest? October 1980, Seattle, WA, Washington State University Research Center.

Forshage, A. & Carter, N.E. 1973. Effect of gravel dredging on the Brazos River. Proceedings of the 27th Annual Conference. *Southeastern Association Game and Fish Commission* (24): 695-708.

Giles, N. 1992. Wildlife after gravel: twenty years of practical research by The Game Conservancy and ARC. The Game Conservancy, Fordingbridge, Hampshire.

Griffiths, G.A. 1979. Recent sedimentation history of the Waimakariri River, New Zealand. *Journal of Hydrology* (18): 6-28.

Keith, S.B. 1969. Sand and gravel. In Mineral and Water Resources of Arizona. Arizona Bureau of Mines Bulletin (180): 424-441.

Klein, R. & Kondolf, G.M. 1994. Lower Eel and Van Duzen Rivers: description of necessary components for development and implementation of river management plan. Report to the County of Humboldt, Planning Department, Eureka, California.

Knudsen, K., Hecht, B. & Holmes, D.O. 1992. Hydrologic and geomorphic factors affecting alternatives for managing the alluvial corridor of the Lower Sisquoc River, Santa Barbara, California. Balance Hydrologics, Berkeley, California, report submitted to S.P. Milling Company.

Kondolf, G.M. 1993. The reclamation concept in regulation of gravel mining in California. *Journal of Environmental Planning and Management* (36): 397-409.

Kondolf, G.M. 1994a. Geomorphic and environmental effects of instream gravel mining. *Landscape and Urban Planning* (28): 225-243.

Kondolf, G.M. 1994b. Environmental planning in regulation and management of instream gravel mining in California. *Landscape and Urban Planning* (29): 185-199.

Kondolf, G.M. 1995. Managing bedload sediments in regulated rivers: examples from California, USA. In *Natural and Anthropogenic Influences in Fluvial Geomorphology* (89): 165-176. Geophysical Monograph, American Geophysical Union, Washington D.C.

Kondolf, G.M. & Swanson, M.L. 1993. Channel adjustments to reservoir construction and gravel extraction along Stony Creek, California. *Environmental Geology and Water Science* (21): 256-269.

Kondolf, G.M. & Wolman, M.G. 1993. The sizes of salmonid spawning gravels. *Water Resources Research* (29): 2275-2285.

Kondolf, G.M., Vick, J.C., & Ramirez, T.M. 1996. Salmon spawning habitat rehabilitation in the Merced, Tuolmne, and Stanislaus rivers, California: an evaluation of project planning and performance. Report No. 90, University of California Water Resources Center, Davis, California.

Lake County 1992. Lake County aggregate resource management plan. Lake County Planning Department, Resource Management Division, Lakeport, California. Draft.

Lehre, A., Klein, R.D. & Trush, W. 1993. Analysis of the effects of historic gravel extraction on the geomorphic character and fisheries habitat of the Lower Mad River, Humboldt County, California. Appendix F to the Draft Program Environmental Impact Report on Gravel Removal from the Lower Mad River. Department of Planning, County of Humboldt, Eureka, California.

Leighton, M.M. 1919. *The road building sands and gravels of Washington.* Washington Geological Survey Bulletin 22.

Leopold, L.B., Wolman, M.G. & Miller, J.P. 1964. *Fluvial processes in geomorphology.* Freeman & Sons, San Francisco.

MacDonald, A. 1988. Predicting channel recovery from sand and gravel extraction in the Naugatuck River and adjacent floodplain. In *Hydraulic Engineering*: 702-707. Proceedings of the National Conference, American society of Civil Engineers Hydraulics Division, Colorado Springs, Colorado.

Marcus, L. 1992. Status report: Russian River resource enhancement plan. California Coastal Conservancy, Oakland.

McBain, S., & Trush, W. 1996. Tuolomne river channel restoration project, special run pools 9 and 10. Report submitted to Tuolumne River Technical Advisory Committee (Don Pedro Project, FERC License No. 2299) by Mc Bain and Trush, Arcata, California.

NRC (National Research Council) 1992. Restoration of aquatic ecosystems. National Academy Press, Washington, D.C.

Parfitt, D. & Buer, K. 1980. Upper Sacramento River spawning gravel study. California Department of Water Resources, Northern Division, Red Bluff.

Parker, G. & Klingeman, P.C. 1982. On why gravel-bed streams are paved. *Water Resources Research* (18): 1409-1423.

Parsons, Brinkerhoff, Gore & Storrie, Inc. 1994. River management study: permanent protection of the San Luis Rey River Aqueduct crossings. Report to San Diego County Water Authority, Parsons Brinkerhoff Gore & Storrie, Inc.

Partee, R.R. & Samuelson, D.F. 1993. Weyco-Brisco ponds habitat enhancement design criteria. Unpublished report, Grays Harbor College, Aberdeen, Washington.

Pauley, G.B., Thomas, G.L., Marino, D.A. & Weigand, D.C. 1989. Evaluation of the effects of gravel bar scalping on juvenile salmonids in the Puyallup River drainage. University of Washington Cooperative Fishery Research Unit Report. University of Washington, Seattle.

Richards, C., Cernera, P.J., Ramey, M.P. & Reiser, D.W. 1992. Development of off-channel habitats for use by juvenile chinook salmon. *North American Journal of Fisheries Management* (12): 721-727.

Sacramento County 1987. Supplemental environmental impact report for Teichert (Aspen IV-A) Land Company use permit. Sacramento County Planning and Community Development Department, Sacramento. Control No. 87-UP-0232.

Schultz, P. 1990. Disused gravel mines provide skiing and boating for Denmark. *Landscape Architecture* (80): 30-32.

Scott, K.M. 1973. Scour and fill in Tujunga Wash – A fanhead valley in urban southern California – 1969. US Geological Survey Professional Paper 732-B.

Sear, D.A. & Archer, D.R. 1995. The effects of gravel extraction on the stability of gravel-bed rivers: a case study from the Wooler Water, Northumberland, UK. paper presented to the Fourth International Workshop on Gravel-Bed Rivers, Gold Bar, Washington, August 1995.

Smith, M.R. & Collis, L. 1993. *Aggregates*. Geological Society Engineering Geology Special Publication No. 9 (second edition). Geological Society, London.

Soil & Water. 1985. Attacks on the Otaki – Gravel or grants? Anonymous article in Soil & Water Magazine, National Water and Soil Conservation Authority. Wellington, New Zealand (21): 2-6.

State Lands Commission, California 1993. California rivers: A public trust report. California State Lands Commission, Sacramento.

Stevens, M.A., Urbonas, B. & Tucker, L.S. 1990. Public-private cooperation protects river. APWA Reporter, September 1990: 25-27.

Svedarsky, W.D. & Crawford, R.D. 1982. Wildlife values of gravel pits, symposium proceedings. University of Minnesota Agricultural Experiment Station. Miscellaneous Publication 17-1982. St. Paul, Minnesota.

Tepordei, V.V. 1992. Construction sand and gravel: annual report. US Department of Interior, Bureau of Mines. Washington, D.C.

Todd, A.H. 1989. The decline and recovery of Blackwood Canyon, Lake Tahoe, California. In Proceedings of International Erosion Control Association Conference, Vancouver, B.C.

Vining, L.J., Blakely, J.S. & Freeman, B.M. 1985. An evaluation of the incubation life-phase of chum salmon in the Middle Susitna River, Alaska. Alaska Department of Fish and Game winter habitat investigations Report No. 5, Volume 1.

WDNR (Washington Department of Natural Resources) 1989. Gravel removal from rivers for reducing flood risk. Washington Departments of Natural Resources and Ecology, Olympia, Washington.

Williams, G.P. & Wolman, M.G. 1984. Downstream effects of dams on alluvial rivers. US Geological Survey Professional Paper 1286.

Woodward-Clyde Consultants 1980. Gravel removal studies in arctic and subarctic floodplains in Alaska. US Fish and Wildlife Service Biol. Serv. Progress Rep., FWS/OBS-80/08.

The coarse aggregate resources of South Africa

J.N. DUNLEVEY & D.J. STEPHENS
University of Durban-Westville, South Africa

1 INTRODUCTION

The objective of this study is to provide an overview of the availability and usage of coarse aggregate in the construction industry in South Africa where a unique blend of First and Third World developments can be found. As South Africa re-establishes its position a period of rapid growth and development to redress the imbalances of the past and build up industrial power is predicted. It is likely that both international agencies and construction organizations will once more establish bases in South Africa to act as regional and continental development centres, and it is intended that this article may assist as a guide to current aggregate usage, availability and quality norms.

2 REGIONAL DEMOGRAPHICS

The population distribution, industrial and commercial activities in South Africa can be directly related to the climate and geology of the country, for certain areas are too arid for development while others have economically important mineral deposits. The country is divided into nine provinces, in which there are five 'urban' growth centres and the rest of the country can be considered as being 'rural' (Fig. 1). These growth areas have been used in this study to evaluate the role of coarse aggregates in South African construction activities. Gauteng is the main industrial and commercial region of South Africa and is based around the many gold mines in the Witwatersrand area; this region also includes Pretoria, where the main government offices and administrative centres are situated. Cape Town and Durban are major ports with typical associated development, but relatively few mineral resources. The national Parliament is located in Cape Town. The Eastern Cape is a largely agricultural district, with a number of light industries, many of which are related to the motor vehicle manufacturing centres located at Port Elizabeth and Uitenhage. The Bloemfontein region is dominated by agriculture and related industrial developments, although it also has importance as a regional centre. The northern and eastern portions of the country, although dominated by agriculture with some forestry activity, contain a few important mining sites. Population densities and construction activities in

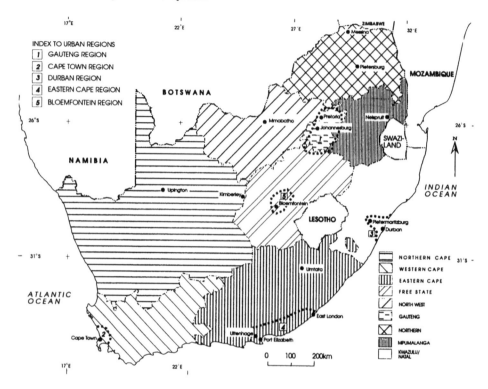

Figure 1. Urban regions in South Africa.

these regions are relatively low. The Karoo, which includes the Northern Cape Province, the western parts of the North West and Free State provinces together with the northern parts of the Western and Eastern Cape provinces, is very sparsely populated, being both arid and lacking mineral resources with most of the population located in small towns.

The South African economy is a peculiar mix of developed and developing sectors, probably mainly due to its colonial development and history. The distribution of aggregate consumption is therefore very complicated because the South African construction industry has to address all facets of this economy and as a result is very diverse. At one extreme, the advanced sector of the industry utilizes high technology and demands high quality materials for its products, while at the other extreme the lower-level sector has only very basic construction material requirements. Within this framework the industry works on many levels in an almost symbiotic relationship, and the criteria used (if any) in the selection of materials depends on the project specifications. In larger-scale projects, stringent requirements are often set that exceed the national standards set by the South African Bureau of Standards (SABS 1995). However, in many small-scale construction projects material selection is left to the builder and is based on experience rather than any scientific evaluation of the physical properties of materials. In addition, in the 'informal' sector this differentiation is even more marked with, for example, aggregates being extracted from the material available on the actual building sites.

The supply of aggregate plays an essential role in the building and construction industries and makes a significant contribution to both employment and the Gross National Product. Data supplied by the South African Minerals Bureau (Van Zyl & Schreuder 1994) indicates that the value of natural sand and coarse aggregate supplied to the construction industry exceeds R200 million (US$56 million) per annum (1990 to 1993).

3 NATIONAL PRODUCTION

Mainly due to the complex commercial considerations described above, accurate national aggregate supply and production figures are difficult to obtain. However, valid estimates can be based on the data from major aggregate producers and the sales of cement around the country. Using these criteria, it is estimated that the annual consumption of concrete aggregate produced from crushed rock is approximately 37.2×10^6 tonnes, while some 3.5×10^6 tonnes of crushed rock are used as railway ballast, with a further 17.3×10^6 tonnes used for road aggregate, giving a grand total of some 58×10^6 tonnes (1990 to 1993).

As construction aggregate has a low price per unit volume, transport costs form a significant portion of the price structure. In rural areas, small quarries tend to serve each local community while in the urban areas a small number of large quarries each serve a specific district (unless a particular material is required from another quarry). Transport cost considerations ensure that local rock types are utilized unless proven unsuitable, and therefore different lithologies within a single quarry can be used as aggregate. Thus the rock type utilization data (Table 1) reflects more the availability of acceptable aggregate sources to the growth areas than the physical properties and quality of each rock type. Table 2 represents the relative intensity of construction activity in the major development centres with Gauteng accounting for almost 50% of the total aggregate consumption in South Africa.

The five major urban growth areas of South Africa shown in Figure 1 are widely dispersed and located in very different geological terrains. The lithologies used in each area (Table 2) vary significantly, and it must be borne in mind that there can be significant differences even where the same lithological description is used. For example, the granite and quartzite used in the Gauteng region were formed approximately 2000

Table 1. Lithologies used as coarse aggregate in South Africa (1990-1993).

Lithologies	% usage	Million tonnes per year
Quartzite	36	20.9
Dolerite	22	12.8
Granite	12	6.9
Hornfels/graywacke	9	5.2
Diamictite (tillite)	5	2.9
Sandstone	4	2.3
Dolomite	3	1.7
Andesite	2	1.2
Others	7	4.1

Table 2. Coarse aggregate usage in major growth areas (1990-1993).

Growth area	Lithologies	% of total used
Gauteng	Andesite Dolomite Gabbro/norite Granite/felsite Quartzite	48
Cape Town	Granite Hornfels/graywacke Quartzite	19
Durban	Diamictite Quartzite Dolerite Granite	15
Eastern Cape	Dolerite Quartzite	12
Bloemfontein	Dolerite Quartzite	6

million years ago while those in the Cape Town region are only 400 to 600 million years old. Rocks in these two regions have consequently had very different diagenetic histories. Therefore, the physical differences between the lithologies utilized in the five urban growth areas discussed here are such that the aggregates have been considered primarily on a regional rather than a lithological basis. It is beyond the scope of this paper to provide a detailed description of complex South African geology and for further information on this topic readers are referred to the geological maps and publications of the South African Council for Geoscience and the Geological Society of South Africa as well as the books on stratigraphy by Tankard et al. (1982) and SACS (1980).

4 AGGREGATE PROPERTIES

Aggregate produced from crushed sedimentary and metamorphosed sedimentary rocks generally display significantly more variation than those derived from igneous rocks. This variation is normally the result of primary differences in the layering of the strata. Igneous bodies may contain significant variations, particularly in small bodies such as sills, but the larger masses, notably the gabbros, norites and granites, are usually quite consistent. Many high-grade metamorphic rocks can be crushed to produce a quality aggregate but foliated material will normally have a very poor shape.

Although many of the physical characteristics of an aggregate can be predicted from the mineralogy, the properties quoted in most specifications must be measured by physical tests. Therefore, emphasis is placed on measured physical properties (Davis & Alexander 1989) rather than the mineralogy and petrology of the various lithologies. The specification quoted in Table 3 is 'SABS 1083-1976 Standard specification for aggregates from natural sources' (SABS 1979). While the SABS codes are generally the most used in construction in South Africa, other specifications have been popular,

Table 3. Summary of SABS 1083 (1979) specifications.

Property	Unit	Concrete aggregate	Base and subbase	Surfacing Surface dressing	Rolled in chips	Bitumi-nous
Flakiness index	%		35 max			
53.0-26.5 mm diameter	%			35 max		35 max
19.0-26.5 mm	%			25 max	20 max	25 max
9.5 mm	%			30 max		30 max
10% FACT						
Arenaceous:						
– Non-siliceous	kN		140 min			
– Siliceous	kN		110 min			
Diamictite (tillite)	kN		160 min	220 min	220 min	170 min
Argillaceous	kN		180 min			
Other rock types	kN		110 min	210 min	210 min	160 min
Surface abrasion conditions exist	kN	110 min				
No surface abrasion	kN	70 min				
ACV, dry	%	29 min				
Arenaceous:						
– Non-siliceous	%		27 max			
– Siliceous	%		29 max			
Diamicitite (tillite)	%		25 max	20 max	20 max	24 max
Argillaceous	%		24 max			
Other rock types	%		29 max	21 max	21 max	25 max
Voids content	%	48 max				
Drying shrinkage (as % of reference)						
Where shrinkage is important	%	150 max				
Where shrinkage is not important	%	200 max				
Water absorption	%	1.0				
Abrasion resistance (Los Angeles)	%loss			40 max	40 max	40 max
Concrete subject to abrasion	%loss	40 max				
Concrete not subject to abrasion	%loss	50 max				
Aggregate for road surfaces	%loss		40 max			
Aggregate for base courses	%loss		50 max			
Railway ballast:						
– Class A	%loss		22 max			
– Class B	%loss		28 max			
– Class C	%loss		34 max			
Durability (wet 10 % FACT as % of dry)	%		75 min	75 min	75 min	75 min
For general use	%	75 min				
Water retaining structures	%	75 min				
Polished Stone Value (PSV)						
Continuously graded bituminous	%			50 min		50 min
Gap-graded bituminous mixture	%			45 min		45 min
Rolled-in chips	%				55 min	
< 200 vehicles/lane/day	%			45 min		45 min
200-600 vehicles/lane/day	%			50 min		50 min
600-1000 vehicles/lane/day	%			55 min		55 min

particularly in road construction. Two such specifications are the CSRA (Committee of State Road Authorities) 'Standard Specification for road and bridge works' (CSRA 1987) and 'TRH14 Guidelines for road construction materials' (NITRR 1985). The sections of the CSRA and NITRR specifications dealing with aggregates are virtually identical to those of SABS 1083 and are therefore not discussed in detail.

5 GAUTENG REGION

Gauteng (Fig. 1) is the largest conurbation and fastest-growing urban region in South Africa, consuming almost 50% of its entire coarse aggregate production. Although not particularly large, this region has a diverse geological setting (SACS 1980, Tankard et al. 1982) which provides a greater variety of rocks suitable for use as crushed aggregate than any of the other regions (Tables 2 and 4). As Gauteng is a major mining centre, in addition to rock from quarries specifically producing aggregate, crushed rock is available from stope and mine development.

Quartzite
Several quartzite bodies are utilized for coarse aggregate in the Gauteng region, however, those from the Pretoria area in the north have somewhat different characteristics to those quarried in the south. Although both of these quartzites are suitable for use as concrete aggregate, those from the northern area are of a somewhat better quality, with lower drying shrinkages and a higher wet/dry ratio for 10% FACT (Fines Aggregate Crushing Test, SABS 1979). The reserves of the Magaliesberg Quartzite (previously known as the 'Pretoria quartzite') in the northern area are very limited and consequently usage of Witwatersrand Quartzite produced by the numerous gold mines in the south of the region is increasing. The Witwatersrand Quartzite has a higher shale content which is responsible for the poorer shape and generally slightly lower quality. Deleterious sulphide minerals such as marcasite, pyrite and chalcopyrite are present in the Witwatersrand Quartzite. Although no problems, other than staining and slight spalling, have been encountered in concrete, the oxidation and subsequent production of ferric salts and sulphuric acid have caused serious complications when the aggregate has been used as fill. Difficulties with sulfate salt migration have also been reported in road construction by Weinert (1980).

Dolomite
The dolomite available in the Gauteng region is considered a good coarse aggregate as concretes made with this aggregate have high test prism crushing values, low drying shrinkage and fairly low abrasion values (Table 4). However, the quality of concrete made with dolomite aggregate is strongly influenced by the siliceous minerals, particular chert. The irregular or flaky shape of the silica-rich bodies makes selective mining of dolomite very difficult and the material produced by a single quarry may vary significantly as the working face advances. Dolomite is a popular aggregate in the manufacture of concrete works exposed to acids and sulfates. In these circumstances both the aggregate and cementitious material are eroded and this reduces the overall rate of erosion. The dolomite used in the manufacture of concrete pipes intended for corrosive environments should contain a minimum of non-

Table 4. Typical physical properties of aggregates used in Gauteng region.

Property	Unit	Andesite	Dolomite	Felsite	Gabbro/ norite	Granite	Magaliesberg quartzite	Wit/rand quartzite
10% FACT:								
– Dry	kN	470	268	238	170	148	222	195
– Wet	kN	442	266	222	132	136	208	141
– Wet/dry	%	94	99	93	87	92	94	72
Los Angeles abrasion loss	%	8	20	18	24	38	25	22
Drying shrink-age	%	0.070	0.051	0.073	0.062	0.060	0.057	0.075
Water absorp-tion	%	0.28	0.12	1.00	0.13	0.29	0.54	0.23
Relative den-sity	kg/m^3	2880	2820	2620	2920	2610	2630	2698
Loose bulk density	kg/m^3	1492	1498	1281	1538	1397	1456	1360
Voids	%	48	47	51	47	47	45	49
Flakiness in-dex		23	26	34	15	23	21	35
Soundness (15 cycles)	% loss	1.7	0.2	6.9	4.1	4.2	4.4	2.3
Polished stone value		38	50	44	52	48	37	52

carbonate material in order to maximize the acid resistance (Addis 1986). Although dolomite adheres well to bitumen (Weinert 1980) it polishes easily and is therefore not suitable as road surfacing aggregate. The high crushing value makes dolomite suitable for use in all classes of railway ballast.

Andesite
Andesitic lavas of the Ventersdorp Group are quarried for aggregate in the southern part of the Gauteng region (south of Johannesburg). The same stratigraphic unit extends through the North West Province and into the Northern Cape Province where it is also worked for aggregate. However, in these rural areas quarrying activities are only carried out on a small scale, and reliable data are not available. The very high crushing strength of andesite (as indicated by the 10% FACT values in Table 4) makes this a good aggregate for concrete (Ballim 1983) and tests have shown that the use of andesite sand can increase the compressive strength of a concrete by up to 25% (Davis 1973). Andesite has been used in all aspects of road construction (Weinert 1980), but tends to polish and is therefore not well suited to road surfacing.

Granite
The fresh granite available in the Gauteng region is an acceptable concrete aggregate, and despite the fact that the water absorption is relatively high for an igneous rock, the drying shrinkage is low (Table 4). Although suitable for concrete subject to surface abrasion, the polished stone value of granite may be too low for high traffic areas. Similarly, the crushing strength of granite is too low for use as rolled-in chips.

Because of the kaolin content, weathered portions of granite bodies are not suitable as concrete aggregate. This also applies to the biotite-rich varieties, although these are often good and durable material for gravel roads and base courses (Weinert 1980).

Felsite

Felsite is a fine-grained equivalent of granite and has a high crushing strength and low abrasion. However, felsite tends to have a poor shape (Table 4) and the lack of strength between the stone and matrix at early stages of concrete setting may be detrimental. This delayed bonding can cause difficulties when sawing joints within 24 hrs after casting, but in normal structures the delay in bond development does not affect the final strength of the concrete (Weinert 1980). Felsite is suitable for most aspects of road making, but stripping of the bitumen and polishing may occur.

Gabbro and norite

Gabbro and norite are available from the Bushveld Igneous Complex (SACS 1980, Tankard et al. 1982) in the northern part of the Gauteng region (around Pretoria) and make good concrete aggregate provided they are unweathered. Both gabbro and norite have low drying shrinkages and are used extensively in road construction. However, both lithologies are generally too soft for surface dressing or use as rolled in chips or railway ballast (Weinert 1980), as shown by the 10% FACT and Polished Stone Value in Table 4. Weathered material should be avoided as the decomposition of pyroxene, a major component of these rocks, can produce undesirable phyllosilicate minerals. In addition, the relatively high density of this aggregate increases separation problems and may promote bleeding in concrete (Addis 1986).

6   CAPE TOWN REGION

The Cape Town region (Fig. 1) is the second greatest consumer of aggregate in South Africa. Although the geological structure of the region is quite complex (Tankard et al. 1982), the variety of lithologies available for use by the construction industry (Table 5) is much less than in Gauteng. The reactivity of certain aggregates with high alkali cements has produced several detailed studies of this problem (Addis 1986, Davis & Coull 1991, Oberholster 1981, Oberholster et al. 1978).

Hornfels and graywacke

The largest proportion of aggregate production in the Cape Town region comes from hornfels and graywacke of the Malmesbury Group. However, the lenses of siltstone, phyllite and slate in these rocks can cause significant variations in batch quality if adequate care is not exercised during mining.

The data in Table 5 shows that the Malmesbury Group hornfels and graywacke are generally good quality materials with high strength and medium to high abrasion resistance. However, the relatively high drying shrinkage, related to the phyllosilicate material in the rock matrix, can be problematic. Laboratory and direct measurement data indicate that a high alkali cement concrete made with Malmesbury Group hornfels and graywacke may undergo deleterious expansion due to the 'alkali-aggregate reaction' (Oberholster et al. 1978, Oberholster 1981, Davis & Coull 1991). The physical properties of Malmesbury Group hornfels and graywacke (Table 5) show that the material is sufficiently strong and durable to be used as road aggregate and all classes of railway ballast.

Granite
There are a number of granite bodies in the vicinity of Cape Town each with their own physical characteristics which explains the variation of the data in Table 5. The soundness and flakiness index of the granites are generally good and the aggregate is well suited for use in concrete and road construction, although bitumen bonding problems have been experienced with some of the more biotite-rich rocks (Dunlevey 1988). The granitic rocks have significant potential as they are present in large reserves, are generally of good quality and have little potential for reaction with high alkali cements.

Quartzite
Quartzite aggregate was originally quarried directly from the Table Mountain Group outcrops, but urban encroachment has sterilized these quarries. Currently only small quantities of this material are still recovered, mainly from Quaternary-age river gravels. The physical properties of quartzite aggregate produced from these gravels are generally poorer than those of material quarried from outcrops, due to both weathering and their mixed stratigraphic source.

Although coarse aggregate produced by crushing rocks of the Table Mountain Group has proved popular, the results of physical tests display significant variability. The data in Table 5 indicates that while the quartzite can be a good aggregate for concrete, some material has low strength, high drying shrinkage and poor shape. Although most of the poorer quality aggregate is produced from terrace gravels, some of the

Table 5. Typical physical properties of aggregates used in the Cape Town region.

Property	Unit	Malmesbury Group hornfels/ graywacke	Cape Granite	Table Mountain Group quartzite		
				Run of quarry	Selected	River terrace gravels
10% FACT:						
– Dry	kN	230-355	175-355	124	208	151
– Wet	kN	206-340	110-322	110	193	116
– Wet /dry	%	86-98	70-100	89	93	76
Aggregate crushing value:						
– Dry	%	11.7-21.5	15.2-24.4		19.6	23.5
– Wet	%	14.6-15.5	16.3-25.3		19.4	23.1
Los Angeles abrasion loss	%	15-28	16-33		24	48
Drying shrinkage	%	0.059-0.108	0.051-0.076	0.012	0.058	0.082
Water absorption	%	0.20-0.89	0.12-0.89	1.42	0.66	1.40
Relative density	kg/m³	2690-2760	2600-2750	2640	2660	2680
Loose bulk density	kg/m³	1363-1462	1420	1360	1374	1432
Voids	%	43.2-50.0	46.2-49.7	48		47
Flakiness index		23	16-31	33		17
Soundness (15 cycles)	% loss	1.7	0.0-22.0	64.0		28.00
Polished stone value		38	47-56		56	57

variability in the rock is due to its sedimentary nature. Within the hard, well-indurated and recrystalline beds, lenses and thin horizons of softer sandstone can be found. Even though the best quality aggregate from the Table Mountain Group is suitable for most general concrete work not subject to exposure and severe conditions, it is not recommended for road pavements, thin reinforced concrete or structures designed for water retention. In certain circumstances aggregate from the Table Mountain Group has proved to be reactive with high-alkali cement. This does not, however, appear to be a general condition and it is still unclear whether rock from a specific stratigraphic unit or tectonic province was responsible for those cases (Davis & Coull 1991, Oberholster 1981, Oberholster et al. 1978).

7 DURBAN REGION

Although Durban is the second most populous metropolitan area in South Africa (Fig. 1), it ranks only third in aggregate consumption (Table 2) which can be related to both political factors and the lack of mineral resources. The geological structure of the area is relatively simple, the predominance of argillaceous sedimentary rocks of the Karoo Sequence greatly limits the materials that are suitable for use in construction. Diamictite from the Dwyka Group is the most commonly used aggregate, but quartzite of the Natal Group is also popular for many applications (Forbes et al. 1986).

Diamictite
Diamictite of the Dwyka Group, being a sedimentary rock, shows significant variability in physical properties (Table 6). Weathered diamictite has a low strength, high water absorption and high drying shrinkage, although fresh rock is acceptable for usage in virtually all types of concrete (Dunlevey & Stephens 1995, Paige-Green1975). Providing the 10% FACT is acceptable, diamictite has proved popular as a road making material. The rock adheres well to bitumen and is the least likely to polish of all South African road building materials. Slightly weathered diamictite can be used in crushed stone and gravel road bases.

Quartzite
Quartzite from the Natal Group is a popular concrete aggregate and is generally of acceptable quality (Table 6), but may have poor shape characteristics (Forbes et al. 1986). As is the case with all sedimentary rocks, there is significant variability in the material available from different quarries. Variability can be accentuated by the localized effects of contact metamorphism caused by dolerite intrusions (Dunlevey & Stephens 1994). Use of the Natal Group quartzite for road surfacing may be limited due to occasional unacceptably low 10% FACT values, and (particularly in high traffic density situations) borderline Polished Stone Values may prove unsatisfactory.

Dolerite
Karoo-age Mesozoic dolerite is used in the Pietermaritzburg portion of the Durban region as it is often the only 'hard rock' available. Although the quality of dolerite is quite variable (Table 5), most bodies have high values for both 10% FACT and the

Table 6. Typical physical properties of aggregates used in the Durban area.

Property	Unit	Natal Group quartzite	Dwyka Group diamictite	Karoo dolerite	Natal granite
10% FACT:					
– Dry	kN	170	291	445	167
– Wet	kN	111	243	429	156
– Wet/dry	%	65	84	97	95
Los Angeles abrasion loss	%	17	15	12	–
Drying shrinkage:	%	0.098			
– Mortar	%	–	0.109	0.093	0.072
– Concrete	%	–	0.033	0.023	0.024
Water absorption	%	0.45	0.52	0.48	0.44
Relative density	kg/m^3	2650	2690	2880	2620
Loose bulk density	kg/m^3	1445	1330	1520	1420
Voids	%	48	51	48	46
Flakiness index		35	43	23	18
Soundness (15 cycles)	% loss	5.1	11.7	3.4	22.4
Polished stone value		54	58	55	55

ratio between the wet and dry 10% FACT. They are suitable for use in concrete, road construction and as railway ballast (Dunlevey & Stephens 1996). Weathered dolerite is generally unsuitable for construction except in the sub-base and base courses of tar roads or gravel surfaced roads (Weinert 1980).

Granite
Unfoliated granite from the Natal basement is relatively little used, although it provides good quality aggregate for use in concrete and as a sub-base and base for road construction. However, the relatively low typical 10% FACT values makes this material only marginal at best for road surfacing applications. Material with a high mica content is not regarded as suitable for use in bituminous pavements.

8 EASTERN CAPE REGION

Quartzite from the Table Mountain Group (very similar to that utilized in the Cape Town region) is the most popular aggregate in Port Elizabeth and East London, the two major towns of the Eastern Cape region (Fig. 1). However, significant quantities of dolerite are also used, generally in the smaller towns of the region and for road making (Table 7), but as the vast majority of these quarries are either temporary or very small-scale enterprises, accurate data is difficult to obtain.

Quartzite
The data in Table 7 indicates that the quartzite aggregate produced in the Eastern Cape region generally has a high crushing value, low drying shrinkage and good shape. Quartzite recovered from river gravels tends to be of poorer quality due to the effects of both weathering and mixed source origins. Unfortunately, certain of these

Table 7. Typical physical properties of aggregates used in the Eastern Cape.

Property	Unit	Table Mountain Group quartzite	
		Mean quarry values	River terrace gravels and boulders
10% FACT:			
– Dry	kN	201	243
– Wet	kN	188	214
– Wet/dry	%	94	88
Los Angeles abrasion loss	%	31	23
Drying shrinkage	%	0.067	0.050
Water absorption	%	0.51	0.42
Relative density	kg/m^3	2654	2620
Loose density	kg/m^3	1441	1420
Voids	%	46	–
Flakiness index		15	–
Polished stone value		59	56

aggregates have proven to be reactive with high-alkali cements, in a similar manner to their stratigraphic equivalents in the Cape Town region.

Dolerite
The dolerite utilized as aggregate in the Eastern Cape region is equivalent to that found in the Durban region and has similar physical properties (Table 6). Eastern Cape dolerite is particularly popular in rural road construction and large volumes are used in the maintenance of gravel roads. Dolerite weathered under certain climatic conditions (Weinert 1980) has proven unreliable as a road construction material due to non-oxidizing alteration of the ferro-magnesian minerals to chlorite and the development of extensive fractures in the primary mineral grains.

9   BLOEMFONTEIN REGION

The Bloemfontein region (Fig. 1) is the smallest of the five South African major growth areas in terms of aggregate consumption (Table 1). Being situated close to the centre of the Karoo sedimentary basin, the dominant lithologies are shale and fine-grained sandstone which are generally unsuitable for use as aggregate. However, certain dolerite intrusions, usually in the form of sills, provide rock suitable for use as aggregates and have in some cases metamorphosed the surrounding sediments to the extent that a quartzite aggregate can be produced. Not all the dolerite bodies produce good aggregate, weathered rock is of inferior quality, and dolerite in which primary olivine has undergone hydration to phyllosilicates is unstable.

Dolerite
Properties of typical dolerites are shown in Table 8. The lack of certain standard test results such as Polished Stone Values is due to the fact that these are rarely carried out on this aggregate which has proved itself highly usable in road surfacing over the years (Loubscher 1995).

Table 8. Typical physical properties of dolerite aggregate used in the Bloemfontein region.

Property	Unit	Dolerite
10% FACT:		
– Dry	kN	294
– Wet	kN	326
– Wet/dry	%	90
Aggregate crushing value	%	11.4
Los Angeles abrasion loss	%	14.6
Drying shrinkage	%	–
Water absorption	%	0.3
Relative density	kg/m^3	2970
Loose bulk density	kg/m^3	1570
Voids	%	–
Flakiness index		–
Polished stone value		–

Quartzite
The Welkom area, 200 km north-east of Bloemfontein, is anomalous in that although rural in character there are several large gold mines. Quartzite aggregate, similar to that in the southern Gauteng region, is available as a by-product of mining and this distorts the local market to the extent that in exceptional circumstances this material has been used for building as far afield as the Bloemfontein municipality.

10 CONCLUSIONS

Many different lithologies are used as aggregate in South Africa, and although the availability of certain rock types varies greatly between the five major urban areas there is generally an adequate supply of rock suitable for use as coarse aggregate. Although development has sterilized some reserves in urban areas, both high-quality aggregates with well-documented physical properties and lower-quality, general-purpose aggregates are readily available. However, as no single lithology produces aggregate ideally suited for all construction applications, the physical criteria of the material required for each project must be carefully considered during project planning and the selection critically balanced against cost and availability. As a means of providing a quick way of ascertaining the general suitability of the aggregates in each of the major growth areas for various engineering applications, Table 9 has been generated from the combined details of Tables 3 to 8 in a semi-qualitative approach that uses four categories of acceptability.

In the rural areas of South Africa there is generally very little variety in the lithology of commercially available aggregate, and although material is readily available, documentation of the physical properties is much less extensive. When high-quality aggregate or material with specific physical properties is required on major projects such as interprovincial highways, this may necessitate the opening of new quarries especially for those undertakings.

Table 9. Suitability of coarse aggregate for specific usage in major growth areas (A = good, B = adequate, C = marginal, D = not recommended).

Growth area	Lithologies	Concrete	Roadstone			
			Base and subbase	Surface dressing	Rolled-in chips	Bituminous surfacing
Gauteng	Andesite	A	A	C	C	C
	Dolomite	A	A	D	D	A
	Gabbro/norite	A/B	A	D	D	D
	Granite	B	A	B	C	C
	Felsite	A/B	B	B	B	C
	Quartzite	B	B/C	B/C	B/C	B/C
Cape Town	Granite	A	A	A	A	B
	Hornfels/graywacke	A/B	A	A	A	A
	Quartzite	A/B	A/B	D	D	D
Durban	Diamictite	A	A	A	A	A
	Quartzite	A	A	B	B	B
	Dolerite	A	A	A	A	A
	Granite	A	A	C/D	C/D	C/D
Eastern Cape	Dolerite	A	A	A	A	A
	Quartzite	A/B	A/B	C/D	C/D	C/D
Bloemfontein	Dolerite	A	A	A	A	A
	Quartzite	B	B/C	B/C	B/C	B/C

ACKNOWLEDGEMENTS

We wish to express our thanks to the University of Durban-Westville for funding the basic research associated with this project, Mr G F Viljoen, Director-General of the Department of Regional and Land Affairs for permission to use his department's technical data and our many colleagues in the South African quarrying and construction industry for their valuable contributions.

REFERENCES

Addis, B.J. (ed.) 1986. *Fulton's Concrete Technology, Sixth Edition*. Portland Cement Institute: Midrand South Africa.

Ballim, Y. 1983. The concrete-making properties of the Andesite Lavas from the Langgeleven Formation of the Ventersdorp Supergroup. MSc. (Eng.) Thesis, Univ. Witwatersrand: Johannesburg.

Committee of State Road Authorities. 1987. Standard specification for road and bridge works: Pretoria.

Davis, D.E. 1973. The concrete-making properties of South African aggregates. PhD. Thesis, Univ. Witwatersrand: Johannesburg.

Davis, D.E. & Alexander, M.G. 1989. *Properties of Aggregates in Concrete (Part 1)*. Hippo Quarries Technical Publication: Sandton.

Davis, D.E. & Coull, W.A. 1991. *Alkali-Aggregate Reaction*. Hippo Quarries Technical Publication: Johannesburg.

Dunlevey, J.N. 1988. The Saldanhian orogeny and the evolution of the Cape Granite Suite. *S. Afr. J. Sci.* 84: 565-568.

Dunlevey, J.N. & Stephens, D.J. 1994. Crushed Natal Group sandstone used as coarse aggregate. *J. S. A. Inst. Civ. Eng.* 36(3): 11-18.

Dunlevey, J.N. & Stephens, D. J. 1995. The use of Dwyka Group diamictite as crushed aggregate in Natal. *J. S. A. Inst. Civ. Eng.* 37(2): 11-16.

Dunlevey, J.N. & Stephens, D. J. 1996. The use of Karoo dolerite as crushed aggregate in Natal. *J. S. A. Inst. Civ. Eng.* 38(4): 33-40.

Forbes, J., Dunlevey, J.N., Oberholster, R.E. & Dobson, D. 1986. *Report on the coarse aggregate industry in the Durban/Pietermaritzburg region.* Contract Report (0090539) for Physical Planning Division, Department of Constitutional Development and Planning. National Building Research Institute, CSIR: Pretoria.

Loubscher, E. 1995. Personal communication.

National Institute for Transport and Road Research. 1985. TRH14: *Guidelines for Road Construction Materials.* CSIR: Pretoria.

Oberholster, R.E. 1981. Alkali-aggregate reaction in South Africa – a review. *Proc. 5th Int. Conf. Alkali-aggregate Reaction in Concrete.* Cape Town.

Oberholster, R.E., Brandt, M.P. & Weston, A.C. 1978. Cement-aggregate reaction and the deterioration of concrete structures in the Cape Peninsula. *Civ. Engr. in S.A.* 20(7): 161-166.

Paige-Green, P. 1975. The geotechnical properties of Dwyka tillite in the Durban area. MSc Dissertation, Univ. Natal: Durban.

South African Committee for Stratigraphy (SACS). 1980. *Stratigraphy of South Africa. Part 1.* (Comp. L.E. Kent). Lithostratigraphy of the Republic of South Africa, South West Africa/Namibia, and the Republics of Bophutatswana, Transkei and Venda. Handb. Geol. Surv. S. Afr., 8.

South African Bureau of Standards. 1979. SABS 1083-1976: *Standard Specification for Aggregates from Natural Sources* (as amended 1979). Pretoria.

South African Bureau of Standards. 1995. SABS 1995 *Yearbook*: Pretoria.

Tankard, A.J. Jackson, M.P.A. Eriksson, K.A., Hobday, D.K., Hunter, D.R. & Minter, W.E.L. 1982. *Crustal Evolution of Southern Africa: 3.8 Billion Years of Earth History.* Springer-Verlag: Berlin.

Van Zyl, L. & Schreuder, C.P. 1994. *South Africa's Mineral Industry.* S.A Dept. Mineral and Energy Affairs Minerals Bureau: Braamfontein.

Weinert, H.H. 1980. *The Natural Road Construction Materials of Southern Africa.* Academica: Pretoria/Cape Town.

The utility of Quaternary thematic maps in the exploitation and the preservation of the natural environment

FRIEDA BOGEMANS

Toegepaste Geologie (Applied Geology), Vrije Universiteit Brussel, Belgium

1 INTRODUCTION

The conflicts between those who exploit natural resources and those who strive to preserve the natural environment have existed for several decades now and are becoming more common place. Indeed, the natural environment is already heavily disturbed by reckless interference of all sorts of human activity. However, natural resource extraction is very important to our society. For example, house construction and infrastructure work require large amounts of building materials, including mineral aggregate products. A growing society needs both aggregate extraction and natural environment preservation. A compromise can be achieved between these contrary interests by using a balanced land use planning. The fundamentals of such land use planning must include geological knowledge of the region. In fact, the natural environment is dominantly defined by its geological history. The morphology of the landscape and the properties of the soil are all the result of the evolution of the sedimentary depositional environments within a geological time scale. Moreover, surface and near surface layers affect socio-economic conditions. The quality of agricultural land, the suitability of the land for construction and the availability of natural resources are directly related or defined by the geological and geotechnical conditions present. Most layers which have a direct impact on society were deposited during the Quaternary, or the last 2.4 million years.

2 PROFILE-TYPE MAPS

Geological maps remain one of the most effective instruments in representing geological knowledge of a region and to convey that particular knowledge to a large public audience. Not only the geological history of the surficial sediments has to be known, but more importantly information on vertical and lateral variations of both the surficial and the deeper lying sediments is needed for a balanced land use planning. Unfortunately, this information can not be presented on traditional geological maps, which only show geological features at the surface (Visser 1980). Information on the third dimension needs to be added to geological maps, notwithstanding the fact that geological maps remain a 2D representation. This objective can be obtained by creating profile-type maps, with map units consisting of vertical sequences of sediments, called special profile-types, present in the surveyed area.

Although Lüttig (1988) is of the opinion that a profile-type map is not easily read or understood by non-geologists, this type of geological map can reach a large public audience if some considerations are taken into account during legend development. For example, geological nomenclature does not form the key to the map but material types like peat, sand, silt, clay and gravel, in relation to their sedimentary depositional environments do. People involved in engineering, natural resources extraction or agriculture are well-informed about the terms mentioned above. Of course, a positive attitude of the non-geologist and the geologist is essential in this situation.

Displaying stratigraphic information is feasible within the conceptual basis of such a geological map. The lithostratigraphy is defined according to the sedimentological and lithological properties. In this context the pure geological information evolves into applied geological information which forms the basic information needed by non-geologists and is of direct interest to a large group of people.

The profile-type map was created by de Jong & Hageman (1960) in order to represent the Holocene coastal deposits. de Jong & Hageman (1960) have defined a profile-type map as a compilation of vertical sequences entirely defined in the field. From the very beginning, the vertical sequences represented on the map are limited to a depth that is attainable by a hand auger. Today this mapping method is used throughout the Netherlands.

3 APPLICATION IN FLANDERS

The basic mapping principle as described by de Jong and Hageman (1960) has been adapted in Flanders profile-type maps. Major differences are that the entire Quaternary sequence is represented and although the units are still defined in the field, additional properties were used to define them. Indeed, by definition sediment genesis, texture, sedimentary environment, the occurrence of paleosols, etc. may be used to define mapping units or components. The criteria for determining map units should be chosen so that the most complete representation of the field geology is presented in a form applicable to the user's needs.

The first time that the entire Quaternary sequence was represented on profile-type maps was in the central part of Belgium, north of Brussels (topographic maps Boom – Mechelen, 23/3-4 and Vilvoorde – Zemst, 23/7-8, scale 1:25,000, Fig. 1a). Here socio-economic impact is very high and the Quaternary record is quite variable (Bogemans 1988, 1991, 1993, 1994). These Quaternary deposits are fluvial, eolian and mass wasting in origin. The deposits are approximately 12-15 m thick and were deposited mainly during the Upper-Pleistocene. Within this area a branch of a paleo-valley, the so called Flemish valley, is present (Fig. 1b). de Moor (1969, p. 338) defined the Flemish valley as an erosional form, mainly of fluvial origin that was largely eroded below the present sea level before the Eemian transgression. The valley is filled predominantly with fluvial deposits, however differentiation within these deposits is possible. Changes in sediment properties are the result of deposition in both meandering and braided river systems. Outside the paleo-valley eolian and mass wasting deposits predominate. Fluvial deposits occur also, but as remnant terraces over a small part of the area and always at the base of the Quaternary sequence. The eolian deposits vary in texture within the test area. In the north, coversand has accumulated; in the south, loess covers the landscape. Between these two areas, sandy and silty sediments were deposited in an area that is known as the transitional area (Paepe & Vanhoorne 1969, Fig. 1c). To date, maps have been produced by assignment of the Ministry of the Flemish Community, Administration of Natural Resources and Economy, all at a scale of 1:50,000. These maps are of areas near the test area and of northeastern Flanders, located north and northeast of Antwerpen (Campine area, see Fig. 2) where both the geological history and the thickness of the

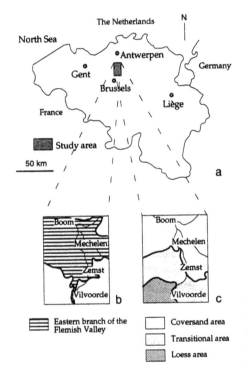

Figure 1. a) Geographic map of Belgium with the location of the test area. b) Delineation of the eastern branch of the Flemish valley in the test area. c) Subdivision of the test area into three eolian sedimentation areas.

Figure 2. Location of the Campine area in Belgium.

Quaternary sediments is quite different. The Campine area is underlain by estuarine sediments in combination with pure continental sediments, over 60 m thick and deposited during the last 2.4 million years.

4 CONSTRUCTION OF A PROFILE-TYPE MAP

The profile-type mapping system is completely based on sedimentological properties of the sediments. These properties are directly defined in the field, determined from bore hole descriptions, or extracted from former geological descriptions which are stored in the archives of geological institutes. Since a profile-type map is composed of a succession of different sedimentary facies deposited during the Quaternary, it also has a paleo-geographic meaning.

The mapping system consists of two hierarchic levels, a higher level that is defined as a compilation of 'complexes' (Barckhausen et al. 1977) and a lower one defined by 'special profile-types' which form the real mapping units.

Quaternary deposits are known for their great variability both laterally and vertically over small distances. Continuous changing properties indicate successive changes in the depositional environments, which appear on both small and large scale. An example of small scale changes is the alternating braided and meandering river deposits in the Flemish valley. Another example, is the alternating estuarine and fluvial deposits in the Campine area formed as a consequence of falling and rising sea level. This particular situation is very sensitive, even to the slightest changes, because two depositional environments (pure estuarine – fluvial estuarine) are interfingering. Large scale variations are the consequence of drastic sedimentary changes, tectonics, etc.; for example, the transition from fluvial into eolian depositional environments.

In the profile-type mapping system, the large scale changes are registered in complexes, whereas, the small scale changes are incorporated in the special profile-types. Furthermore, the creating of the complexes depends on the sedimentological and lithological characteristics of the survey area, the duration of the depositional history and the mapping scale. To illustrate the latter statement an example is taken from the transitional and coversand area near Brussels. In the area, Upper-Pleistocene mass wasting products are only found in restricted places. Because these sediments are still representable on a scale of 1:25,000 a complex was introduced to include all sequences containing Upper-Pleistocene mass wasting deposits. On a map with a scale of 1:50,000 no specific complex is created because the distribution of these deposits is too small. The complexity of the map would increase tremendously in order to display such a complex without necessarily increasing the usefulness of the map. However, the representation of the Holocene mass wasting deposits are maintained because of their direct impact on society.

Although a profile-type map is constructed to reach a large public audience, it remains a geological map with stratigraphic meaning. The components of both the complexes and the special profile-types are commonly represented as lithostratigraphic units, however belonging to different stratigraphic levels according to the classification of Hedberg (1976). A component of a complex will always be higher or equal in rank than a component of special profile-types. A genetic unit cor-

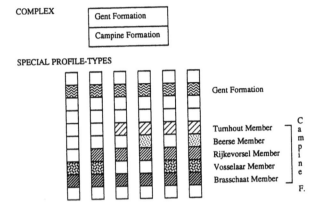

COMPLEX

Gent Formation

Campine Formation

SPECIAL PROFILE-TYPES

Gent Formation

Turnhout Member
Beerse Member
Rijkevorsel Member
Vosselaar Member
Brasschaat Member

C
a
m
p
i
n
e
F.

Figure 3. One complex defined on the level of Formation and the associated special profile-types, present in the survey area. Note that the special profile-types are composed of a succession of Members in exception of the top unit, which has a higher lithostratigraphic rank. Formation is used due to a lack of detailed data all over the survey area.

responds in general to a Formation, Group or Complex (Fig. 3) whereas a Member stands for the depositional environment, usually the working unit of the special profile-types (Fig. 3).

Special profile-types are the key to profile-type maps. It is important that the number of components per special profile-type stays limited, in order to keep the map readable and understandable. Mengeling & Vinken (1975) opted for a maximum of eight components, four to five components are perfect to keep the map optimal, readable and clear. In general, special profile-types are composed of a succession of components each related to a specific depositional environment. Each environment has specific sedimentological properties which are of interest to both geologists and non-geologists. When there are too many components, the data do not allow detailed interpretation or the scale of the map is not suitable, the map is constructed so that each component no longer stands for a depositional environment but becomes a unit higher in rank (Fig. 3).

5 REPRESENTATION OF THE LEGEND OF THE MAP

A special profile-type is defined in the legend and on the map by means of a number. The special profile-type number is given in relation to the complexity of the special profile type. The simplest special profile-type is part of the first complex and will be number one. The more complex the special profile-type is, the larger the number.

The legend is designed in such a way that the special profile types are schematically represented by columns; each built from several components. Each component is represented in the same way by means of a square filled with a specific pattern (Fig. 5). Each square has an unchangeable position in the special profile-type columns. Its position reflects its stratigraphic position in the field. For the ease of the map reader each pattern is briefly explained on the bases of its sedimentological properties (Fig. 4) and is incorporated into a regional stratigraphic scheme where chronostratigraphic and lithostratigraphic positions are shown (Fig. 4). Blank squares occur in special profile-types, if one or more components are missing due to non-deposition or erosion in a particular place (Figs 3-5).

CHRONOSTRATIGRAPHY		LITHOSTRATIGRAPHY			
Series	Stages				Code
LOWER - PLEISTOCENE	Bavelian	St. Lenaarts Formation			(symbol)
	Menapian				
	Waalian				
	Eburonian				
	Tiglian	Campine Formation	Turnhout Member		(symbol)
			Beerse Member		(symbol)
			Rijkevorsel Member		(symbol)
			Vosselaar Member		(symbol)
			Brasschaat Member		(symbol)
	Pretiglian	Merksplas Formation			(symbol)
PLIOCENE					

(symbol) Estuarine deposits (clayey-sandy complex, dominance of mica-ceous clayey sediments. Sand facies present in the north. One or more soil horizons possible).

(symbol) Fluvial and eolian deposits (fine to medium fine sand with several soil horizons, periglacial phenomena).

Figure. 4. Illustration of a part of a stratigraphic table used in the legend of a profile-type map, followed by a brief sedimentological description of the lithostratigraphic units. (The patterns used in this figure and the layout are different from the ones used on the real profile-type maps.)

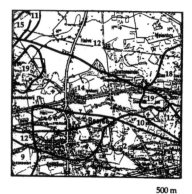

500 m

Explanation of the profile types

| 1 | 2 | 5 | 6 | 9 | 10 | 11 | 12 | 14 | 15 | 18 | 19 |

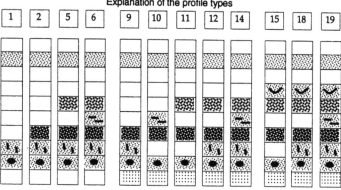

Figure 5. Illustration of a part of a profile-type map. The polygons are on this figure blank, in practice they are coloured. (The patterns used in this figure are different from the ones used on the real profile-type maps.)

Each complex on a profile-type geological map is given a colour or sometimes a colour in combination with a pattern. Within these colours a great diversity of shades is applied to designate each special profile-type.

6 EXAMPLE OF A PROFILE-TYPE MAP

To illustrate a profile-type map, an example is taken from the central part of the Turnhout map, situated north of Antwerpen in the Campine area (Fig. 5). The polygons are not coloured on this figure, they are left blank. On this part of the map three complexes are represented, containing 12 special profile-types.

The first complex is composed of the Gent Formation, consisting of eolian deposits, and the Campine Formation, composed of an alternation of estuarine and fluvial deposits. Four special profile-types belonging to this complex are present in the example. The first two special profile-types consist of the Gent Formation followed by the oldest units of the Campine Formation, whereas the two last ones contain a great number of Members belonging to the Campine Formation or the complete sequence of the Formation.

The second complex contains the Gent and Campine Formations and the Merksplas Formation. The Merksplas Formation is also composed of estuarine deposits, but with different characteristics than those of the Campine Formation. The second complex consists of five special profile-types. The difference in numbering is the result of an increase in the total number of Members present, all belonging to the Campine Formation.

The third complex contains the St. Lenaarts Formation in addition to the Gent, Campine and Merksplas Formations. Three special profile types are represented on this part of the map.

7 THE USE OF THEMATIC APPLIED GEOLOGICAL MAPS

Profile-type maps provide the basic geological information for balanced land use planning. Before fundamental changes in the landscape are carried out, like extraction of natural resources or development, a set of thematic applied geological maps is produced for the local situation, in order to avoid problems, extra costs and to prevent unwanted destruction of the natural environment. These are customized for specific interest groups or problems. In the Netherlands, applied geology maps are standardized and subdivided into engineering geological maps and environmental maps (Geological Survey of the Netherlands and Delft Soil Mechanics Laboratory 1984, Geological Survey of the Netherlands et al. 1986, de Mulder 1986, 1988, 1989).

Since the possible applications of thematic applied geological maps is so large, only maps that meet the specific needs of the authorities or an interest group are produced.

8   EXAMPLES OF THEMATIC APPLIED GEOLOGICAL MAPS

In our society their is great demand for natural resources. Thematic applied geological maps can greatly assist the exploitation for and development of these natural resources. Examples of two thematic applied geological maps are presented. The first showing the properties and the distribution of the sand facies in a branch of the Flemish Valley (Fig. 1b) and the second depicting the distribution and total thickness of clay layers present in the Northern Campine area (Fig. 2) above a depth of 10 m. Ten meters is taken as the cut off point because present day extraction seldom goes beyond this depth. A greater depth of extraction could have serious consequences for the surrounding agricultural lands because the groundwater level may drop tremendously.

It is impossible to image constructional works proceeding without sand. Sand is used as drainage material, foundation material, road material, fill and is used in the production of concrete. Depending on the application, the content of glauconite, shells, organic matter and fine clastic material is regulated, as well as the grain size distribution of the sand itself. Sand is subdivided into fine, medium and coarse sand (Ministry of Public Works 1978). Fine sand contains more than 50% sand between 0.06 mm and 0.20 mm. Medium sand is obtained when the grain size distribution between 0.06 mm and 2 mm makes up more than 50% of the total, however no more than 50% of the material can be between 0.20 and 2 mm. Finally, sand is called coarse when the fraction between 0.20 and 2 mm is greater than 50%. In nature, however, a combination of the different grain size distributions may be present. On the thematic geological map, five grain size distributions are shown: medium sand, coarse sand, fine to medium sand, medium to coarse sand and fine to coarse sand. The thickness and the initial depth of the sand facies are also put on the map. Three subdivisions are used for both thickness and initial depth. Sand facies are subdivided into units of less than 5 m thick, between 5 and 10 m thick and more than 10 m thick. With regard to the initial depth, a distribution is made between the sand facies found above 5 m, between 5 and 10 m and deeper than 10 m. In this framework, the economic feasibility is taken into account. At present it would be unacceptable and unrealistic to start an excavation in an area where the top of the sand facies is situated more than 10 m below the surface and where the sand deposits are less than 5 m thick. In addition to the tremendously high exploitation costs, the development itself, would probably have a negative impact on the natural environment. On the map, the areas of different grain size distributions are outlined in colour.

In the Campine area, clay is a needed natural resource. In this application, sediments are considered to be clay if the material is described as clay, silty clay or clay with sand lenses. Alternatively clay and sand layers and sand with clay lenses or layers are defined as sandy material. The term 'clay' as well as 'sandy' material is used in a practical sense and not in the strict sedimentological sense of the word. The quality of the clay is not shown on the thematic applied geological map. The map shows how much 'clay' is present in the subsoil and what is its distribution. To display the distribution, the number of clay layers are presented. In the example (Fig. 6), the total number of clay layers range from 0-2 within the first 10 m. The total thickness of the 'clay' within the 10 m depth is expressed as a percentage and subdivided into four groups. The number of clay layers is represented on the map by means of a

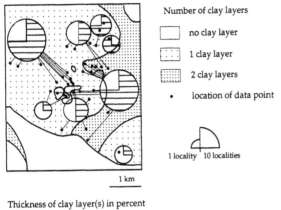

Number of clay layers

▢ no clay layer

▢ 1 clay layer

▣ 2 clay layers

• location of data point

1 locality   10 localities

1 km

Thickness of clay layer(s) in percent

between 0.01 - 24,99 %

between 25 - 49,99 %

between 50 - 74,99 %

between 75 - 100 %

Figure 6. Thematic applied geological map on which the number of clay layers over a depth of 10 m are represented together with the total thickness of the clay layers.

colour, whereas the percentage of 'clay' is shown by using pie charts. The colours in this example are replaced by patterning. Pie charts were chosen because of the large lateral and vertical variations of the sandy and clayey deposits and for the ability to display number of data sites. The diameter of the pie chart is defined by the number of drilling sites where the same percentage group is present.

9 CONCLUSION

Quaternary deposits, which are characterized by their great variability, have a direct impact on our life. The presence of aggregates, the fertility of the soil, the suitability of the land for construction, the behaviour of the ground water all depend on the natural distribution of the Quaternary sediments.

Fundamental geological information is represented by means of a profile-type map, in which the distribution of both the surficial and the Quaternary subsurface layers are represented. Specific information, concerning topics such as those described above is important to ensure wise planning of our natural resources and environment and is shown by means of thematic applied geological maps. Which type of these geological maps is needed, depends on the users needs and the geology of the region. Profile-type maps and thematic applied geological maps containing either fundamental or applied geological information are indispensable in land use planning.

ACKNOWLEDGMENTS

The author is greatly indebted to R. Paepe, as the head of the Belgian Geological Survey for allowing the geological data to be used and, as the head of the Department of Applied Geology at the Vrije Universiteit Brussel for giving the opportunity to develop the different mapping systems used in Flanders. Thanks are also extend to the Administration of Economy and Natural Resources (ANRE) of the Flemish Community who gave the permission to publish some parts of the geological maps. Thanks go also to C. Baeteman, who made some of the drawings. The paper in its final form has benefited from constructive reviews of Peter Bobrowsky and Ed de Mulder.

REFERENCES

Barckhausen, J., Preuss, H. & Streif, H. 1977. Ein lithologisches Ordnungsprinzip für das Küsten-holozän und seine Darstellung in Form von Profiltypen. *Geologisches Jahrbuch* A44: 45-74.

Bogemans, F. 1988. Thematische kwartairgeologische voorstellingen als toepassingsmodellen in de economische ontwikkeling. Brussel: unpublished doctoral thesis.

Bogemans, F. 1991. Quaternary thematic mapping of continental deposits: a tool in land-use planning. *Acta Geologica Taiwanica* 29: 139-147.

Bogemans, F. 1993. Quaternary geological mapping on basis of sedimentary properties in the Eastern Branch of the Flemish Valley. *Toelichtende Verhandelingen voor de Geologische Kaarten en Mijnkaarten van België* 35: 49.

Bogemans, F. 1994. The usefulness of Quaternary thematic maps in land-use planning. In G.W. Lüttig (ed), *Aggregates – Raw Materials' Giant*: 263-275. Erlangen: University Erlangen-Nuremberg.

de Jong, J.D. & Hageman, B.P. 1960. De legenda van de holocene afzettingen op de nieuwe geologische kaart van Nederland. *Geologie en Mijnbouw* 39: 644-653.

de Moor, G. 1963. Bijdrage tot de kennis van de fysische landschapsvorming in Binnen-Vlaanderen. *Bull. Soc. belge d'Etudes Géogr.* 32: 329-433.

de Mulder, E. 1986. Applied and engineering geological mapping in the Netherlands. *Proc. Fifth Int. Congress I.A.E.G., 6.1.5.*: 1755-1759.

de Mulder, E. 1988. Engineering geological maps: A cost benefit analysis. In P. Marinos & G. Koukis (eds), *The Engineering Geology of Ancient Works, Monuments and Historical Sites*: 1347-1357. Rotterdam: Balkema.

de Mulder, E. 1989. Thematic applied Quaternary maps: A Profitable investment or expensive wallpaper? In E. de Mulder & B. Hageman (eds), *Applied Quaternary Research*: 105-117. Rotterdam: Balkema.

Geological Survey of the Netherlands & Delft Soil Mechanics Laboratory 1984. *Ingenieurs Geologische Kaarten van Nederland.*

Geological Survey of the Netherlands, Institute for Applied Geosciences & National Institute for Soil Mapping 1986. *The subsurface uncovered.*

Hedberg, H.D. (ed.) 1976. *International stratigraphic guide.* New York: John Wiley & Sons.

Lüttig, G. 1988. Large scale maps for detailed environmental planning. In *Commission for the geological map of the world, C.G.M.W.* Bulletin 38: 12-17.

Ministerie van Openbare Werken 1978. *Typebestek 150.*

Mengeling, H. & Vinken, R. 1975. Die Profiltypekarte – ein Schritt in der Weiterentwicklung geologischer Karten. *Geologisches Jahrbuch* 29: 65-80.

Paepe, R. & Vanhoorne, R. 1967. The stratigraphy and palaeobotany of the Late Pleistocene in Belgium. *Toelicht. Verhand. Geologische Kaarten en Mijnkaarten van België*: 96.

Visser, W.H. (ed.) 1980. *Geological nomenclature.* Bohn: Scheltema & Holkema.

Sand accumulation in a gravel-bed river

M.C THOMS
Co-operative Research Centre for Freshwater Ecology, University of Canberra, Australia

1 INTRODUCTION

Alluvial deposits of sand and gravel are an important exploitable resource of building aggregate in eastern Australia. Sediment is mined from the active channel (instream mining) or from pits excavated in the floodplain or terraces. The economic potential of these coarse-grained deposits depends largely upon their composition. Grain-size influences the porosity and permeability of gravel deposits and also dictates the degree of secondary treatment necessary in gravel extraction for construction activities. An understanding of the processes that form gravely deposits is also important for: the location of new or the extension of existing deposits; interpretation and evaluation of exploration data; the modeling of aggregate deposits; and, the preparation of appropriate methods of extraction.

Coarse-grained deposits commonly display a bimodal grain-size distribution, with coarse particles (the *framework*) being dominant, on a frequency by weight basis, and finer material (the *matrix*) occupying the interstitial spaces between the framework clasts (Carling & Reader 1982). There is an array of framework-matrix mixtures. 'Open framework gravels' are composed solely of gravel clasts and at the other end of the spectrum there are 'matrix supported gravels' in which the gravel clasts are dispersed in a matrix of fine sediment (Carling & Glaister 1987). A full range of coarse-fine sediment mixtures can be found within any alluvial facies (Miall 1978) or reach of river (Thoms 1987).

While something is known of the mechanisms that control the deposition of coarse framework gravels very little is known of matrix deposition in gravel-bed rivers. The fine matrix component is important because it can influence the texture, porosity and permeability of coarse-grained deposits. Carling & Glaister (1987) suggest three possible ways in which matrix sediment can become incorporated into gravely deposits.

1. Sand and gravel mixtures may form with the synchronous deposition of sand and gravel. However, this would imply large and rapid energy fluctuations that are uncharacteristic of natural river discharges (Dyer 1970).

2. Allen (1983) suggested that gravel particles may overpass and eventually become incorporated into a predominantly sand bed. However, the deposition of gravel

clasts occurs at velocities too vigorous to allow the settling of large quantities of fine material (Frostick et al. 1984).

3. The incorporation of matrix material following the deposition of gravel clasts is termed secondary ingress or infiltration. Einstein (1968) and Carling (1984) simulated matrix ingress in a laboratory flume and indicated that the interstitial spaces within framework gravels can become completely infilled with fine sediment as a result of this process.

Secondary ingress/infiltration is an important mechanism that can influence the composition of coarse-grained deposits. However, it has received scant attention. Laboratory studies undertaken by Einstein (1968) and Carling (1984), were limited because of controlled flow and sediment transport rates and uniform gravel sizes employed, conditions rarely seen in natural river channels (Carling & Reader 1982). Field experiments conducted by Frostick et al. (1984) were restricted to intermittent sampling over a six month monitoring period and employed methods that did not collect all the ingressed material (Thoms 1987).

This paper presents the results of an experimental field study conducted in a small upland gravel-bed stream located in the Adelaide Hills region of South Australia. Rates of ingress into an array of different gravel mixtures were monitored continuously over a 12 month period under varying flow and sediment transport conditions.

2 THE EXPERIMENT

2.1 *Field site*

Sixth Creek, a tributary of the River Torrens, drains a 43.4 km^2 catchment of the Adelaide Hills, South Australia (Fig. 1). Conglomeratic sandstone and quartzite dominate the catchment geology and extensive erosion in the headwater reaches are the source of the predominantly coarse-grained bed material in Sixth Creek. The channel is confined in a narrow valley bottom where there is little room for floodplain development. Thus the channel planform, largely determined by the position of local faults, comprises a series of straight reaches punctuated at irregular intervals by marked changes in direction. Experiments were conducted in one of these straight reaches (Fig. 1), characterized by channel widths varying between 3.5 and 5.2 m and bankfull depths of 0.90 to 1.75 m.

The bed material includes cobble, gravel and sand sized material with the median grain-size (D_{50}) of the surface 'armour' layer varying from -7.5 to -8.5 ϕ ($\phi = -\log_2$ (mm)). This distinguished it from the sub-surface material which has a D_{50} between -4.2 and 7 ϕ (Thoms 1992). The sub-surface material had a bimodal grain-size distribution with the framework component (-3 to -7 ϕ) being separated from the matrix (-2 to $<+4$ ϕ) by a saddle frequency positioned in the -2 to -3 ϕ interval.

Flows in Sixth Creek were monitored by the South Australian Engineering and Water Supply Department (EWS) approximately 1 km upstream of the Torrens confluence (Fig. 1). Mean monthly discharges for the period 1980-89 ranged between 0.005 and 1.68 m^{3}s^{-1} with peak discharges exceeding 3.5 m^{3}s^{-1}. Runoff from the catchment produced a typical 'flashy' hydrograph characterized by steep rising and falling limbs. High flow events were frequent during the winter months of June-

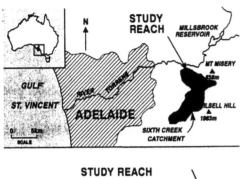

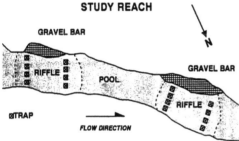

Figure 1. Location of study site.

September while in contrast, the summer months were characterized by long periods of low flow.

2.2 *Matrix traps*

Traps designed to simulate the natural framework gravels of the river-bed were installed at two riffles between January 1989 and February 1990. Each trap consisted of two circular fenestrated wire baskets 30 cm high and 24 and 25 cm in diameter, with the smaller basket fitting inside the larger. The inner basket was filled with natural framework gravels, with a size distribution equivalent to that of the natural river-bed. This basket was placed in a plastic bag and lowered into the outer basket and the entire assembly was buried in the river-bed, flush with the surface of the substratum. The plastic bag was forced to the bottom of the space between the baskets and a plastic collar inserted into the top of space between the two baskets to prevent bed material from filling this space. Thus traps allowed both the vertical and lateral ingress of fine sediment.

2.3 *The experiment*

A total of 16 traps were installed at two riffle sites. Four traps were placed equidistant across the channel on the riffle head and tail, at each site, in order to reveal spatial differences in matrix ingress. At Site 1 the surface and sub-surface framework gravels had the same grain-size distribution but at Site 2 one half of the traps had a coarse surface layer in comparison to the sub-surface while the other half had the reverse. Two traps of each sequence i.e. fining or coarsening upwards, were placed on the riffle head and tail at this site. This was done to allow an investigation of the influence of the vertical structure on matrix ingress.

The traps were emptied after 30 days. During retrieval, flow was isolated across the top of the traps with a cylindrical drum, the plastic collar removed and the plastic bag pulled up to enclose the inner basket. The bag and inner basket could then be removed from the river-bed without the loss of any matrix material. The framework clasts from each trap were washed in a bucket to remove the ingress matrix and detritus then placed back in the trap. Matrix samples were returned to the laboratory for granulometric analysis. Initially, each sample was passed through a -1 ϕ sieve, air dried, weighed, and then sieved to half ϕ intervals.

2.4 *Sediment transport*

Suspended sediment samples were taken on a regular basis every fourteen days at Site 1. Samples were also collected at half hourly intervals during ten high flow events, in 1989. Each sample was passed through a 0.22 μm membrane filter and the dry weight recorded along with its colour (using a Munsell colour chart).

Bed material movement was assessed at Site 1 during the 12 month monitoring period. Seventy-five surface gravel clasts (size range: 5 ϕ to 8.5 ϕ) collected from the site were painted and placed across the channel to determine movement of the coarse bed material. After each high flow event the painted clasts were recovered and replaced across the channel.

3 RESULTS

3.1 *Sediment transport*

Mean monthly discharges for Sixth Creek ranged from 0.02 to 1.05 m^3s^{-1} and peak discharges from 0.12 to 2.49 m^3s^{-1} during 1989. Suspended sediment concentrations increased with discharge (Fig. 2a) in a similar manner as that documented for other Australian streams (e.g. Loughran 1977, Olive & Reiger 1985). There is, however, a considerable degree of scatter in this positive relationship with the variance being greater during the summer months (Nov-Apr) compared to the winter months (May-Oct). For example, at a discharge of 0.0021 m^3s^{-1} the range of suspended sediment concentrations recorded was 1887% greater during summer (0.93-138 mg 1^{-1}) than winter (2.3-9.3 mg 1^{-1}).

Rating curves for suspended sediment concentration *versus* discharge were constructed for the summer and winter months by fitting Model II logarithmic regression curves. Both curves had significant slopes (ANOVA: $p < 0.05$) and the slopes of each curve were significantly different from one another (ANCOVA: $p < 0.05$). For any given discharge, suspended sediment concentrations tended to be higher during summer in comparison to the winter months.

Figure 2b illustrates the behaviour of suspended sediment concentration and discharge during two typical high flow events in June and September 1989. All events, with the exception of those in September, displayed a clockwise hysteresis where peak sediment concentrations occurred before peak discharge. By comparison, high

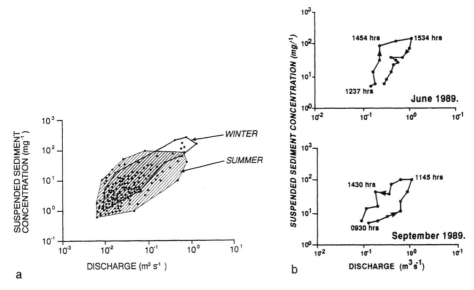

Figure 2. Suspended sediment concentrations recorded in Sixth Creek during 1989. a) Suspended sediment concentration versus discharge. b) The behaviour of suspended sediment concentrations during floods in July and September.

flow events during September were characterized by anti-clockwise hysteresis. Delayed sediment input results in peak suspended sediment concentrations preceding peak discharge, thus increased concentrations on the falling of the hydrograph (Wood 1977).

Variations in the colour of suspended sediment samples were noted during storm events in 1989. For example, the brownish yellow colour (HUE 10YR 6/6) of sediment on the falling limb of hydrographs in September was distinctly different to the reddish brown (HUE 5YR 5/3) colour of the rising limb. The reddish brown coloured sediment was similar to soils found in the upper head-waters and higher slopes of the catchment. In contrast, the colour of the sediment sampled on the rising limb of events in September and during all other storm events (brownish yellow) was characteristic of material found within the channel and on the lower slopes of the catchment. Sediment source areas, distinguished on the basis of sediment colour, have a unique set of characteristics which determine the quantity of material in storage and its release to the channel during a rainfall event. Differential supply conditions from the various sediment source areas in Sixth Creek result in exhaustion and delayed sediment input causing distinct hysteretic effects during storm events.

Movement of the sediment tracers was recorded during June-October. All 75 gravel clasts were recovered after each month of movement. However, a number of the clasts were partially buried during July and August indicating vigorous bed load transport. Further evidence of bed load movement was noted in the experimental reach, with the reduction in the size of a small bar immediately downstream of Site 2. No estimate of the depth of bed disturbance was made.

3.2   *Rates of matrix accumulation*

A two-way analysis of variance (Zar 1984) was used to assess temporal and spatial trends in the rates of matrix accumulation at Site 1. On a monthly basis there was no significant difference ($p < 0.05$) in the spatial variation of matrix accumulation. Distinct temporal variations in the mean rate of accumulation were evident and associated to variations in discharge. Positive relationships were recorded between the mean rate of accumulation and the mean monthly discharge ($r^2 = 0.68$) and peak monthly discharge ($r^2 = 0.74$). Figure 3 shows the variation in the mean rate of accumulation (data were averaged for the whole site) and discharge. Rates of accumulation varied between 0.06 and 0.54 kg m^2d^{-1} and were over 300% greater during the winter (mean = 0.33 kg m^2d^{-1}) in comparison to summer months (mean = 0.08 kg m^2d^{-1}).

The percentage of the available void-space filled by deposition each month was variable. Over the 12 month period the average value was 51.27% with a standard deviation of 12.41%: values ranged from 12.36% during February to 78.17% in August.

Despite there being no statistically significant difference spatially in the rate of matrix accumulation, cross-channel and longitudinal patterns at Site 1 could be detected in the monthly data set. These spatial variations became clear when data for the individual traps were averaged for the 12 month period (Fig. 4). During 1989 the deposition of matrix was 142% greater at the riffle head (mean = 0.29 kg m^2d^{-1}) than the riffle tail (mean = 0.12 kg m^2d^{-1}) despite a distance of only 5 m separating the two. This pattern was observed for all the individual months. At a reduced scale this decrease in matrix accumulation along the riffle corresponds to 'Sundborgs Law' of decreasing sedimentation from a source (Sundborg 1967).

Rates of matrix accumulation increased across the channel from the right to left bank at both the riffle head and tail. This uneven cross-sectional distribution was more pronounced at the riffle tail, as noted by an average 675% difference between the right and left bank traps in comparison to 89% recorded at the riffle head. Cross-channel variations were noted during all months but there was a tendency for these to decrease during the winter months from June to August.

Distinct channel filaments of increased flow velocities were identified at this riffle site for discharges up to 1 m^3s^{-1} (i.e. just below bankfull conditions). At these dis-

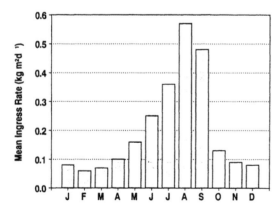

Figure 3. Mean rates of matrix accumulation and discharge variations for Sixth Creek.

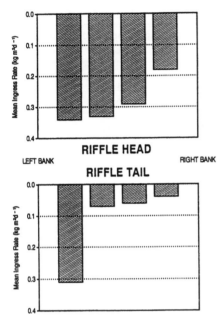

Figure 4. Spatial variations in matrix accumulation recorded at Site 1.

charges, flow over the riffle became concentrated towards the left bank approximately half way along the riffle resulting in increased velocities. However, at discharges in excess of 1.2 m^3s^{-1} cross-sectional data indicate velocities were evenly distributed across the riffle. Velocity patterns can dictate the pattern of cross-channel transport of bed-load at-a-site (Carling & McCahon 1985) and the results of this study indicate they may also influence rates of matrix accumulation.

Table 1 presents data on the accumulation of matrix sediment in the gravel substratum's of different vertical compositions monitored at Site 2. Rates of accumulation were consistently greater in the fining upward sequence by an average 121% over the 12 month period. Cross-sectional variations in flow velocity at this riffle site were negligible, therefore, differences between the traps were assumed to be a direct result of variations in the vertical composition of the traps. Inspection of the traps after retrieval from the river-bed revealed that in the coarsening upward sequence there appeared to be clogging of the interstitial spaces approximately 100

Table 1. Rates of matrix deposition into fining and coarsening upwards gravel-bed deposits. Data are 12 month averages for each trap.

Trap no. and position		Mean, minimum, maximum ($kg.m^2d^{-1}$)			Mean, minimum, maximum ($kg.m^2d^{-1}$)		
		Fining upward			Coarsening upward		
1	Riffle head	0.26	0.14	0.78	0.12	0.06	0.46
2	Riffle head	0.26	0.20	0.81	0.12	0.04	0.40
3	Riffle tail	0.22	0.18	0.86	0.10	0.05	0.42
4	Riffle tail	0.22	0.20	0.80	0.10	0.04	0.49

Table 2. Textural composition of ingressed sediment. Data are means, with range in *italics.*

	D50 (ϕ)	Coarse sand (%)	Medium sand (%)	Fine sand (%)	Silt/clay (%)
Site 1					
All traps	2.50	4.58	29.08	61.50	4.92
N = 8	*1.68-2.83*	*1-15*	*17-54*	*26-75*	*1-18*
Riffle head	2.10	6.75	41.01	47.25	5.29
N = 4	*1.68-2.62*	*3-10*	*28-51*	*36-61*	*3-8*
Riffle tail	2.39	5.23	34.02	53.75	7.25
N = 4	*1.65-2.77*	*3-10*	*25-52*	*35-62*	*3-10*
Site 2					
Coarsening	2.35	5.21	35.11	50.18	6.25
Upwards	*2.23-2.48*	*3.5-11*	*28-54*	*32.-57*	*4-8*
N = 4					
Fining	1.94	11.21	53.12	35.41	4.11
Upwards	*1.68-2.22*	*8-16*	*32-58*	*30-58*	*2-6*
N = 4					

mm below the surface layer which prevented further movement into the substratum. Matrix material in the fining upwards sequence was more uniformly distributed within the interstitial spaces of the gravel traps.

4 DISCUSSION

There are few measurements of infiltration rates of fine sediment into gravel-bed river sediments. Rates of deposition, recorded by Carling and McCahon (1985) i.e. 0.001 to 1.43 kg m^2d^{-1}, in a comparable upland gravel-bed stream were consistently lower than that recorded in this investigation during base flow and storm events. However, Welton (1980) and Frostick et al. (1984) both measured deposition rates in lowland gravel-bed rivers, albeit experiencing high sediment loads, which were approximately an order of magnitude greater (> 10 kg m^2d^{-1}) than that recorded in Sixth Creek. Differences between these studies may reflect variations in sediment supply and/or variations in the composition of the river-bed substratum.

The procedure employed in this and other studies, provides a measure of the maximum potential ingress of matrix only. Removal of the matrix sediment from the traps every 30 days, artificially 'cleans' the gravel substratum. Under 'normal' field conditions gravel bed rivers contain a quantity of matrix and the actual flux of this material will depend not only on the supply of matrix to the bed (as measured in this study) but also the removal or flushing of matrix with the disruption/turnover of the gravel bed material in high flow events. Movement of the surface tracers during the winter months at this site may suggest that an unmeasurable quantity of matrix was removed from the gravel substratum during high flow events. However, the results of this study do suggest that a number of supply processes may influence the accumulation of matrix within a gravel bed substratum.

The infiltration of fine matrix sediment into coarse river-bed sediments is controlled at two levels (Beschta & Jackson 1979).

4.1 *Processes acting at and/or above the bed surface*

Infiltration of fine sediment is governed by the size and shape of the ingressing particles with respect to the surface voids and turbulence at the sediment-water interface (Carling 1984). However, the dominant control on rates of deposition is suspended sediment concentration (Einstein 1968), although silting of gravels can be rapid even at low suspended sediment concentrations (Thoms 1987).

Marked seasonal variations in discharge and suspended sediment concentration were recorded in Sixth Creek during 1989 along with similar variations in the rates of deposition. The rates of matrix deposition may be a result of differential supply conditions. Increased sediment concentrations and loads, during the winter months, resulted in elevated deposition rates. However, there is an important anomaly when the rates of deposition for July and September are compared. The mean rate of deposition in July was less than September despite the former experiencing more high flow events, higher discharges and elevated suspended sediment concentrations. Data indicate the behaviour of sediment transport during the July and September high flow events differed and that this may have influenced rates of matrix deposition.

The dynamics of sediment transport during high flows are important in understanding the conveyance of fine sediment within a river channel. Mobilization and transport of sediment occurs on the rising limb of the hydrograph and deposition on the receding limb. High flows in July had a pronounced clockwise hysteresis whilst those in September were characterized by anti-clockwise hysteresis. Thus the relative supply of matrix available to be ingressed during high flows was greater in September than July. Figure 5 provides a schematic illustration of the influence of differential sediment transport behaviour for matrix accumulation during high flows. Sources of sediment transported in Sixth Creek during most of the year (with the exception of September) were in-channel deposits and the lower catchment slopes. These sedi-

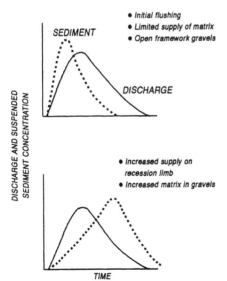

Figure 5. Schematic diagram of the implications of elevated suspended sediment concentrations on the rising and falling limbs of the hydrograph.

ment source areas have a limited budget and will become depleted with prolonged rainfall near the end of the winter flood season. Sediment exhaustion contributes to lower sediment concentrations on the falling limb of the hydrograph in comparison to the rising limb, thus limiting the potential supply of matrix to be ingressed into a gravel-bed substratum. By comparison, sources of sediment transported during high flows in September were from the headwater areas of the catchment. The increased distance of this source from the experimental reach, results in a sediment lag and elevated sediment concentrations on the falling limb of the hydrograph. Anti-clockwise hysteresis, therefore, may promote an increase in the potential supply of matrix to the river-bed.

The conveyance of sediment during a high flow event can influence the storage of material along a river channel system. From their work on the Sandusky River in Ohio, USA, Verhoff et al. (1979) suggested that suspended sediment moves through a river channel in a discontinuous manner. Each storm event entrains sediment from the river-bed, banks, floodplain, transports it some distance downstream and deposits it within the river channel. A model, based upon the difference between the flood wave celerity and the velocity of water flow, which assumes the flood wave moves kinematically, was derived to calculate average travel distances of sediment pulses (waves or sedographs) during a flood event. This model was applied by Lambert & Walling (1986) to the River Exe in Devon (UK) noting that derived travel distances corresponded to known areas of sediment accumulation. Elevated sediment concentrations on the rising limb of the hydrograph were associated with long travel distances, resulting in a natural flushing through the river channel system and reduced supply of matrix for ingress. Anticlockwise hysteresis relationships displayed the converse.

Suspended sediment concentrations are an important control on rates of matrix deposition (Einstein 1968, Carling 1984, Frostick et al. 1984). This study indicates that the dynamics of sediment transport during high flows are also an influencing factor. The importance of whether or not the hydrograph and sedograph are in- or out of phase has not previously been considered. This has a number of important implications concerning the composition of gravel-bed substrata. For example, rivers which are dominated by a 'first flush' of sediments, such as in urban catchments (Thoms 1987), may have a limited supply of fine sediment for ingress regardless of high suspended sediment concentrations. Thus the river-bed may be characterized by relatively open framework gravels. Similarly other catchment disturbances which result in a change to the quantity and timing of sediment delivered to a channel may alter the supply of matrix along the channel. Attenuation of the hydrograph and sedograph will occur downstream but variations in sediment sources and roughness along a channel will alter the speed and size of the water and sediment wave (Pickup et al. 1983). This may result in distinct zones of either elevated or limited matrix supply depending on the relative timing of each and therefore zones of relatively open framework and matrix supported gravels along a channel.

4.2 *Mechanisms acting within the substratum*

Once a particle has moved through the surface layer its further movement into the sub-surface sediment is controlled by the relationship between its size and the size of pores which it approaches. In the absence of significant lateral movement within the substratum, each grain will settle through a deposit until it reaches a pore too small to allow it to pass. Thus the vertical structure of the substratum is an important factor. Frostick et al. (1984) noted that the presence of sub-surface framework material reduced mean rates of matrix accumulation over a three month period by up to 87%.

It is evident from the results obtained at Site 2 that the vertical composition of a gravel substratum does influence rates of deposition. Matrix accumulation was greater in the fining upward sequence by 121% in comparison to the coarsening upward sequence. In a fining upward deposit, matrix sediment which passes through the surface layer will meet progressively larger void-spaces. In this situation the gravel deposit fills from the bottom up in a manner similar to that described by Einstein (1968).

Gravely river-bed deposits commonly develop a surface armour layer which is coarser than the sub-surface framework (Carling & Reader 1982). As a result the size of void-spaces get progressively smaller with depth in such a deposit. Matrix sediment once it has passed through the larger, surface pores of the armour may then be too large to pass down through lower pores. Because these particles are prevented from further movement they bridge the interstitial pores causing clogging of the framework just below the surface layer and reduce the rate of ingress. This leads to the development of patchy matrices with unfilled void-spaces at depth. The likelihood of this occurring is relatively high because when floods recede, coarser matrix particles will be the first to ingress, increasing the probability of clogging of the near-surface pores.

5 CONCLUSIONS

Modern and ancient alluvial gravel deposits have long been utilized as sources of building aggregate. Successful modeling and the ultimate exploitation of these deposits require a knowledge of their formation processes and composition. Whilst there are abundant data on the geomorphic features, strata type and stratification sequences of aggregate deposits, experimental studies have not contributed greatly to gravel deposit models. This is because of problems of scale in flume work, and the difficulty of observing active transport and deposition in high energy environments.

This experimental study has indicated that the accumulation of fine matrix sediment into framework gravel deposits is controlled by two broad factors: 1. The rate of sediment supply and the nature of the sediment and water discharge waves during a high flow event, and 2. Controls within the river-bed. The interaction of these variables dictates the complex pattern of matrix development and may explain the wide range of gravel fabrics found.

Matrix sediments are derived from suspended and finer bed load material. Varying catchment conditions which alter the character of sediment carried as either suspended or bed load will influence the matrix content of and eventual storage in the channel bed. The increased supply of sediment to a channel will encourage the infilling of interstitial spaces. Alterations to the dynamics of water and sediment discharges may either elevate or limit the supply of matrix to be ingressed.

Knowledge of the distribution of matrix development within framework gravels is also of significance to their economic potential. Matrix supported deposits are likely in fining upward gravel sequences and patchy development is encouraged in coarsening upward sequences. The presence and nature of impermeable layers in gravel deposits that act as reservoirs can also determine sites of liquid and gas accumulation.

ACKNOWLEDGMENT

Part of the research was funded by the Royal Society of South Australia.

REFERENCES

Allen, J.R.L. 1983. Gravel overpassing on humpback bars supplied with mixed sediment: Examples from the Old Red Sandstone, Southern Britain. *Sedimentology* 30: 285-294.

Beschta, R.L. & Jackson, W.L. 1979. The intrusion of fine sediments into a stable gravel bed. *Journal of Fisheries Research Board Canada* (36): 204-210.

Carling, P.A. 1984. Deposition of fine and coarse sand in an open-work gravel bed. *Canadian Journal of Fisheries and Aquatic Sciences* (41): 263-270.

Carling, P.A. & Glaister, M.S. 1987. Rapid deposition of sand and gravel mixtures downstream of a negative step: The role of matrix infilling and particle overpassing in the process of bar front accretion. *Journal of the Geological Society, London* (144): 543-551.

Carling, P.A. & McCahon, C.P. 1985. Natural siltation of brown trout (*Salmo trutta* L.) spawning gravels during low-flow conditions. In J.F. Craig & J.B. Kemper, J.B. (eds), *Regulated Streams: Advances in Ecology*: 229-244. Plenum Press, New York.

Carling, P.A. & Reader N.A. 1982. Structure, composition and bulk properties of upland stream gravels. *Earth Surface Processes and Landforms* (7): 349-365.

Dyer, K.R. 1970. Grain size parameters for sandy-gravels. *Journal of Sedimentary Petrology* 40(2): 616-620.

Einstein, H.A. 1968. Deposition of suspended particles in a gravel bed. *Journal of the Hydraulics Division*, ASCE (94): 1197-1205.

Erskine, W.D., Geary, P.M. & Outhet, D.N. 1985. Potential impacts of sand and gravel extraction on the Hunter River, New South Wales. *Australian Geographical Studies* (23): 71-86.

Frostick, L.E., Lucas, P.M. & Reid, I. 1984. The infiltration of fine matrices into coarse-grained alluvial sediments and its implications for stratigraphical interpretation. *Journal of the Geological Society, London* (141): 955-965.

Lambert, C.P. & Walling, D.E. 1986. Suspended sediment storage in river channels: a case study of the river Exe, Devon, UK. In R.K. Hadley (ed.), *Drainage basin sediment delivery*, IAHS publication 159: 303-315.

Loughran, R.J. 1977. Sediment transport from a rural catchment in New South Wales. *Journal of Hydrology* (34): 357-375.

Miall, A.D. (ed.) 1978. *Fluvial Sedimentology*. Canadian Society of Petrology and Geology Memoir 5.

Olive, L.J. & Reiger, W.A. 1985. Variation in suspended sediment concentration during storms in five catchments in southeast New South Wales. *Australian Geographical Studies* 23(1): 38-51.

Pickup, G., Higgins, R.J. & Grant, I. 1983. Modeling sediment transport as a moving wave: The transfer and deposition of mining waste. *Journal of Hydrology* (60): 281-301.

Sundborg, A. 1967. Some aspects on fluvial sediments and fluvial morphology, 1. General views and graphic methods. *Geografiska Annaler* 49, A(2): 333-343.

Thoms, M.C. 1987. Channel sedimentation in the urbanised River Tame, UK. *Regulated Rivers: Research and Management* 1(3): 229-246.

Thoms, M.C. 1992. A comparison of grab- and freeze-sampling techniques for the collection coarse-grained channel sediments. *Sedimentary Geology* (78): 191-200.

Verhoff, F.H., Melfi, D.A. & Yaksich, S.M. 1979. Stream travel distance calculations for total phosphorous and suspended materials in rivers. *Water Resources Research* 15(6): 1354-1360.

Welton, J.S. 1980. Dynamics of sediment and organic detritus in a small chalk stream. *Archive Hydrobiologia* (90): 162-181.

Wood, P.A. 1977. Controls of variation in suspended sediment concentration in the River Rother, West Sussex, England. *Sedimentology* (24): 437-445.

Zar, J.H. 1984. *Biostatistical Analysis*, 2nd Edition. Prentice-Hall, New Jersey.

Planning for aggregate extraction: Using an integrated resource management approach

DOUGLAS BAKER

Faculty of Natural Resources and Environmental Studies, UNBC, Prince George, Canada

1 INTRODUCTION

In most cases, aggregate extraction involves the mining of sand, gravel or bedrock in an urban or suburban context. Although aggregate mining is similar to most mining methods, requiring excavation and on-site processing, the location of aggregate operations makes it a very different type of mining. Aggregate extraction takes place in a landscape that includes a myriad of other land uses that often are not compatible with this type of industrial use. Yet, aggregate extraction is a necessary mining activity for any form of urban development.

Due to the unique character of this type of mining, the regulation of aggregate mining in Canada has followed two general approaches: to regulate sand and gravel activities as a resource at the provincial level, or to plan for aggregate extraction as a land use within a municipality. At the municipal level, aggregate extraction is frequently controlled as a land use through soil removal by-laws or extractive-industrial zones. This type of regulation is often restrictive and attempts to control the negative impacts of aggregate mining due to competing land uses and public sensitivity to the noise, dust, and traffic associated with extraction and processing. Aggregate mining in many communities is viewed as a negative land use and there is little provision at the local level to identify and protect aggregate sources as a resource. When this is the case, high quality sand and gravel deposits are overlooked or ignored as a valuable resource and urban development is allowed to sterilize high quality deposits.

At the provincial level, aggregate deposits are recognized as a provincially significant resource and inventories are created to locate prime mining deposits. Within this context, the regulation of aggregate resources tends to focus on the singular sectoral aspect of mining and controlling the site development. However, this approach fails to recognise the dynamic land use environment in which urban mining takes place, and the emphasis of regulation tends to be on the singular aspect of mining to the detriment of other land uses.

The focus of this paper is to examine a different approach to the planning and regulation of aggregate resources based on integrated resource management principles. Because the character of aggregate resources is unique, and not easily regulated using the present planing processes, an integrated resource management approach may provide a better means to develop and reclaim this resource. Integrated resource management provides a method to pull together provincial, municipal, public, and mining interests to provide a more efficient and effective means to plan for aggregate extraction.

2 THE CURRENT REGULATION PROCESSES

The control of sand and gravel resources in most provinces in Canada is shared between municipal and provincial interests. The relative control of each level of government varies between provinces. Two examples of where the control of aggregate extraction varies significantly between provinces occurs in the cases of Ontario and British Columbia. Ontario has developed comprehensive legislation since 1971 (Pits and Quarries Control Act) to regulate aggregate extraction at a provincial level. In British Columbia, there is very little provincial guidance for the management of aggregate resources, and the bulk of the regulation occurs at the municipal level of government. Each of these regulatory regimes will be examined with respect to the strengths and weaknesses of the management of aggregate resources.

2.1 *Ontario*

In Ontario, the extraction of sand, gravel, and bedrock resources is regulated under the Aggregate Resources Act (1989). The Ministry of Natural Resources (MNR) administers the Aggregate Resources Act in conjunction with the Mineral Aggregate Resources Policy Statement (1986) which is issued under the authority of Section 3 of the Planning Act (1983). The Statement (MARPS) provides direction as a policy statement to the Province and municipalities for the planning of aggregate resource extraction.

2.1.1 *Aggregate Resources Act (1989)*
The Aggregate Resources Act, proclaimed on January 1990, replaces the former Pits and Quarries Control Act (1971). The Act provides a means to control pit development and rehabilitation by a licensing system that is administered by MNR. A Class 'A' license (to excavate annually more than 20,000 tonnes) or a Class 'B' license (less than 20,000 tonnes) is required for aggregate extraction. For a Class 'A' license, the operator must submit a detailed site plan that outlines features such as topography, mining activity, adjacent properties, the water table and surface water, the sequence of development, and the progressive rehabilitation. The plan is to be prepared in at least three separate drawings outlining: existing features, the operational plan, and the progressive, and final reclamation stages. The site plan for a Class 'B' license requires similar feature descriptions, but with less detail.

In addition to a site plan, an applicant for a Class 'A' license, must submit a report that details (Section 9):
 – The suitability of the reclamation plans with regard to the adjacent lands,
 – The environment that may be expected to be affected by the pit operation and any remedial measures that might be required,
 – The social and economic effects that may be expected,
 – The main haulage routes and proposed truck traffic to and from the site,

– Potential hydrology and hydrogeology problems on site,

– The location of on-site overburden, top soil, and aggregate stockpiles, and

– Any other planning or land use considerations or information respecting the site.

To obtain a license, the operator must first apply to MNR by filing a site plan, and then give public notice of the application. Every application for a license is circulated internally within MNR and externally to conservation authorities, municipalities, the public, and other concerned agencies. An environmental impact assessment can be requested in those areas considered environmentally sensitive. Any objections to the application, by the municipality or public, can be filed with MNR. Objections may require the Minister to refer the issuing of a license to the Ontario Municipal Board (OMB). The OMB is a semi-judicial tribunal that is provincially appointed and conducts hearings into grievances and objections to activities under the Planning Act. The OMB is not bound by precedent, but rather the legislation and policy of the day generally defines the parameters within which judgements are made.

Under the Aggregate Resources Act, an annual fee is payable for the operation of a pit or quarry and the site is subject to inspection by the MNR. A production fee of six cents per tonne of aggregate is levied against the operator, and of this, four cents goes to local municipalities, one cent to the Province, one half cent to counties/regions, and one half cent to the abandoned pit and quarry rehabilitation fund.

2.1.2 *The Planning Act (1983)*

Municipal control of aggregate resource extraction is provided in the Planning Act (Section 34) through zoning by-laws and the Official Plan. Municipalities cannot regulate established pits and quarries through zoning by-laws; however, they have location control over new or proposed pit and quarry development. No license under the Aggregate Resources Act can be issued unless the conditions comply with the Official Plan and the respective zoning. In the event that municipal by-laws or an Official Plan conflicts with the Aggregate Resources Act, the Act takes precedence and the municipal regulations are inoperative to the extent of the conflict with the Act (Section 66:1-4).

Further control of pit operation is granted to municipalities by the Municipal Act. Under this act, the municipal council may regulate hours of operation, dust control, types of machinery used, setbacks, and grades.

2.1.3 *Mineral aggregate resources policy statement (MARPS)*

The Ontario government has issued a total of seven policy statements, under Section 3 of the Planning Act (1983). One of these is MARPS, which establishes aggregates as a matter of provincial interest at the highest level of policy (i.e. approved by Cabinet). All levels of government are required to have 'due regard' to policy statements issued under the Act. The objective of MARPS is to ensure that aggregate resources remain available to local producers, that existing licenses are protected from incompatible uses, and that designated reserves (through the Aggregate Resources Inventory Program) are available for future mining.

The following planning principles from the basis for the policy:

– Aggregates as a non-renewable resource should be recognized as an important component to any comprehensive land use plan,

– Aggregates should be available to the consumers of Ontario at a reasonable cost and as such it is necessary to maintain sources of supply as close to markets as possible,

– All parts of Ontario that contain aggregate deposits share a responsibility for meeting the provincial demand,

– It is essential to ensure that extraction is carried out with minimal social and environmental cost,

– Other land uses may in 'specific instances', take precedence over aggregate extraction, and

– Municipalities have an important role in the planning of this resource and 'should encourage the concept of extraction as an interim land use activity'. Mineral aggregate sources should be identified and protected from other incompatible land uses.

The policy embodies an attempt to place emphasis on the resource management of aggregates with direction at the provincial level. As well as provincial guidance, the Province's interests are also implied in terms of supplying the provincial demand, and maintaining a reasonable cost to the consumers of Ontario.

2.1.4 *Problems with regulation*

Ontario has designed some of the most comprehensive regulations in the country to control aggregate extraction. However, despite the approach taken, there remains considerable conflict within the province over aggregate mining. Since 1971, over 150 Ontario Municipal Board (OMB) hearings have convened to settle disputes involved with objections to aggregate mining. The OMB is a quasi-judicial, provincially appointed tribunal that conducts hearings into grievances and objections to activities under the Planning Act. These hearings can be expensive and time consuming, for example, the Puslinch OMB hearings that extended from 1988-89, cost a total of $8 million and lasted 161 days.

An analysis of the conflict generated by OMB hearings (Baker 1992) indicates that 68% of the hearings centred on objections to license applications by individuals or municipalities; from 1980 to 1990 over 50% of new license applicants required an OMB hearing (MNR 1992). The strong objections to aggregate mining by individuals, groups, and in many cases, municipalities has led to a very expensive and time consuming licensing process. The primary issue for most of the disputes centred around the negative impacts of pit and quarry development on local residents and their environment. With respect to municipalities, the conflict frequently involved disagreements with the Province over Official Plan designations and zoning violations. Municipalities would attempt to restrict aggregate extraction with zoning by-laws in contravention to MARPS and provincial policy. The OMB often facilitated the 'battle ground' for these jurisdictional disputes.

2.2 *British Columbia*

Regulations of aggregate mining in British Columbia has been primarily controlled by soil removal by-laws at the municipal level. The Province has taken a laissez-faire approach to sand and gravel resources, having no specific acts dealing with aggregate extraction. Rather, the legislation that has historically regulated this resource at a provincial level has been fragmented among a variety of statues and regulations. Hora & Basham (1981) documented the proliferation of regulations that controlled aggregate extraction prior to 1980. No less than 13 different agencies and ensuing legislation and regulations affected the extraction of sand and gravel, with no coordinating lead agency. There has been some minor changes since this study, particularly with mining legislation, but no substantive legislation has been enacted that changes the extraction of this resource. The primary legislation that controls aggregate extraction is divided between the Mines Act and the Municipal Act, with each act having separate regulations and administration enforcing different aspects of aggregate mining.

2.2.1 *Mines Act (1989)*

The Mines Act defines aggregate extraction as a mine and provides direction for the permitting, reclamation, and inspection of any mining operations. It is administered by the Ministry of Employment and Investment. The primary requirement for an aggregate source to be developed consists of a permit (Section 10:1-8) which requires the following work system information requirements as defined under Part 6.1.3 of the Health, Safety and Reclamation Code of the Mines Act:

– A regional map showing the location of the mine property,

– A plan at a scale of 1:10,000 or less showing topographic contours, claims, leases or licences, lakes, streams, buildings, roads, railways, power transmission lines, pipelines, and the locations of all proposed mining undertakings and related facilities,

– Descriptions, design data, and details of the geology and ore reserves, surface mining, roadways, material handling, overburden, waste rock dumps, stockpiles, processing plant and facilities, buildings, tailings transportation and impoundment, and water systems and storage facilities,

– The methods to be followed in the construction of haulage roads,

– A traffic control plan,

– In addition, for underground developments, detailed maps of present and proposed underground workings, and plan of surface installations.

Reclamation is governed under Part 10 of the Health, Safety and Reclamation Code of the Mines Act. The reclamation standards for aggregate mining are very generally defined and are excluded from the more rigorous standards for coal and hardrock mineral mines (Part 10.1.4). General reclamation standards are determined in Part 10.6 (1-17) and to a large degree, rely on the discretion of the mining inspector. For example:

– Land Use 10.6.3. The land surface shall be reclaimed to an acceptable use that considers previous and potential use,

– Productivity 10.6.4. The level of land productivity to be achieved on reclaimed areas shall not be less than existed prior to mining on an average property basis unless the owner, agent, or manager can provide evidence which demonstrates to the satisfaction of the chief inspector the impracticality of doing so,

– Revegetation 10.6.6. Land shall be revegetated to a self sustaining state using appropriate plant species.

The standards lack definition and detail, especially when applied to valuable urban and sub-urban land that requires reclamation after aggregate extraction. There is neither incentive for progressive reclamation nor are there standards to guide the final rehabilitation of the pit or quarry.

2.2.2   *The Municipal Act (1979)*

Municipalities are given control to regulate aggregate extraction through soil removal by-laws, however, the process by which control is allocated is disjointed. At a general level, the Official Community Plan (Section 944) is directed to contain statements and map designations respecting 'the approximate location and area of sand and gravel deposits that are suitable for future sand and gravel extraction'. However soils, including sand and gravel, are not considered a land use and therefore are not to be controlled by zoning by-laws within the Official Plan. Rather, each municipality has the option to implement soil removal by-laws under Section 930.1. Within this section:

2. The council may, by by-law, regulate or prohibit a) the removal of soil from, and b) the deposit of soil or other material on any land in the municipality or in any area of the municipality, and different regulations and prohibitions may be made for different areas.

3. A provision in a by-law under Subsection 2 that prohibits the removal of soil has no effect until the provision is approved by the minister with the concurrence of the Minister of Employment and Investment.

Thus, the municipality unilaterally has the ability to implement soil removal by-laws, but only to the point where the by-law restricts soil removal. Subsection 3 stipulates that the Minister of Employment and Investment must approve those by-laws that attempt to 'prohibit' soil removal.

In addition, municipalities can impose fees on permits or activities relating to the removal or deposition of soils within the municipality, and the fees may vary between different municipalities. As a result, there are considerable differences between municipalities in the structure of rates affecting aggregate operations. In areas such as the Lower Mainland, this differentiation between rates can create unfair competitive advantages for producers that have pits in municipalities with lower royalties.

In addition, municipalities can impose fees on permits or activities relating to the removal or deposition of soils within the municipality, and the fees may vary between different municipalities. As a result, there are considerable differences between municipalities in the structure of rates affecting aggregate operations. In areas such as the Lower Mainland, this differentiation between rates can create unfair competitive advantages for producers that have pits in municipalities with lower royalties.

3   PROBLEMS WITH REGULATION

Municipal regulation provides a comprehensive regulatory framework to control aggregate extraction but often the zoning by-laws are enacted to 'zone out' aggregate mining rather than provide a resource management strategy to effectively develop local sources. At the municipal level, in most cases, there is limited expertise to develop a comprehensive inventory to identify local sand and gravel resources. Other competing land uses for the same land base, such as housing, provide more revenue and do not have the 'operational' negative impacts. The end result of not protecting and conserving this resource is that many municipalities, particularly in the Lower Mainland, will be facing shortages of local supply.

Poor provincial regulations and a lack of provincial direction to identify and inventory the resource has essentially left the management of aggregate resources in the hands of the municipalities. Other than providing safety regulations and limited excavation and reclamation standards, the Mines Act is inadequate in its regulation and standards for urban mining. Reclamation standards are poorly defined, there is no requirement for progressive reclamation, and the Act does not control off-site impacts such as noise or visual effects. The administration of aggregate resources is further confounded by numerous agencies with differing levels of responsibility and mandates. The Province has not coordinated the management of aggregate resources to provide direction to municipalities, agencies, and producers. As a result, there are few effective inventories that guide the development of sand, gravel, and bedrock within municipalities and crown land.

4 INTEGRATED RESOURCE MANAGEMENT AND AGGREGATE MINING

Integrated Resource Management (IRM) is a relatively new concept whose application still remains relatively ambiguous (Lang 1986). IRM is used by many resource sectors in a variety of countries, but the application of this method is irregular and diffused in the literature. Mitchell (1986, p. 22) defines IRM as involving the following elements: multiple participant strategies, blending of various resource sectors, using resource management as a mechanism for social and economic change, and striving for accommodation and compromise. The primary purposes of IRM are to enhance communication between parties and to provide a means to carry out group decision-making. Similar to other planning methods, IRM is process driven, focusing on adopting multiple perspectives and integrating interests. Although process is an important component to IRM, equally important is that it is strategic and can focus on specific problems.

Walther (1987) has cautioned against idealistic beliefs in the problem solving capacity of IRM, suggesting that there are considerable obstacles to this form of consensus decision-making. Primary impediments consist of historically rooted differences (and different interests), economic competition, and entrenched bureaucracies. He further suggests that IRM needs to be based in a clear political and legal commitment, so that the decision making process can be enforced. Indeed, the problem solving capacities of IRM will not work in all situations and certain resource sectors will offer more favourable opportunities for the application of IRM than other sectors.

Recent work by Born & Sonzongni (1995) reduces the conceptualization of IRM into four principal characteristics: 1. Comprehensive, 2. Inconnective, 3. Strategic, and 4. Interactive/coordinative. The first three dimensions deal with the 'what' issues and the final one pertains to 'how'. Within the context of IRM, *comprehensive* refers to the breadth of the overview in which all the relevant variables are included. These variables may include the physical ecological system of which a particular resource is part, the human dimensions, and the economic and institutional aspects. A comprehensive approach effectively outlines the relative spatial characteristics that are affected by a resource use.

The second concept of *interconnective* addresses the need for interrelationships and linkages within the physical and social systems. The connections between the ecological and human environments must be acknowledged to recognize the interdependencies within social and natural relationships. For example, an interconnective approach is required for watershed planning where diverse elements need to be integrated such as the physical linkages within the basin ecosystem, the social dimensions of resource use within that watershed, and the history of involved multiple interests. Connecting these different aspects of natural resources is complex and requires a multi-disciplinary approach.

A *strategic* element is required for IRM in order to scope to the relevant problem variables. Strategic planning provides a method to reduce and bound the complexity of resource issues. As Born & Sonzongni (1995, p. 171) note 'a strategic-reduced approach aims to make integrated environmental planning and management adaptive, anticipatory, and more attuned to the realities of the political decision area'.

The final dimension is defined as *interactive/coordinative* and refers to the planning process as to how IRM should be carried out. Interactive defines the relationship between stakeholders. Lang (1986) suggests that interaction is made necessary by at least five factors: dispersed information, shared action space, conflict, legitimacy, and the need for behavioral change with respect to resource use. The coordinative aspect involves the decision-making process and includes a multiple stakeholder approach through such methods as mutual learning, negotiation and conflict resolution strategies, and involving all affective interests. The coordination and sharing of information and values is an integral part of the planning process in IRM.

5   APPLYING IRM TO AGGREGATE MINING

Aggregate extraction provides a good opportunity to apply IRM principles because many of the failures identified in the provincial and municipal management regimes in Ontario and British Columbia can be addressed by using an integrated framework. The purpose here, is to provide a conceptual framework to guide future attempts to regulate aggregate mining. The unique character of aggregate extraction, as primarily an urban form of mining, requires different management approaches compared to other forms of mining. Applying IRM to aggregate extraction provides a means to analyze present regulations and provide an alternative method to manage this valuable resource. An overview of the regulation and management issues affecting aggregate mining will be examined using the Born & Sonzongni (1995) framework for integrated resource management. Each of the four principle characteristics of IRM will be assessed with respect to aggregate extraction.

5.1   *Comprehensive*

As detailed earlier, comprehensive refers to the completeness and breadth of analysis with reference to the resource system. With respect to aggregate resources, an essential component in defining the resource system is an inventory of the potential reserves. Integral to the success of the provincial policies and legislation in Ontario was the initial inventory of the mid-seventies, termed the Aggregate Resources Inventory Program (ARIP), which identified different classes of reserves throughout the province. The ARIP is easily understood by the public and municipal regulators and provides an accessible guide to planners at the local level for land use decisions.

This approach is needed in municipalities throughout British Columbia. Inventories for aggregate resources exist for the Ministry of Transportation and Highways, which has one of the most comprehensive inventories in the province. However, the inventories are not available to most municipalities and are not normally used in land use decisions at the municipal and regional levels. A comprehensive aggregate resources inventory, which is readily understood and available to the public and the planners, is essential for future planning and land use considerations.

However, an inventory program alone is inadequate as a means to comprehensively plan for aggregate resource extraction on a long term basis. The negative impacts associated with sand and gravel mining require that adjacent land uses also be considered. Other land uses and other resources, such as environmentally sensitive areas, need to be integrated into an inventory of aggregate reserves to evaluate what is accessible and what cannot be mined. The aggregate resource inventory must be fitted into a comprehensive resource use survey to determine the relative trade-offs to be made if the resource is to be developed (or not developed). Community land use priorities and aggregate development need to be coordinated in order to provide opportunities for both land uses. This is best achieved by integrating an aggregate inventory into the community official plans and integrating different land uses on a long time horizon.

5.2 *Interconnective*

Interconnective, as applied to aggregate mining, involves the biophysical and social landscape that is physically affected by this resource use. Both the direct environmental and social impacts of mining and the shadow effects (Marshall 1983) or secondary impacts have to be accounted for in an assessment of how aggregate resources are interconnected with other resource uses. This may not require a formal environmental assessment, but it at least requires scientific awareness of the cumulative impacts associated with aggregate mining.

Of course, the antithesis is also required: when other resource uses are being considered, their impact must also be considered on mining operations and the potential availability of sand and gravel for future use. Aggregate operations are often forced out of a neighbourhood when urban encroachment finds it an undesirable land use. The State of the Resource Study (MNR 1993) identified the primary accessibility constraints to aggregate resource areas in southern Ontario as: environmental, resource quality, social and political considerations, transportation, planning considerations, and urban growth. Certainly within this context, what appears available for mining and what can actually be mined are considerably different once the accessibility constraints are factored in.

Much of the land use conflict in Ontario, with respect to aggregate development, originated as a result of the province's failure to recognize other resource values when licensing aggregate operations (Baker & Shoemaker 1995). Cumulative environmental effects on groundwater resources and wetlands provide an example where aggregate mining must be assessed relative to other resource values. Aggregate mining can negatively affect environmental quality and other biophysical resources as well as produce unwanted social impacts. It is essential to recognize that sand and gravel sources, such as eskers and kames, are also sources for groundwater and the basis for other natural systems. For example, the eskers in the western parts of the Northwest Territories are highly prized for their aggregate sources but are also crucial habitat for species that den in the loose soils such as the arctic fox. The mining of this material removes it permanently from an ecosystem that may depend heavily on the water or shelter these landforms provide. If sources are to be developed, then the alternative and adjacent resource uses need to be considered.

5.3 *Strategic*

A strategic approach to planning for aggregate mining ensures that there is a future resource base to meet the economic demand. Inevitably, any resource management strategies that focus on aggregate resource extraction do so in order to secure an adequate future supply for surrounding markets. The high expenses involved in the transportation of aggregates create a much more costly product if it has to be hauled long distances. For example, at a $10.00 per tonne delivered price, a 10 km transport price may take approximately 25% of the cost, however, if the distance is increased to 80 km, then the total cost of transport will increase to 60% (MNR 1992). Development of sources close to markets can considerably reduce the delivered price of the aggregate, increase an operator's competitive edge, and cut construction costs for products such as houses or highways.

Strategic planning for aggregate mining ensures a timely development plan for local sources that can keep pace with irregular market demands. Long delays in licensing operations inevitably mean that the extra expenses are passed on to the consumers. A timed sequence for mining within a region or municipality should be incorporated into the Official Plans of communities both to ensure an adequacy of supply and to identify those areas that will be slated for extraction. This type of information will inform future developers and landowners and reduce potential land use conflicts around aggregate sources.

A second aspect of strategic development of aggregate sources deals with the allocations of resource rent. The taxing structure of aggregate resources needs to reflect a fair tax on the operator and a reasonable tax for the provincial and municipal governments. The six cents per tonne levy that Ontario generates through the Aggregate Resources Act provides a payback to the municipalities of four cents per tonne. This rent provides a basis for compensation for such items as road damage along haul routes. A tax such as this also provides an incentive for municipalities to treat sand, gravel, and rock as valuable local resources, and to plan for these resources.

A final component to strategic development of aggregate sources is determining the reclamation landscape before the pit is depleted. Depending on the volume of the deposit and life of the pit, a reclamation strategy should be in place as soon as possible. McLellan (1985) has concluded that a progressive reclamation strategy is less expensive for the operator (compared to leaving reclamation to the end), and provides environmental benefits in the interim. Property in the urban and suburban fringe is valuable in a variety of ways, from the economic land development potential to a greenspace opportunity for parks or wildlife habitat. A reclamation strategy that integrates the final landscape into the community provides means to increase community acceptance of aggregate operations and gives local people an opportunity to participate in the decision-making. As Bauer (1993, p. 693) suggests: 'To continue as an urban land user the aggregate producer must re-evaluate its activity in the light of being surrounded by an ever increasing number of people. The industry must pursue a more sensitive and organized approach to mining, site improvement, environmental awareness, and land shaping'.

Considerable success has been documented (Baker & McLellan 1992) where provincial agencies, conservation authorities, and wildlife groups have come together to formulate reclamation strategies for aggregate operations that benefit both the producer and surrounding land uses.

5.4 *Interactive/coordinative*

This final component of IRM defines the 'how' in terms of implementing strategies to provide an integrated approach to managing the resource. An interactive approach defines the relationship between parties: information sharing, value exchange, and mutual learning. An interactive approach provides the opportunity for all parties to educate each other and to share in the decision-making process.

With respect to aggregate resources, many local communities and municipalities need to be educated in the need to conserve and protect aggregate reserves. The average consumer does not have an awareness of the importance of aggregate resources for local construction needs. In conjunction to this awareness is a need to implement aggregate reserves in municipal Official Plans with phased time lines for development. In British Columbia, the Province needs to adopt a more progressive strategy for aggregate resource policies to direct municipal governments to protect potential reserves. The present vacuum in policy and legislation at the provincial level is only facilitating an inevitable depletion of reserves around rapidly growing urban centres.

Thus, an interactive approach includes integrating different interests at four primary levels: the public, industry, municipal government, and provincial government. A primary problem in the Ontario planning process is the separation of provincial and township (municipal) interests. There continues to be poor coordination of the different interests at each level of government. The same can be said for British Columbia, where there is very little coordination and communication between municipal and provincial interests.

A coordinated approach provides a means to bring parties together. Within Ontario, the State of the Resource Study recommended that Aggregate Advisory Committees be formed to integrate interests in areas where there is continuous aggregate mining. This approach would also be beneficial in British Columbia. For example, the city Prince George has recently developed a Soil Removal By-law Committee consisting of provincial, municipal, industry, and public interests to advise the city on regulations for aggregate extraction as a land use. This type of forum is useful for education, an exchange of ideas, and a sharing of values.

6 CONCLUSION

Integrated resource management provides a planning approach that is designed to facilitate resource development. It is not a panacea to be applied to all resource issues. Aggregate extraction provides a good opportunity to apply the methods of IRM to a context that has generated considerable conflict in the past between the public, interest groups, and different levels of government. IRM provides a sound conceptual framework to analyze regulations and planning policies that control aggregate mining. It has been argued in this paper that the institutional arrangements presently being implemented in Ontario and British Columbia can be improved by adopting IRM methods. The primary challenge remains to integrate the institutional interests at municipal and provincial levels to provide a suitable planning framework. The second challenge is to bring together local residents and producers to integrate aggregate mining as an integral land use within the community.

REFERENCES

Baker, D.C. 1992. Conflicting justifications and claims to property rights: Planning for aggregate resource extraction in southern Ontario. Unpublished Ph.D. Thesis. University of Waterloo.
Baker, D.C. & McLellan, A.G. 1992. Substantive techniques for conflict resolution: Aggregate extraction in southern Ontario. In P. Fenn & R. Gameson (eds), *Construction Conflict Management and Resolution*: 161-171. London: Chapman and Hall.
Baker, D.C. & Shoemaker, D. 1995. *Environmental assessment and aggregate extraction in southern Ontario: The Puslinch Case*. The Environmental Assessment and Planning in Ontario Project. Dept. of Environment and Resource Studies, University of Waterloo.
Bauer, A. 1993. Site planning elements for aggregate mining operations. In *The Challenge of Integrating Diverse Perspectives in Reclamation*: 624-641. Proceedings of the American Society of Surface Mining and Reclamation.
Born, S.M. & Sonzogni, W. 1995. Integrated Environmental Management: Strengthening the Conceptualization. *Environmental Management* 19(2): 167-181.
Hora, Z.D. & Basham, F.C. 1981. *Sand and Gravel Study 1980 – British Columbia Lower Mainland*. British Columbia Ministry of Energy Mines and Petroleum Resources. Mineral Resources Branch, Paper 1980-10.
Lang, R. 1986. Achieving integration in resource planning. In R. Land (ed.), *Integrated Approaches to Resource Planning and Management*: 27-50. University of Calgary Press.
Marshall, I.B. 1983. *Mining, land use and the environment 2: A review of mine reclamation activities in Canada*. Land Use in Canada Series No. 23 Ottawa: Lands Directorate, Environment Canada.
McLellan, A.G. 1985. Government Regulation Control of Surface Mining Operations: New Performance Guideline Models for Progressive Rehabilitation. *Landscape Planning* 12: 15-28.
Mitchell, B. 1986. The Evolution of Integrated Resource Management. In R. Lang (ed.), *Integrated Approaches to Resource Planning and Management*: 13-26. University of Calgary Press.
Ontario 1993. *Aggregate Resources of Southern Ontario: A State of the Resource Study*. Planning Initiatives Ltd., Ministry of Natural Resources. Queen's Printers for Ontario.
Walther, P. 1987. Against idealistic beliefs in the problem-solving capacities of Integrated Resource Management. *Environmental Management* 11(4): 439-446.

Issues affecting development of natural aggregate near St. George and surrounding communities Washington County, Utah, USA

ROBERT E. BLACKETT & BRYCE T. TRIPP
Utah Geological Survey, Salt Lake City, Utah, USA

1 INTRODUCTION

1.1 *Background*

As seen throughout the United States, urban growth depletes an area of the natural aggregate resources necessary for continued development (Beeby 1988, Mikulic 1995). With growth, increasingly large amounts of aggregate are required for building roads, parking lots, houses, and other structures. As new structures are built atop potential aggregate resources, access to those resources is lost, and aggregate is effectively removed from the resource base. This situation inevitably requires communities to haul aggregate long distances, which increases the cost of future development.

In the early 1970s, California's construction aggregate industry experienced backlash from its own success as explosive urban growth created land-use pressures that caused the premature closure of pits and quarries at the urban fringe. Following the release of an urban-geology master plan by the California Division of Mines and Geology (Alfors et al. 1973), the State of California inventoried aggregate materials in metropolitan regions of California. The findings startled land-use planners. The California Division of Mines and Geology estimated that $17 billion worth of resources, primarily construction aggregate, would be excluded from mining by the year 2000 if existing land-use practices were continued. The agency also estimated that 90% of this loss could be prevented if economic geologic data, compiled in a systematic, resource-deposit inventory, were used in the local planning process. Consequently, California's Surface Mining and Reclamation Act of 1975, incorporating the master-plan recommendations, was passed into law. Under this act, nearly 50 billion tons of high-quality aggregate resources in 15 California regions were identified by 1988 and designated 'regionally significant', giving those deposits a level of protection from urbanization previously unavailable (Beeby 1988).

Utah is now experiencing this same type of explosive growth. In order to avoid future shortages of low-cost aggregate for construction, local and state government agencies will need to coordinate efforts to conserve sand and gravel resources, particularly in resource-poor regions.

1.2 *Purpose and scope*

Rapid urban expansion has resulted in significant loss of aggregate resources in the communities surrounding St. George, Utah (Fig. 1). These communities had relatively little high-quality aggregate available initially, and exploitable reserves are further restricted by building and environmental regulations. In the summer of 1995, the Utah Geological Survey began a study of the availability of natural aggregate in the St. George region, starting with informal meetings with personnel from the Utah Department of Transportation, Washington County Planning Department, and US Bureau of Land Management (BLM). We found that workers in other government agencies had also recognized conflicts created by rapid urban growth, protected biological areas (under the Endangered Species Act of 1973 as amended), lands withdrawn for wilderness or other preservation, and the continuing need for construction aggregate in the St. George area.

The first phase of our study involves compiling information on individual sand and gravel pits, surficial geology, land-status issues, and protection zones for both endangered and candidate species. From this phase of the study, we intend to identify target areas for later, detailed geologic field mapping, surveying, and deposit sampling. Initially, the study area included all of Washington County. However, to keep the work focused and of manageable scope, we limited the first phase to the area covered by the St. George and Hurricane 15-minute quadrangles (37°-37°15'N lati-

Figure 1. Regional location of St. George, Utah.

tude, 113°15'-113°45'W longitude). The goal of the study is to delineate all known deposits of good-quality aggregate, and identify areas where potential high-quality resources may occur.

1.3 *Location, physiography, and climate*

St. George is a city of about 40,000 people located in extreme southwestern Utah. With surrounding communities, the population of the area exceeds 65,000. Interstate Highway 15 connects St. George with Las Vegas, Nevada, 188 km (117 miles) to the southwest, and Salt Lake City, Utah, 488 km (303 miles) to the north. The St. George basin, the area included in the first phase of the study, comprises a low-lying region surrounding the Santa Clara and Virgin River valleys (Fig. 2). The study area is near the west margin of the Colorado Plateau, just southeast of the Basin and Range-Colorado Plateau Transition Zone (Stokes 1977). Geographic boundaries are the Pine Valley Mountains and Bull Valley Mountains to the north, the Beaver Dam Mountains to the west, and the Hurricane Cliffs to the east. The Utah-Arizona state line forms the southern boundary about 10 km (6 miles) south of St. George.

The Virgin River is the principal drainage of the Kolob Terrace, a highland region north and east of the Hurricane Cliffs that includes part of Zion National Park. The main stem of the Virgin River flows into the study area from the east, cutting through the Hurricane Cliffs. Many tributary streams drain the southeast flanks of the Pine Valley Mountains, flowing southward into the Virgin River. The Virgin River flows southwest out of the study area and has cut a deep gorge (Virgin River Canyon) through sedimentary rock formations southwest of St. George. From there, the river flows into Lake Mead along the Arizona-Nevada border.

The Santa Clara River drains the west and southwest flanks of the Pine Valley Mountains, the southeast flank of the Bull Valley Mountains, and the west flank of the Beaver Dam Mountains. This river flows into the study area from the northwest and joins the Virgin River just south of St. George.

2 SOCIOECONOMIC CONDITIONS

The area's warm climate (mean annual temperature of 61.1°F [16.2°C]), clean environment, and surrounding natural beauty, which includes desert ecosystems, spectacular landforms, and vistas, have led to changes in the economy of the area. The traditional mining/agriculture-based economy has changed to a more trade/service-based economy over the past decade, supporting retirement communities and recreation activities.

Washington County's population, centered around St. George, increased nearly 80% between 1980 and 1993 making it the fastest growing county in Utah (Duffy-Deno & Brill 1995). The Governor's Office of Planning and Budget (1994) estimated the population at 65,885 and has projected additional growth by more than 250% over the next 25 years.

Economic sectors in Washington County have shifted from resource-based to service-based employment between 1980 and 1993 (Duffy-Deno & Brill 1995). Presently, the largest sectors of Washington County's economy are services (legal,

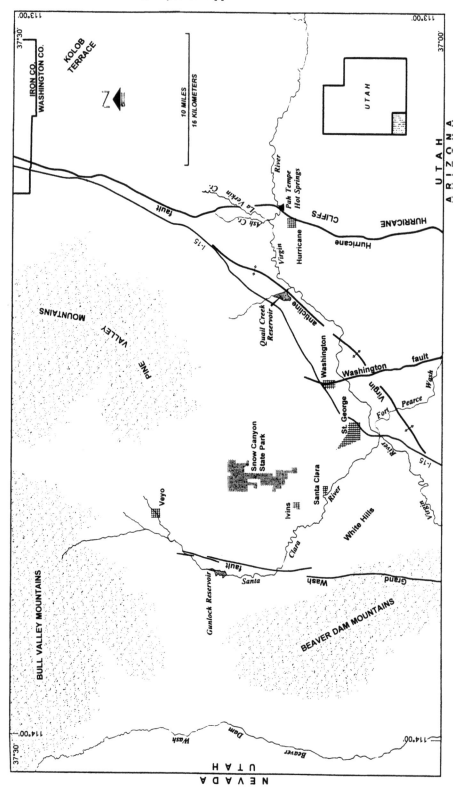

Figure 2. Map of the study area showing geographic features, principal rivers, and geologic structures.

personal, professional, and others) and trade (wholesale and retail), whereas in 1980 the economy was mostly agriculture-based. The services and construction sectors account for the largest percentage increases in economic activity over the same 13-year period. The construction sector increased by 11% and now accounts for nearly one-quarter of the total county economy. The construction sector consists of building construction (by general contractors or operative builders), heavy construction (other than building by general contractors and special trade contractors), and construction activity by other specialized-trade contractors (Duffy-Deno & Brill 1995)

3 GEOLOGIC SETTING

3.1 *Regional structure*

The St. George area lies within a zone of structural transition between generally flat-lying sedimentary rocks to the east, typical of the Colorado Plateau, and fault-bounded mountain blocks to the west, characteristic of the Basin and Range Province. Within the St. George basin, sedimentary rocks are gently folded along northeast axes. The Virgin anticline (Fig. 2) trends northeastward across the basin, extending from about 8 km (5 miles) south of St. George northeastward for more than 32 km (20 miles).

The basin is bounded on the east by the prominent west-facing Hurricane Cliffs (Fig. 2), which form the hanging wall of the Hurricane fault. The Grand Wash fault (Fig. 2), mapped by Hammond (1991) west of the St. George basin, is en echelon with the Hurricane fault and was suggested by Longwell (1952) to be the west edge of the Colorado Plateau in Arizona. The block between these two faults dips gently to the northeast (Peterson 1983).

3.2 *Sedimentary stratigraphy*

Paleozoic sedimentary rocks are exposed in the Beaver Dam Mountains, Virgin River Canyon, and the Hurricane Cliffs. In the Beaver Dam Mountains, more than 3962 m (13,000 ft) of Cambrian through Permian quartzite, shale, and carbonate formations rest unconformably on Precambrian schist, gneiss, and pegmatite (Hammond 1991). The Permian Kaibab Formation is exposed extensively southeastward from the Beaver Dam Mountains through the Virgin River Canyon. The Permian Toroweap and Kaibab Formations crop out along the Hurricane Cliffs.

Sedimentary rocks exposed in the study area are mainly Mesozoic units and have a combined thickness of about 5800 m (19,000 ft) (Fig. 3). These rocks include the Triassic Moenkopi, Shinarump, and Chinle Formations; and the Jurassic Moenave, Kayenta, Navajo, Temple Cap, and Carmel Formations (Cook 1960, Hintze 1988, Higgins & Willis 1995, Hamblin, unpubl. maps). Triassic and early Jurassic (Moenave and Kayenta) formations are primarily fine-grained clastic deposits of terrestrial origin. Upper Jurassic rocks consist of massive eolian sandstone (Kayenta Formation and Navajo Sandstone) and marine sediments (Carmel Formation).

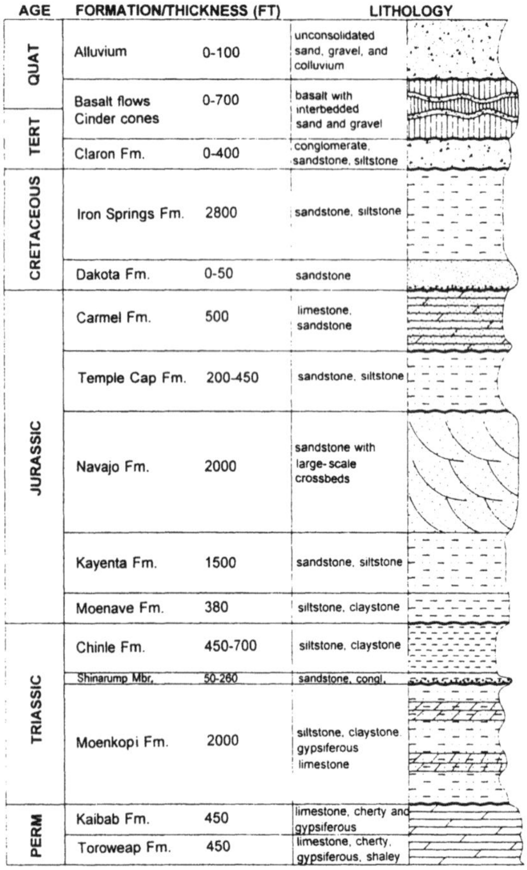

AGE	FORMATION/THICKNESS (FT)		LITHOLOGY
QUAT	Alluvium	0-100	unconsolidated sand, gravel, and colluvium
TERT	Basalt flows Cinder cones	0-700	basalt with interbedded sand and gravel
TERT	Claron Fm.	0-400	conglomerate, sandstone, siltstone
CRETACEOUS	Iron Springs Fm.	2800	sandstone, siltstone
CRETACEOUS	Dakota Fm.	0-50	sandstone
JURASSIC	Carmel Fm.	500	limestone, sandstone
JURASSIC	Temple Cap Fm.	200-450	sandstone, siltstone
JURASSIC	Navajo Fm.	2000	sandstone with large-scale crossbeds
JURASSIC	Kayenta Fm.	1500	sandstone, siltstone
JURASSIC	Moenave Fm.	380	siltstone, claystone
TRIASSIC	Chinle Fm.	450-700	siltstone, claystone
TRIASSIC	Shinarump Mbr.	50-260	sandstone, congl.
TRIASSIC	Moenkopi Fm.	2000	siltstone, claystone gypsiferous limestone
PERM	Kaibab Fm.	450	limestone, cherty and gypsiferous
PERM	Toroweap Fm.	450	limestone, cherty, gypsiferous, shaley

Figure 3. Stratigraphic column for the St. George basin and surrounding region (generalized from Hintze 1988, Charts 94 and 96). Unit thicknesses are in feet.

3.3 *Cenozoic igneous rocks*

Oligocene and Miocene calc-alkaline ash-flow tuffs, erupted from calderas in southern Nevada, are widespread in the northwestern part of Washington County in the Bull Valley Mountains, and in the northern part of the Pine Valley Mountains (Best et al. 1987). Grant (1991) described the igneous mass of the Pine Valley Mountains, separating it into an upper extrusive latite and a lower intrusive monzonite, both of Miocene age. The Pine Valley latite is a complex of flows and domes that achieves thicknesses of 488 m (1600 ft). The Pine Valley monzonite, thought to be either a sill or a laccolith, is about 305 m (1000 ft) thick (Grant 1991). Blank (1959) showed that a similar pluton occupies the core of an eruptive center in the Bull Valley Mountains.

Cenozoic volcanic rocks in the study area are mostly Pliocene dacite and basalt, and Pleistocene and Holocene basalt. Hamblin (1970) provided a classification and estimated ages of these units based on relative elevation of flows and potassium-argon dating. Basalt flows originating from vents in and around the southern flanks of the Pine Valley Mountains flowed south down paleovalleys toward the St. George basin. Eventually, they cooled and solidified. The solidified flows diverted drainage causing erosion and downcutting in less resistant sedimentary rock. This process was repeated several times over the past two million years, resulting in many long, narrow, basalt-capped sinuous ridges called inverted valleys (Hamblin 1970). Successively younger basalt flows filled the new valleys, and now older flows lie topographically above younger flows.

3.4 *Unconsolidated and semi-consolidated material*

Cook (1960) described Quaternary sediments in general terms throughout Washington County and categorized them as old pediment gravels and young alluvial channels. He described older gravels as generally coarse, poorly sorted, and commonly occurring at higher elevations than younger alluvial deposits. He described younger deposits as forming 'narrow alluvial strips, bars, and benches in modern valleys'. Nevertheless, he lumped both deposit types into one map unit.

Christenson & Deen (1983) separated Quaternary sediments on the basis of grain size and age, and discussed the availability of construction materials (sand and gravel) in the St. George area. Channel and flood-plain deposits of the Santa Clara and Virgin Rivers and their tributaries are principally sand with varying percentages of gravel, silt, and clay. Eolian sand deposits are scattered throughout the study area. Christenson & Deen (1983) delineated older Quaternary gravels in terraces at several levels above modern channels, mostly associated with the Santa Clara and Virgin Rivers, and Fort Pearce Wash.

Hamblin (unpubl. maps) prepared relatively detailed surficial geologic maps of the St. George and Hurricane 15-minute quadrangles (Fig. 4). He divided surficial deposits into four categories based on relative age, elevation, and association with the present drainage system (Table 1). The two youngest of Hamblin's mapped units comprise stream-channel, flood-plain, and low-level river-terrace deposits in active alluvial channels and alluvial fans. Hamblin's two older mapped units consist of Pleistocene and Pliocene alluvial-terrace gravels positioned at levels high above present stream channels. The older deposits commonly do not appear associated with present drainage systems and may be equivalent in age to the older, inverted valleys capped by Hamblin's (1970) Stage I, basalt flows. Higgins & Willis (1995) mapped the St. George 7.5 minute quadrangle in detail, separating the surficial deposits into 19 map units based on relative age and deposit type. For ease of presentation, surficial deposits discussed herein mostly follow the simpler descriptions by Hamblin (unpubl. maps).

4 SUITABILITY OF AGGREGATE MATERIALS

4.1 *Known aggregate sources*

The Utah Department of Transportation (1966) prepared an inventory of aggregate pits and quarries for Washington County that included material test data for representative samples. This study provided a basis for assessing the suitability of material mainly for highway construction. Considerable quantities of aggregate suitable for roadbed construction are present in the St. George area, but sources of concrete aggregate are much less common. Most source areas contain either silicic volcanic

EXPLANATION

Qa — Holocene alluvium

Qb — Holocene basalt

Qt — Holocene and late Pleistocene alluvial terrace deposits

QTt — Pleistocene and Pliocene alluvial terrace deposits

Tb — Tertiary basalt

R — Pre-Tertiary bedrock (mostly Mesozoic and Paleozoic sedimentary rocks)

⊠ — Sand and gravel

⚙ — Volcanic cinders or riprap

Figure 4. Surficial geologic map of the study area. Symbols for the various geologic units are explained in Table 1. Geologic contacts are based upon the mapping of Hamblin (unpublished maps). Some of Hamblin's units are combined for presentation purposes.

Table 1. Description of surficial deposits and basalt flows in the St. George and Hurricane 15 quadrangles (after Hamblin, unpubl. maps). Map symbols correspond to units shown on Figure 4.

Qa: Holocene alluvium. Sand and minor gravel and mud deposited in stream channels and adjacent flood plain.
Qb: Holocene basalt flows and cinder cones. Dense, black olivine basalt flows that retain original flow structures. This basalt comprises the Santa Clara flow, which originated at two cinder cones just northeast of Snow Canyon and flowed down (southward) Snow Canyon to the Santa Clara River. Although there are no definitive age-estimates, Hamblin (1970) suggests that the Santa Clara flow may be as young as 1000 years.
Pleistocene basalt. Medium-grained basalt flows extruded onto a pediment. Basalt slightly modified by weathering and erosion.
Qt: Holocene and late Pleistocene alluvial terraces. Low-level, approximately 7 m (23 ft) above the present drainage. Sand and gravel deposited in stream channels, and in alluvial fans. Most deposits are in strike valleys eroded into the Chinle, Moenave, and Moenkopi Formations, and in depressions on downthrown blocks. Includes minor colluvium.
QTt: Pleistocene and Pliocene high-level alluvial terrace deposits. Gravel and sand preserved in stream terraces up to 60 m (200 ft) above present stream channels. Deposits are as much as 30 m (100 ft) thick.
Alluvial terrace deposits (Early Pleistocene and Pliocene). Gravel and sand capping the highest terraces not obviously associated with present drainage systems. These deposits are probably equivalent in age to the oldest and highest inverted valleys capped by Tertiary basalt.
Tb: Younger Tertiary basalt. Black to medium gray, vesicular basalt flows which form inverted valleys as high as 70 m (230 ft) above the present drainage. Most flows were extruded near the base of the Pine Valley Mountains and flowed southward toward the Virgin River. Upper flow surfaces are smooth, flat, and covered by well-developed soils.
Older Tertiary basalt. Dense, black, vesicular basalt flows preserved as segments of dissected inverted valleys as high as 200 m (660 ft) above the adjacent drainage. Original margins and surface features are destroyed by weathering and erosion.
R: Mostly pre-Eocene sedimentary rock formations. Units in this category include: Permian – Toroweap and Kaibab Formations; Triassic – Moenkopi and Chinle Formations; Jurassic – Moenave, Kayenta, Navajo, Temple Cap, and Carmel Formations; Cretaceous – Dakota and Iron Springs Formations; Paleocene – Claron Formation.

rocks, or Mesozoic sedimentary units with soft or soluble minerals. The quality of gravel deposits along the Santa Clara and Virgin Rivers, and Fort Pearce Wash reportedly varies, although most deposits are suitable for use in roadbeds and asphalt. Clasts in these deposits tend to be coated with calcium carbonate and the deposits commonly contain gypsum or other soft minerals making them less desirable for use in concrete. The deposits along Fort Pearce Wash, however, are reportedly used as concrete aggregate (Larry Gore, BLM, verbal comm.).

Gravel from terraces along the Virgin River and tributaries southwest of St. George was used in the construction of Interstate 15. The Utah Department of Transportation (1966) reported that this gravel was derived from Quaternary basalt and Mesozoic sedimentary rock and was of poor quality. Because they are older, and have been subjected to carbonate soil-forming processes over a longer time period, higher level terrace deposits generally contain more calcium carbonate than younger deposits nearer the present stream level (Christenson & Deen 1983).

Christenson & Deen (1983) reported that the better quality gravel in the area is found in younger (generally Qa and Qt on Fig. 4), lower terraces along the Virgin

River east of St. George. Older, higher terrace deposits (generally QTt and some Qt) are carbonate cemented and of less quality. They also reported that the best sources of concrete aggregate are in the lower terraces located north of the Virgin River. With the exception of Fort Pearce Wash, previous studies indicate that the better sources of aggregate are in young, lower terrace deposits along the Virgin River east of St. George.

Holocene (Qa and Qt) terrace deposits along Fort Pearce Wash are mined extensively for a variety of uses, including concrete aggregate. Most of the aggregate supply for the area comes from these terrace deposits because they are: 1. Close to St. George, 2. Relatively thick, and 3. Have a more desirable clast-size distribution than most other deposits. Also, Fort Pearce Wash deposits contain relatively fewer deleterious soft and soluble clasts than are found in other areas. The source areas for Fort Pearce Wash deposits are outcrops of Paleozoic carbonate and quartzite located mainly to the southeast in Arizona.

Basaltic cinders or crushed basalt from sources in and around St. George (Qb and Tb on Fig. 4) are used in road metal and could also be used in light-weight aggregates. Crushed volcanic cinders in bituminous pavements can give desirable (non-skid) characteristics to road surfaces, and can provide good drainage for unpaved roads. The vesicular nature of some young basalt flows can be desirable for light-weight aggregate uses such as making cement (cinder) blocks.

4.2 *Potential aggregate sources*

We consider deposits derived from the Kaibab and Toroweap Formations potential aggregate sources. These carbonate rocks could also be used as bedrock sources of aggregate. The Kaibab and Toroweap are primarily hard, Permian limestones that crop out in the Beaver Dam Mountains, Virgin River Canyon, and along the Hurricane Cliffs. During recent reconnaissance, we observed deposits of alluvium and colluvium along the Hurricane Cliffs (Fig. 5) that contain clasts of mostly Kaibab limestone and some clasts of quartzite and basalt. We consider these deposits to have potential for additional development. At least two pits currently produce aggregate from these deposits.

The Kaibab and Toroweap Formations contain chert and gypsum horizons that could be detrimental to Portland cement concrete. Detailed studies would be needed to determine the extent of deleterious materials in talus, colluvium, and alluvial-fan deposits shed from these rock units.

Besides the Kaibab and Toroweap, Christenson & Deen (1983) suggested that the Shinarump Member of the Chinle Formation, and Tertiary/Quaternary basalt are other potential bedrock sources of aggregate. Bedrock sources require increased excavation and crushing costs, and also require separate sources of sand. An advantage is that crushed bedrock can provide more uniform grades of coarse aggregate.

The coarse-grained igneous rocks (quartz monzonite) of the Pine Valley intrusive body reportedly make excellent aggregate for road base (Utah Department of Transportation 1966). Clasts of the Pine Valley intrusive are in alluvial fans and stream channels extending downslope on the southeast and southwest flanks of the Pine Valley Mountains. However, these deposits also contain clasts of the less desirable Pine Valley latite and soft Mesozoic sandstone and siltstone formations, which crop

Figure 5. Southward view along the Hurricane Cliffs from a point south of the town of Hurricane. Early morning (low-angle) sunlight shows the surface trace of the Hurricane fault where recent movement along the fault (between arrows) has truncated alluvial fans and talus deposits at the base of the cliffs. The cliffs are composed mostly of the Permian Kaibab Formation. Alluvial material (included in Unit Qt on Fig. 4) extending from the base of the cliffs out onto the valley floor may comprise an relatively untapped source of aggregate. The valley floor is underlain by Triassic and Jurassic sedimentary rocks.

out on the south flank of the range. These deposits may also contain detrital gypsum or other undesirable, soluble minerals derived from the Mesozoic units.

Extensive pediment gravels up to 100 m (300 ft) thick accumulated from the late Pliocene to the Pleistocene (?) along the west slope of the Beaver Dam Mountains (Beaver Dam Slope) (Hintze 1985). These deposits consist of silt, sand, gravel, and boulders derived mostly from Precambrian metamorphic and Paleozoic sedimentary rocks of the Beaver Dam Mountains. They also include clasts of volcanic rocks from the Bull Valley Mountains. Large-scale development may be precluded, however, because the deposits locally contain siliceous material in volcanic rock clasts (detrimental as concrete aggregate). Protective regulations for the desert tortoise along the Beaver Dam Slope may also preclude mining of these deposits.

Older alluvial gravels cap hilltops and cover hillslopes, mostly along the south slope of the Bull Valley Mountains and the southwest slope of the Pine Valley Mountains, and attain maximum thicknesses of about 30 m (100 ft). Commonly, these deposits are capped by basalt. The deposits are unconsolidated and poorly sorted, containing clasts as large as boulder size. Lithologies suggest that these deposits are derived from local bedrock units.

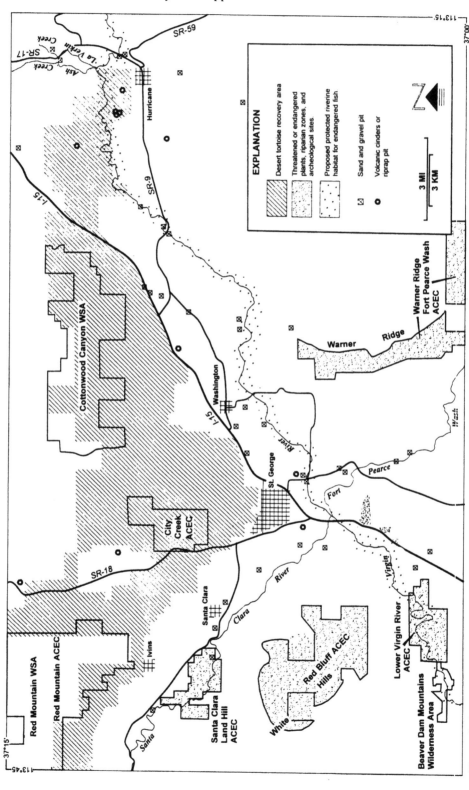

Figure 6. Areas protected for threatened and endangered species, wilderness study areas (WSAs), BLM proposed Areas of Critical Environmental Concern (ACECs), construction-material pits/quarries, and principal drainages in the study area. Compare to Figure 7.

5 LAND MANAGEMENT ISSUES AFFECTING AGGREGATE DEVELOP-
MENT

5.1 *Public lands administration*

The majority of lands within Washington County are public lands administered by the BLM, US Forest Service, and the National Park Service. Within the study area, the majority of lands are administered by the BLM, whereas the remainder are roughly equal proportions of state and private lands. The south part of the Dixie National Forest extends into the study area. The BLM administers public lands in the Dixie Resource Management Area in accordance with the Resource Management Plan and Environmental Impact Statement (Bureau of Land Management 1990). In its Resource Management Plan, the BLM presents its management strategy for the entire Dixie Resource Area and includes many considerations such as mineral materials, grazing, visual resources, cultural resources, wilderness, recreation, riparian systems, soil conservation, and wildlife habitat.

Natural aggregate is considered by the BLM as saleable mineral material. Mineral material on BLM land is available for extraction at the fair-market value, either from established 'community pits' or other areas where mineral permits are obtained. The BLM's Dixie Resource Management Area has over 194,000 ha (480,000 ac) open for mineral-material sales pending site-specific review. Part of this acreage is subject, however, to restrictions identified in the Dixie Resource Management Plan. About 57,000 ha (140,000 ac) in the Dixie Resource Management Area are closed to mineral-material sales (Bureau of Land Management 1990).

5.2 *Areas of critical environmental concern*

Outside of designated wilderness areas, the BLM applies its most intensive conservation management to Areas of Critical Environmental Concern (ACEC). ACEC designations help protect environmentally sensitive areas from activities otherwise permitted under the Resource Management Plan. ACECs are designated to protect scenic value, threatened and endangered species, archaeological sites, riparian habitat, sensitive soils, and other attributes. Six ACECs enclosing 7300 ha (18,000 ac) fall within the study area (Table 2, Figs 6 and 7). All of these ACECs are 'closed to mineral material sales'.

5.3 *Wilderness areas and wilderness study areas*

A small part (less than 500 acres) of the Beaver Dam Mountains Wilderness Area lies about 16 km (10 miles) southwest of St. George along the Virgin River Canyon (Fig. 6). This tract is an outlier of a much larger wilderness area that extends northward from Arizona. The area was designated wilderness as part of the Arizona Wilderness Act of 1984.

Two Wilderness Study Areas (WSAs), the Red Mountain WSA (7405 ha [18,290 ac]) and the Cottonwood Canyon WSA (4587 ha [11,330 ac]) lie within the study area (Fig. 6). Until Congress either designates these areas wilderness, or releases them from wilderness, the BLM manages the WSAs according to an Interim Man

Table 2. Areas of critical environmental concern (ACEC) within the phase one study area. The outlines of the ACECs are shown on Figure 6 (after Bureau of Land Management, 1990, 1995).

ACEC name	Location	Concerns	Hectares (acres)
Red Bluff	8 km (5 mi) SW of St. George	Dwarf Bearclaw poppy, saline soils	2433 (6010)
Warner Ridge-Fort Pearce Wash	8 km (5 mi) ESE of St. George	Dwarf Bearclaw poppy, Siler cactus, Spotted bat, saline soils, riparian zones	1494 (3690)
Santa Clara River – Land Hill	Just west of Santa Clara	Riparian zones, archeological sites, Virgin spindace	717 (1770)
Lower Virgin River	10 km (6 mi) SSW of St. George	Riparian zones, woundfin, Virgin River chub, archeological sites	591 (1460)
Red Mountain	6.5 km (4 mi) N of Santa Clara	National scenic resources	2219 (5480)
City Creek	5 km (3 mi) N of St. George	Desert tortoise, community watershed	1051 (2595)

agement Policy. The general rule of this policy is that the only activities permissible are temporary uses that create no new surface disturbance, nor involve permanent placement of structures. Exceptions to the general rule include grand-fathered uses such as livestock grazing, mining, and leases in-place on October 21, 1976 (approval date of the Federal Land Policy and Management Act). The WSAs are closed to road and trail construction, establishment of permanent right of ways, oil and gas leasing, and post-Federal Land Policy and Management Act mining claim exploration (Bureau of Land Management 1995).

5.4 *Protection for endangered species*

The region around Beaver Dam Wash and the St. George basin encompass the Utah portion of the Mojave Desert. The variety of desert soils and landforms, and the warm, arid climate make this region a unique environment in Utah. These factors, combined with urban expansion, have resulted in loss of habitat for a number of rare plant and animal species. The greatest number of threatened and endangered species in Utah are in Washington County (Jane Perkins, Utah Division of Wildlife Resources, verbal comm.). Mojave Desert animal species either now protected, or soon-to-be protected through critical habitat designation under the Endangered Species Act include the desert tortoise (*Gopherus agassizii*) and the Southwestern willow flycatcher (*Empidonax trailli extimus*). The main stem of the Virgin River is protected habitat for the Virgin River chub (*Gila seminuda*) and the woundfin (*Plagopterus argentissimus*). Protected plant species in the region include the dwarf bearclaw poppy (*Arctomecon humilis*) and the siler pincushion cactus (*Pediocactus sileri*). In addition, the virgin spinedace (*Lepedomeda mollispinis mollispinis*), chuckwalla (*Sauromalus obesus*), banded gila monster (*Heloderma suspectum cinctum*), and at least 35 other reptiles, amphibians, small mammals, insects, and birds are considered candidate species. Maddux et al. (1995) suggested that many of the candidate species will benefit from critical-habitat designation for endangered species.

5.4.1 *Desert tortoise*

The US Fish and Wildlife Service (USFWS) (1994) listed the desert tortoise as a threatened species on April 2, 1990. After several years of study, the USFWS designated desert tortoise (Mojave population) critical habitat on February 8, 1994 and established rules for the desert tortoise's recovery. Although the tortoise is described as having delayed maturity and long life, the USFWS estimates that pre-reproductive adult mortality approaches 98%. If the USFWS recovery criteria are met, the earliest that delisting of the species could take place would be in the year 2019 (US Fish and Wildlife Service, 1994).

In Washington County, desert tortoise protection zones, known as Desert Wildlife Management Areas, occupy the west slope of the Beaver Dam Mountains, most of Beaver Dam Wash (designated the Beaver Dam Slope Unit by the BLM), and the southern slopes of the Pine Valley Mountains (Upper Virgin River Unit) (Figs 2 and 6). Within the Desert Wildlife Management Areas, many activities, particularly grazing and off-road vehicle use, are prohibited. Activities such as hiking, camping, and scientific data gathering are allowed. Mining, including sand and gravel extraction, is allowed but reviewed by the USFWS on a case by case basis. The desert tortoise recovery plan states that the cumulative effects of all mining activities may not significantly impact the habitat of the desert tortoise, and any potential effects on desert tortoise populations must be carefully mitigated during the operation. Mined lands must also be restored to their pre-disturbed condition.

5.4.2 *Southwestern willow flycatcher*

The southwestern willow flycatcher was listed in the USFWS final rule as an endangered species (Federal Register, February 27, 1995, p. 10694, FR Doc. 95-4531). The willow flycatcher is a small bird approximately 15 cm (6 in) in length that occupies riparian zones and wetlands in the southwest United States. The willow flycatcher's habitat typically contains dense growths of willows and other plants with a scattered overstory of cottonwood. Presently, this type of habitat is rare and widely separated by arid lands in the desert southwest.

Reduced populations of the willow flycatcher in the southwest United States is reportedly due to: 1. Loss and modification of habitat, 2. Predation, and 3. Brood parasitism by the brown-headed cowbird (Marshall 1995). The cited causes of the loss of habitat in Utah include urban expansion along the Virgin River, inundation by Lake Powell along the Colorado and San Juan Rivers, livestock grazing in riparian zones, and the encroachment of tamarisk throughout the region (Marshall 1995). Studies summarized by Marshall (1995) indicate that increased predation on willow flycatcher eggs and hatchlings may be indirectly due to habitat fragmentation. The brown-headed cowbird, a transplant from the northern Great Plains, removes eggs from the nests of other birds, laying cowbird eggs for the host bird to hatch and rear. The Cowbird now commonly invades willow flycatcher nests.

No recovery plan for the willow flycatcher had been prepared at the time of this writing. Presumably, areas managed in the recovery plan would include the riparian zones along the Virgin and Santa Clara Rivers.

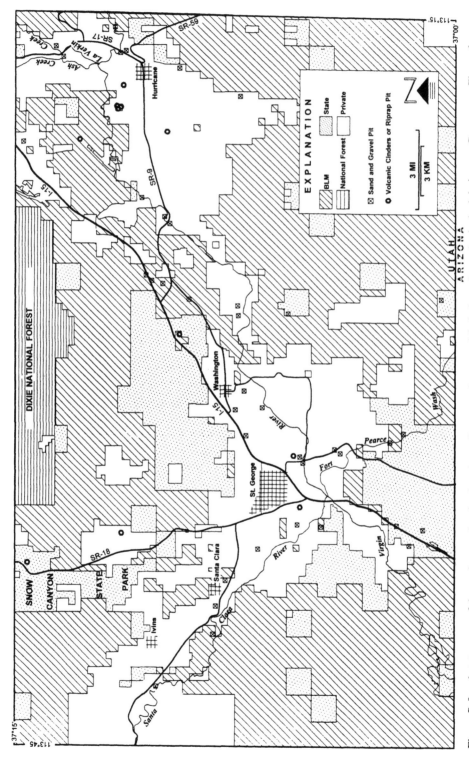

Figure 7. Land status map of the study area showing locations of construction-material pits/quarries and principal drainages. Compare to Figure 6.

5.4.3 *Endangered fish*

The USFWS has recently proposed the designation of critical habitat for three fish species endemic to the Virgin River basin (Federal Register, v. 60, no. 65, April 5, 1995). The fish include the Virgin River chub, the woundfin, and the virgin spinedace. The woundfin and Virgin River chub are listed as endangered under the Endangered Species Act. The virgin spinedace has been proposed for listing as threatened. Maddux et al. (1995) suggested that populations of these fishes have declined as a result of the loss of habitat due to lower stream flows caused by water diversions, proliferation of non-native fish, and alterations to natural flow, temperature, and sediment regimes of rivers and streams.

The USFWS originally proposed approximately 330 km (207 miles) of overlapping critical habitat along the main stem of the Virgin River and its tributaries in parts of Utah, Arizona, and Nevada for the three listed fishes. In Utah, the overlapping habitat included the main stem of the Virgin River plus portions of Beaver Dam Wash, the Santa Clara River, Ash Creek, La Verkin Creek, and the North and East Forks of the Virgin River (Maddux et al. 1995).

A revised plan withdraws the critical habitat protection of the virgin spinedace from the listing. The withdrawal would result in the designation of only the main stem of the Virgin River, from Pah Tempe Hot Springs (near Hurricane) downstream to Lake Mead, as critical habitat for the other species (Henry Maddux, USFWS, verbal comm.).

5.4.4 *Endangered plants*

The Utah Native Plant Society (1989) believes that roughly one-sixth of Utah's native vascular plant species are rare. From these hundreds of species, the dwarf bearclaw poppy is reportedly in the greatest danger of becoming extinct. In Utah, the dwarf bearclaw poppy grows only within a few kilometers surrounding St. George and Bloomington, mostly on state-owned land. The dwarf bearclaw poppy was officially listed in 1979 as endangered under the Endangered Species Act. It is one of only three species in the genus *Arctomecon*, which are all found in the western United States. All are gypsum loving (gypsophiles) and rare. The rapidly growing urban region of St. George has made the poppy vulnerable to disturbances from off-road vehicles, residential/commercial construction, grazing, and mining (Utah Native Plant Society 1989). Protected habitat for the dwarf bearclaw poppy includes the northeast slopes of the White Hills (Red Bluff ACEC), the west face of Warner Ridge, and several small parcels of land south of St. George (Fig. 6).

Recovery actions initiated by the BLM have improved the status of the siler pincushion cactus such that it was proposed to be reclassified from endangered to threatened in March 1993. The siler pincushion cactus is protected within the Warner Ridge-Fort Pearce ACEC.

6 AGGREGATE RESOURCE CONSERVATION

As demonstrated in the urbanized regions of California and elsewhere, protection of sand and gravel deposits is necessary to ensure a low-cost supply of construction aggregate. Presently, however, there is no coordinated effort to conserve aggregate resources in the region around St. George. Knowledge of the location, composition, and volume of deposits is required in order to develop long-term, comprehensive land-use plans. A combination of conditions in the St. George basin contributes to the likelihood of future shortages of low-cost, high-quality aggregate materials needed for continued development (Fig. 8). These conditions include, but are not limited to:

1. High-population growth rate and resulting high-demand for building sites and infrastructure which restricts access to aggregate resources.

2. Relatively little high-quality natural aggregate available due to the geologic setting of the region.

3. Large percentages of federal land ownership with some lands having ACEC or other land-use restrictions.

4. Protection of habitat for endangered plants and animals (often included in ACECs).

These conditions may affect future availability of aggregate and require that local planners and developers consider aggregate resources in their planning efforts.

As the natural aggregate of the area is removed from the accessible resource base (by extraction, urbanization, and designation of protected areas), construction costs will likely rise. The main resource areas along the terraces of the Virgin River and Fort Pearce Wash are experiencing increased extraction as urban expansion and infrastructure development continues. Some pit closures, due to encroachment of subdivisions, have already taken place. Moreover, establishment of protected areas for rare plants and animals is an on-going process that may continue to have an effect on aggregate resources. Higher cost alternatives to local sources of aggregate could include transporting aggregate from increasingly distant sources, or crushing stone quarried from favorable bedrock formations.

Figure 8. View southwest across Fort Pearce Wash. Sand and gravel operations in the center of the photo mine young stream-terrace deposits. Eroded outcrops in the upper right are the Shnabkaib (gypsiferous) member of the Triassic Moenkopi Formation. Distant hills are in Arizona and consist of Paleozoic formations.

Potential sources of aggregate in and around the St. George basin are both known and yet to be defined. Excavation of sand and gravel will continue in areas along the Virgin River east of St. George and along Fort Pearce Wash, and resources there will eventually be depleted. New areas will need to be opened for development. Aggregate resources along the Hurricane Cliffs, around the southern slopes of the Pine Valley and Bull Valley Mountains, selected areas outside of protected zones on the western slopes of the Beaver Dam Mountains, and elsewhere might eventually be considered for development. Sources of crushed rock for aggregate may include Paleozoic sedimentary rocks and Cenozoic basalt.

The availability of low-cost construction materials may eventually place constraints on economic expansion in the St. George area unless measures are taken to conserve the resource. These measures could include: 1. Inventorying aggregate resources through reviews of geological and geotechnical data on resources throughout the St. George basin; 2. Ranking of areas for detailed studies based on deposit volume, quality, and accessibility; 3. Detailed mapping and surveying of highly ranked deposits; 4. Representative sampling for deposit characterization; and 5. Determination of in-place tonnages. With this information, land planners, developers, and government officials would be able to make better-informed land-use decisions that recognize natural aggregate as a valued commodity.

ACKNOWLEDGEMENTS

We thank Clark Maxwell of the Utah Department of Transportation's Cedar City office for his assistance in preparing information on sand and gravel resources of Washington County. We also thank Larry Gore of the BLM's St. George office for his help with information on threatened and endangered species, and sand and gravel operations in the area. And, thanks to Jeff Rowe of the Utah School and Institutional Trust Lands Administration for providing digitized land-grid and land status information.

REFERENCES

Alfors, J.T. Burnett, J.L & Gay, T.E. 1973. *Urban geology master plan for California*. California Division of Mines and Geology Bulletin 198: 112.

Beeby, D.J. 1988. Aggregate resources – California's effort under SMARA to ensure their continued availability. In Tooker, E.W. & Beeby, D.J. (eds). *Industrial minerals in California – economic importance, present availability, and future development*. US Geological Survey Bulletin 1958: 118-121.

Best, M.G. Mehnert, H.M. Keith, J.D. & Naeser, C.W. 1987. *Miocene magmatism and tectonism in and near the southern Wah Wah Mountains, southwestern Utah*. US Geological Survey Professional Paper 1433-B: B31-B46.

Blank, H.R. Jr. 1959. Geology of the Bull Valley district, Washington County, Utah. Seattle, University of Washington, unpublished Ph.D. dissertation: 177.

Bureau of Land Management. 1990. Proposed Dixie resource management plan/final environmental impact statement. US Department of Interior, Bureau of Land Management: 240. 2 pts. scale 1-100,000.

Bureau of Land Management. 1995. Dixie resource area resource management plan and environmental impact statement. US Department of Interior, Bureau of Land Management: 512 appendices.

Christenson, G.E. & Deen, R.D. 1983. Engineering geology of the St. George area, Washington County, Utah. *Utah Geological and Mineral Survey Special Studies 58*. Salt Lake City: 52. 2 pts.

Cook, E.F. 1960. Geologic atlas of Utah – Washington County. *Utah Geological and Mineralogical Survey Bulletin 70*. Salt Lake City: 119.

Duffy-Deno, K.T. & Brill, Thomas. 1995. *Utah county economic profiles*. Utah Department of Natural Resources, Office of Energy and Resource Planning: 70.

Grant, S.K. 1991. Geologic map of the New Harmony quadrangle, Washington County, Utah. *Utah Geological Survey Miscellaneous Publication 95-2*. Salt Lake City: 14. 2 pts. 1:24,000.

Greer, D.C. Gurgel, K.D., Wahlquist, W.L. Christy, H.A. & Peterson, G.B. 1981. *Atlas of Utah*: 300. Provo: Brigham Young University Press.

Hamblin, W.K. 1970. Late Cenozoic basalt flows of the western Grand Canyon. In W.K. Hamblin & M.G. Best (eds) Guidebook to the geology of Utah – the western Grand Canyon district. *Guidebook to the Geology of Utah* 23: 21-37. Salt Lake City: Utah Geological Society.

Hammond, B.J. 1991. Geologic map of the Jarvis Peak quadrangle, Washington County, Utah. Utah Geological Survey Open-File Report 212: 52. 2 pts.

Higgins, J.M. & Willis, G.C. 1995. Interim geologic map of the St. George quadrangle, Washington County, Utah. Utah Geological Survey Open-File Report 323: scale 1-24,000.

Hintze, L.F. 1985. Geologic map of the Castle Cliff quadrangle, Utah. US Geological Survey Open-File Report 85-120: scale 1-24,000.

Hintze, L.F. 1988, *Geologic history of Utah*. Brigham Young University Geology Studies Special Publication 7: 202. Provo: Brigham Young University Press.

Longwell, C.R. 1952. Basin and Range geology west of the St. George basin, Utah. *Guidebook to the Geology of Utah* 7: 27-42. Salt Lake City: Intermountain Association of Petroleum Geologists.

Maddux, H.R. Mizzi, J.A. Werdon, S.J. & Fitzpatrick, L.A. 1995. Overview of the proposed critical habitat for the endangered and threatened fishes of the Virgin River basin. US Department of the Interior, US Fish and Wildlife Service: 47.

Marshall, R.M. 1995. Southwestern willow flycatcher. Carnegie Mellon University, Ecological Services State Office, US Fish and Wildlife Service Phoenix, Arizona

Mikulic, D.G. 1995. Uncertain future for Chicago aggregate industry. *Rock Products*. 98(8): 21-23. Chicago: Intertec.

Office of Planning and Budget. 1994. State of Utah economic and demographic projections – 1994. Governor's Office of Planning and Budget: 391.

Petersen, S.M. 1983. The tectonics of the Washington fault zone, northern Mohave County, Arizona. *Brigham Young University Geology Studies*. 30(1): 83-94. Provo: Brigham Young University.

Stokes, W.L. 1977. *Physiographic subdivisions of Utah*. Utah Geological and Mineral Survey Map 43. scale 1-2,500,000.

US Fish and Wildlife Service. 1994. Desert tortoise (Mojave population) recovery plan. US Department of the Interior, US Fish and Wildlife Service report: 73.

Utah Geological Survey. 1995. Central Virgin River ground-water study, southwestern Utah. Utah Geological Survey project proposal submitted in cooperation with the US Geological Survey to the Utah Division of Water Rights: 11.

Utah Department of Transportation. 1966. Washington County materials inventory. Utah Department of Transportation, Materials and Research Division, Materials Inventory Section: 17.

Utah Native Plant Society. 1989. *Endangered – the Dwarf Bearclaw poppy*. Utah Native Plant Society, Inc., information brochure: 2.

Construction materials in the Netherlands: Resources and policy

B. DE JONG
Ministry of Transport and Public Works, Den Haag, Netherlands

E.F.J. DE MULDER
Geological Survey of the Netherlands, Haarlem, Netherlands

1 INTRODUCTION

Geological conditions determine the availability of construction materials in the Netherlands. Outcrops of Mesozoic and even some Paleozoic bedrock, including limestone, exist only in the extreme southeastern part of the country (Fig. 1). Mesozoic limestones occur locally in the east. With a few exceptions only, in the western part of the country, bedrock is not encountered any shallower than 1000 m deep. Ninety-nine percent of the country's surface consists of unconsolidated marine, fluvial, aeolian and glacial deposits of Quaternary age (< 2.5 million years old). These consist of sand, clay, gravel, and peat. For this reason, only sand, gravel, and clay resources are discussed in this paper as construction materials.

The schematic cross-section (Fig. 2) shows that the Pleistocene and Holocene deposits slope from east to west, i.e. towards the centre of the North Sea Basin, which formed some 70 million years ago. Early Pleistocene deposits, exposed in the upstream parts of the delta of the rivers Rhine, Meuse and Scheldt (Fig. 1), generally consist of fine- to medium-grained fluvial sand, whereas the Middle Pleistocene deposits encompass predominantly coarse-grained sand laid down by braided rivers during glacial periods. Such coarse-grained sand is predominately exposed in the up to 120-meters-high ice-pushed ridges in the central (e.g. Veluwe) and eastern parts of the country.

This sand is overlain by generally fine-grained aeolian and periglacial sands and glacial or (peri)marine clays of Middle to Late Pleistocene age. Almost the entire western half of the country is covered by marine and perimarine clays, fine-grained sands and peat layers of Holocene age. The cross-section (Fig. 2) clearly demonstrates that the Holocene beds increase in thickness in a westerly direction. Deep tidal channels filled with fine-grained sand are incised in a thick succession of soft marine clays and even softer peat layers. Where peat beds were dug out by man for fuel or were eroded during stormy periods, lakes were created in the western part of the country, some 1000 years ago. Most of these lakes were milled or pumped dry between the 16th to 20th century, which increased the Dutch land surface area substantially. In the coastal zone along the North Sea up to 60-meters-high sand dunes developed from the 12th century onwards. As in many other areas in the Netherlands, no sand production is allowed from the dunes for reasons of landscape preservation.

Figure 1. Schematic map of the Netherlands.

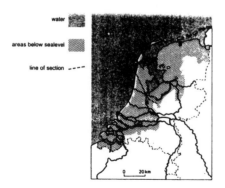

Figure. 2. Schematic cross-section through the central and western parts of the Netherlands.

The high demand for construction materials in the Netherlands is influenced by a number of factors including: 1. Population density which is the second highest in the world (after Bangladesh), 2. GNP per capita which ranks among the highest in the world, and 3. The still growing economy growing in this highly industrialized country. In addition, every construction project undertaken on the very soft soil in the western and central parts of the country requires strong and costly foundations to ensure stability. Consequently large quantities of construction materials are required.

This paper briefly outlines the sand, gravel and clay resources available in the Netherlands, followed by an indication of present and expected future demands. The policy on construction materials of the Dutch Government and its objectives are summarized and special attention is given to alternative construction materials and recycling. The new Dutch Excavation Act is briefly outlined at the end of this paper.

2 RESOURCES

A variety of primary construction materials are extracted in the Netherlands. These are briefly described in the following sections.

2.1 *Sand*

Two types of sand that can be used as construction material can be distinguished:
1. Concrete and mortar sand, consisting of medium- to coarse-grained sand suitable for mortar and concrete production,
2. Fill sand, for landfill, which has to meet fewer grain-size constraints.

Medium- to coarse-grained sand occurrences of the first type are confined to the upstream parts of the delta of the three main rivers, the Rhine, the Meuse and the Scheldt (Figs 1 and 3). Furthermore, this type of sand can be found along former river courses of the Middle and Younger Pleistocene North German rivers, which are mainly located in the subsurface in the central, eastern and northern parts of the country. In addition, medium- to coarse-grained sand occurs in a very small area of the Dutch sector of the North Sea.

Sand that is to be used for landfill or for raising the land surface needs to meet less stringent requirements than 'concrete and mortar sand'. This sand may be fine-grained and might contain small quantities of other components, such as gravel, peat, clay and shells. Large quantities of fine-grained sand can be found at or near surface in many parts of the Netherlands and the Dutch sector of the North Sea. Only in the western and central parts of the country is this sand not generally found at surface because it is buried here by thick sequences of Holocene clay deposits and peat layers. Offshore resources of fine-grained sand are often used for beach nourishment.

2.2 *Gravel*

Gravel has a very limited distribution at or near the surface in the Netherlands (Fig. 3). Gravel that can economically be used for concrete production is confined to the Province of Limburg in the southeastern (most upstream) part of the Netherlands. There, substantial volumes of gravel have been deposited by modern and ancient

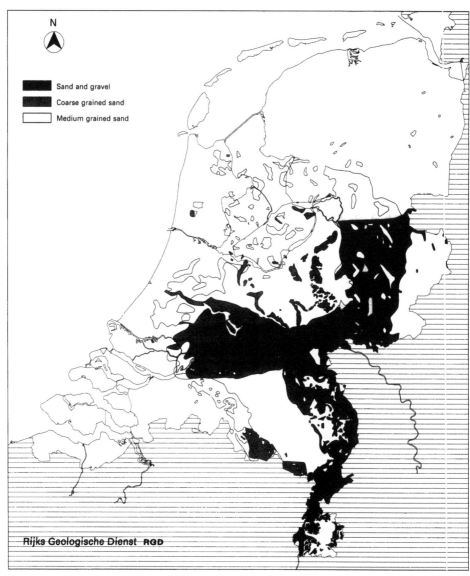

Figure 3. Distribution of coarse-grained sands at or near surface in the Netherlands. Note the gravely sands (in black) are limited to the upstream parts of the River Meuse in the southeast of the country, while the coarse-grained river sands (heavily punctuated) occur mainly in the central and in eastern parts. The predominantly medium-grained sands (lightly punctuated) occur in the more downstream parts of the rivers.

river courses. This gravel is now found in uplifted river terraces and buried in the river beds. Recent studies by the Geological Survey (unpublished) have demonstrated that gravel has a much wider distribution in the southeastern part of the Netherlands at depths of 25 m or more. Under the present economic conditions quarrying these resources is uneconomic.

2.3 *Clay*

Clay is used primarily as a resource for brick and roof-tile production in the Nether-lands. In addition, clay is widely applied for dike construction. Clay resources are lo-cated primarily along the downstream reaches of the main rivers onshore, and very substantial clay resources occur offshore in the Dutch sector of the North Sea (Fig. 4). Clay that is to be used for the brick and tile industry may contain higher quanti-ties of sand than clay used for dike reinforcement.

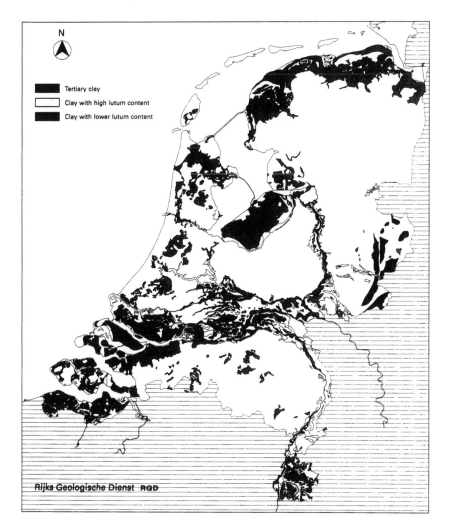

Figure 4. Distribution of clay at or near surface in the Netherlands. Note the three main occur-rences: Holocene river clays with 17.5-35% particles clay (< 2 micron) are displayed lightly punc-tuated in the central parts, Holocene, predominantly marine clays with clay percentages < 17.5 (heavily punctuated) in the western and northern parts, and Tertiary clays (black) situated in the east and southeast.

2.4 *Potentially recyclable waste material*

Construction materials need not necessarily be restricted to sand, clay or gravel. Recyclable waste can also be considered as a potential resource. Several types of waste can be used as construction material, e.g. building and demolition waste (concrete and brick debris), steel slag, pavement debris, blast-furnace slag, phosphor slag, burnt coal residues, cleaned or slightly polluted ground and dredging sludge. Sometimes, re-use is possible only after intensive treatment, for instance in the case of dredging sludge. Storage of dredging sludge presents a large waste-disposal problem in the Netherlands. The possibility of separating dredging sludge into a clean, relatively coarse-grained fraction (sand, mostly of fill sand quality) and a highly polluted fine-grained fraction (silt), is being examined. Pollutants stick mainly to the silt fraction. The sand fraction could then be used for landfill. For the silt fraction an isolated and supervised disposal location would have to be found.

General problems connected with recycling are: 1. The extra costs, 2. The emission of pollutants, and 3. The required constant technical quality. In some cases, problems such as radiation and health hazards are involved in treatment.

Almost all potentially recyclable industrial waste is re-used for construction purposes, and so is up to 68% of building and demolition waste (1994). Cleaned or slightly polluted ground and dredging sludge has, until now, been re-used only to a small extent. Thirty to forty million tonnes of the waste produced annually in the Netherlands is more or less suitable as construction material (Fig. 5).

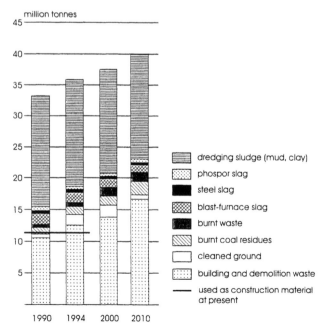

Figure 5. Potential supply of recycled materials period 1990-2010. (ref. Ministry of Transport and Public Works & Ministry of Housing, Physical Planning and the Environment 1994).

3 DEMAND FOR CONSTRUCTION MATERIALS

In 1994, the annual consumption of construction materials in the Netherlands included:
  – 21 million tonnes of concrete and mortar sand (about 80% for concrete production and the rest for mortar),
  – 73 million tonnes of fill sand (used for road and dike construction and for raising the land surface),
  – 20 million tonnes of gravel (about 80% for concrete production and the rest mainly for asphalt paving),
  – 4 million tonnes of clay (about 3 million tons for the coarse ceramic industry and about 1 million tonnes for dike construction),
  – 12 million tonnes of recycled material (mainly building and demolition waste, industrial waste, burnt coal residues and some cleaned and slightly polluted soil. Over 90% of the recycled material is used in road construction.

Studies have shown some correlation between the demand for construction material and investments in buildings, roads, dikes etc. These investments are currently forecast for the next five years. On the basis of the long-term predictions for the national economy it is possible to anticipate investments for a longer period. Short- and long-term predicted demand for construction materials is shown in Table 1. Obviously the predictions, especially the long-term ones, are only tentative.

4 POLICY OBJECTIVES

At the end of the 1970s, the building industry (especially the excavation companies), provinces and environmental groups asked the national government to develop a policy on the excavation of sand, clay and gravel. Each group had different interests and concerns. The provinces wanted a nation-wide survey of the demand for construction materials and desired to be included in decision-making on potential excavation sites. The excavation companies foresaw problems with the supply of construction materials as a result of the increasing time involved in obtaining excavation licenses. The environmental groups asked for reduction of the use of natural construction materials through recycling and also wanted a better planning procedure for deciding where excavation would be permitted.

Table 1. Demand for construction materials in the Netherlands (million tonnes/year). The data in question are the demand of the natural materials sand, clay and gravel or the alternatives (Ref. Ministry of Transport and Public Works & Ministry of Housing, Physical Planning and the Environment, 1995, p. 14).

Material	1994-2000	2000-2011	2011-2020
Concrete and mortar sand	24	24/27	Increasing
Gravel	23	23/26	Increasing
Clay for coarse ceramic industry	4	4	Equal
Clay for dykes and other purposes	5	1	Decreasing
Fill sand	89	86/112	Increasing

Discussion eventually led to the formulation of a national policy on the supply of construction materials. This policy has been in place for the past 10 years. Originally, consultation between the national authorities and the provinces was considered the best solution. Later, however, general consensus grew that the State Government should have a general regulating role. Each province would have the primary responsibility for excavation policy in its own territory, however, the State Government (i.e. the Minister of Transport and Public Works) would be responsible for ensuring a sufficient supply of construction materials for building purposes was available in the Netherlands. Promoting recycling was considered to be primarily a national responsibility.

The main objectives of the current national policy are to:

– Promote the economical use of natural construction materials; e.g. by designing construction projects so that as little material as possible is used,

– Recycle as much as possible. Much attention is given to the prevention of pollution problems caused by recycled materials,

– In some cases encapsulating linings (e.g. to prevent leaching of recycled material used in road construction) are required,

– Ensure a sufficient supply of construction materials for building purposes. More precisely, to ensure that a sufficient proportion of the required natural construction materials can be excavated from within the Netherlands,

– Excavate preferably from the area of application. In this way, transport is limited and consequently energy preserved and road space saved. At the same time this objective promotes regional diversity in land use,

– Achieve a more systematic and coordinated approach to the quarrying policy to ensure a balance with other policy issues, in particular physical planning,

– Develop quarry sites and surrounding areas in a socially responsible manner. When choosing extraction sites, the potential future use of that site should be borne in mind. Consultation with people living near the quarry site is crucial; in many cases a communication plan is desirable,

– Give excavation companies a sufficiently long-term idea of the economically acceptable quantities of construction materials that may be excavated from Dutch soil, or of alternative possibilities.

5   POLICY

5.1   *Instruments to promote economical use and recycling*

To promote the economical use of natural construction materials and recycling, various policy instruments are used:

– A general agreement between three ministries (the Ministry of Housing, Physical Planning and the Environment; the Ministry of Transport and Public Works, and the Ministry of Economic Affairs) and the construction industry to encourage a more economical use of natural construction materials and promote recycling. This agreement was also signed on behalf of the provinces, architects, housing associations and municipalities,

– Research on this issue subsidized by the same three ministries to about two million Dutch guilders a year,

– An agreement between the construction industry, the provinces, the municipal authorities, the environmental groups and the same three ministries to encourage, each in his own field, a growth in the proportion of recycled building and demolition waste used in new construction projects, from 60% in 1990 to 90% by the year 2000,

– An agreement between the Ministry of Transport and Public Works and the provinces to use an increasing percentage of recycled materials in provincial and central government public works,

– A measure to prohibit dumping of all economically recyclable waste. Such a measure is in force now for a great deal of the economically recyclable waste types,

– Taxation on dumping almost all other waste, provisionally excluding dredging sludge. Since 1-1-1995 an Act is in force taxing waste disposal by Dfl 29.20/tonne. The first objective of this Act is to collect more taxes, but in addition it will reduce waste disposal and promote recycling,

– Taxation on newly quarried natural construction materials such as sand, clay, and gravel, both on those quarried from within the Netherlands and on imported materials (in study). The aim of this measure is purely to promote the economical use of natural materials and to stimulate offshore extraction which are excluded from taxation.

As a result of all these measures the nation-wide use of recycled materials is expected to increase to 25 million tonnes a year by the year 2010. This would only meet about 20% of the demand for construction materials. A large quantity of sand, gravel and clay will still be required.

5.2 *Policy regarding concrete and mortar sand*

Currently, the largest volumes of concrete and mortar sand are excavated in the southern and eastern provinces. Part of the Dutch requirement (the coarse-grained concrete sand) is imported from Germany. Mortar-quality sand may even be exported (see Fig. 6).

Increasingly, the distribution of concrete and mortar sand excavation sites throughout the country is regarded as a problem. The problem concerns net national demand, i.e. the quantities of construction materials that are not used in the regions from which they are extracted. The current policy aims at reducing extraction from the southern and eastern provinces and increasing the supply from other provinces. The distribution of resources in the subsurface would permit such a measure to a certain extent. To achieve a fair distribution of excavation sites throughout the country, agreements are negotiated between provincial authorities and the State Government (Ministry of Transport and Public Works). At present, such agreements exist for the period 1989-1998 and provisional for the period 1999-2008.

The possibility of using medium- or even fine-grained sand for concrete production instead of coarse-grained (concrete) sand has been evaluated. Recent studies demonstrate that this might be feasible in some cases. So, in the future more excavation sites would be available because the reserves of medium- and fine-grained sand are much larger than the coarse-grained ones.

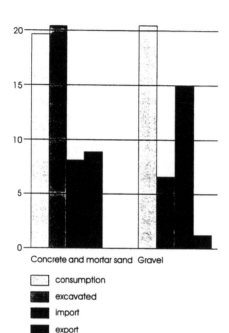

Figure 6. Consumption, excavated in the Netherlands, import and export of concrete, mortar sand and gravel in 1992 (ref. Ministry of Transport and Public Works & Ministry of Housing, Physical Planning and the Environment 1994, p. 187 and 194).

5.3 Fill sand

About 50% of the fill sand required is excavated from onshore locations. The number of scattered small onshore excavation sites available is declining as a result of objections to the ensuing fragmentation of the land. An increasing amount of fill sand is being extracted from a few large central sites. About 25% is extracted from the Dutch sector of the North Sea; the remaining part from large inland waters such as Lake IJssel (Fig. 1). Extraction of fill sand is preferably combined with public works such as widening and deepening of fairways (for example in the North Sea and Lake IJssel). In many shallow waters, extraction has had to be stopped in order to protect the (water) environment.

The current policy is to promote the use of fill sand extracted from the Dutch sector of the North Sea, especially in the very densely populated western provinces. The policy strives to prevent excavation from new onshore sites that would create new water bodies. The Dutch State receives a compensation from extraction when sand is extracted from the Dutch sector of the North Sea. A similar compensation has to be paid when construction materials are extracted from large Dutch inland waters (the State owns these waters and the underlying strata). To promote the use of fill sand from the North Sea, compensation to be paid by the extractors for this resource is at present about 1/3 (i.e. Dfl. 0.30/tonne) that of the compensation for sand from the large inland waters (about Dfl. 1.0/tonne). The use of North Sea sand in the coastal provinces is also encouraged by an agreement with the provinces. The neighbouring eastern provinces pursue a restricted policy with regard to supplying fill sand to the coastal provinces.

5.4 *Gravel*

Gravel reserves in the Netherlands are very limited. Extraction of gravel results in the formation of large expanses of water in the small areas from which it is excavated. For this reason extraction encounters much opposition. Because of this opposition it was decided some years ago to stop excavation of gravel within about a 15 year period. Imported crushed rock from surrounding countries is a good, sustainable alternative for the Netherlands. Investigations have shown that the supply that can be expected from imports is sufficient for the long term. Figure 6 shows data of the gravel demand and supply for 1992.

5.5 *Clay*

The supply of clay will present no problems for the next 10 to 20 years. A sufficient amount of clay will be available as a result of the creation of new nature reserves in the flood plains along the rivers in the Netherlands. These nature reserves are mostly developed by lowering the flood plain through excavation of the upper one or two meters of clay.

6 THE EXCAVATION ACT

The current 'Ontgrondingenwet' (Excavation Act) is being updated and is currently under consideration in Parliament. In January 1996 the 'Tweede Kamer' (the Dutch House of Commons) adopted the updated Act. According to the new Act, the Minister of Transport and Public Works will have to develop, together with the Minister responsible for Physical Planning, a 'Structuurschema Oppervlaktedelfstoffen' (National Plan for Resources at Surface). This plan should cover both the supply of construction materials and physical planning. Such a plan will have to be developed every five years for the next 25 years. The issues addressed and published in the first draft plan are the:
 – Supply of natural construction materials, including import and export,
 – Economical use of construction materials and recycling,
 – Physical planning aspects of excavating natural construction materials.
This national plan specifically outlines the role to be played by the provinces. They will have to comply with the general guidelines of this national plan when developing their own policies on extraction. The national plan only provides very general criteria regarding the locations where excavation will be permitted. Decisions about the exact extraction sites will remain the responsibility of the provincial authorities.

According to the updated Act, the Minister of Transport and Public Works and the Minister responsible for Physical Planning, will be authorized to force provinces to give permission for excavating a sufficient quantity of natural construction materials, if strictly necessary.

Other important issues that are addressed by the Act include:
– Improved (physical planning) instruments for the provinces,
– Some more possibilities to claim money from excavation license holders to ensure redevelopment of quarry sites and surrounding areas upon abandonment, enabling efficient subsequent use, and
– A major reduction of the time involved in the legal procedures required before excavation is allowed to start.

7 CONCLUSIONS

1. As a result of specific geological conditions in the Netherlands no significant resources of rock aggregates occur. Gravel reserves are very limited and confined to the southeastern part of the country, coarse sand reserves are limited, fine sand is distributed widely over the country and clay occurrences are abundant. 2. Thirty to forty million tonnes of the potentially recyclable waste products in the Netherlands is more or less suitable as construction material on an annual basis. 3. In 1992 more than 90% of the recycled material was used in road construction. 4. The main policy objectives of the Dutch government concerning the supply of construction materials include a more economical use of natural construction materials and encouragement of recycling. In addition, the supply of sufficient construction materials will be ensured. The use of offshore construction materials will be promoted. 5. Agreements between three ministries, provinces and the construction industry have been signed to promote more economical use and recycling. Some taxation measures on primary materials and waste dumping should further contribute to implementation of the policy objectives. This should lead to in increase of the use of recycled material by 25 million tonnes/year in 2010 to a total of approximately 20% of the demand of construction materials in the Netherlands.

REFERENCES

De Jong, B. 1989. Sediment resources in the Netherlands. *Proceedings of the 4th International Symposium on River Sediments, June 5-9, 1989, Beijing.* Vol II: 1563-1570. ISBN:7-5027-0551-1/Z.119.
De Mulder, E.F.J. 1984. A geological approach to traditional and alternative aggregates in the Netherlands. *Bulletin I.A.E.G.* 29:49-57.
De Mulder, E.F.J. 1990. Engineering Geology in the Netherlands. *Proceedings of the 6th International Congress of I.A.E.G. 1990, Amsterdam,* Balkema Rotterdam: 3-20
De Mulder, E.F.J. & Hillen, R. 1994. Construction materials in The Netherlands. In G.W. Lüttig (ed.), *Aggregates – Raw Materials' Giant*: 49-60. Erlangen, Germany, ISBN: 3-9801716-7-1
Ministry of Transport and Public Works & Ministry of Housing, Physical Planning and the Environment 1994. Structuurschema Oppervlaktedelfstoffen (draft, in Dutch)
Ministry of Transport and Public Works & Ministry of Housing, Physical Planning and the Environment 1994. Wetsvoorstel Herziene Ontgrondingenwet (Proposal Updated Excavation Act, in Dutch)
Ministry of Transport and Public Works & Ministry of Housing, Physical Planning and the Environment 1995. Structuurschema Oppervlaktedelfstoffen, view points of the State Government (in Dutch).

Economics of recycled aggregates

RICHARD POULIN & W.S. MARTIN
*Departement de mines et metallurgie, Universite Laval, Quebec, Canada, Department of Mining
and Mineral Process Engineering, University of British Columbia, Vancouver, Canada*

1 INTRODUCTION

Crushed stone together with sand and gravel constitute the two main sources of natural aggregates, the vast majority of which are used in the construction industry. Together they constitute the largest, by tonnage, non-fuel mineral commodities currently produced in North America. Further exploitation of this type of resource, however, has been significantly restricted by increasing urbanization and growing public concerns over environmental issues (Thomson 1980). The growth of populated areas has put pressure on aggregate producers to maintain supply whilst being inconspicuous. However, production economics require that quarry sites and their related producing facilities be located in or near population centers. Herein lies the paradox of this industry: a constant, predictable need for products and the community's desire that mining operations be conducted far from its boundaries (Carter 1975). Urban sprawl means that existing aggregate producers face increasing environmental costs as they are engulfed by development. These producers will relocate at the fringe of urban areas (Carter 1981) when this cost is greater than the extra transport cost of the new location.

Recycled aggregates are a substitute for conventional aggregates in many uses. They are differentiated by cost as well as by perception of quality. This substitution is favored by the internalization of environmental costs related to the exploitation of virgin resources for aggregates. An organization internalizes a cost by assuming responsibility of a cost imposed on others. Recycled aggregate become more competitive if additional costs are accounted in the production of conventional aggregate. Also recycling sites are often located in an urban setting close to demand. This is enhanced by the exodus of conventional aggregate producer forced by urbanization to relocate farther from consuming centers.

Land use conflicts involving aggregate producers are not new. Concerns about the sterilization of aggregate resources in the United States through uses that indefinitely render resources inaccessible have been reported for some time (Goldman 1961, Hogberg 1970). Sterilization of proximal resources has an influence on the level of recycling. This paper intends first to review the basic principles of land use planning and then examine the issue of land use planning for aggregates. The economics of recycled aggregate is analyzed in the framework of multiple land use. Examples from the Vancouver area are presented prior to concluding.

2 LAND USE PLANNING

Land use policy will be influenced by three principal issues that will affect the mineral aggregate industry. The first is the identity of the parties advocating different land-use policies and their motives for doing so, the second is the changing nature of benefits derived from land, and the third is the change of property regime which is devised to accommodate the (changing) benefits from using land and its associated natural resources. These three issues influence the outcome of the land use planning process, which can be seen as a form of contracting (Libecap 1989).

The parties involved in the land use planning process are, following Libecap (1989), the private claimants to rents and to managerial rights (existing or prospective owners), the politicians who can maintain or modify property rights arrangements, and the bureaucrats to whom politicians delegate management power but who also have their own interest to preserve. The benefits from using land varies over time depending on the attributes (including location and endowments with minerals and other environmental resources) of the area in question, and the economy's demand for each of these. This means that land use planning must not only take into account the current demand for aggregate but also the existence of other resources which might or might not be affected by aggregate production.

Because of the changing benefits some type of government intervention is needed to achieve optimum use of the natural resources and protection of the environment. Governments establish goals in the public interest in order to control the use of public land and influence the use of land owned by individuals and corporations (McDonald 1989). However, little consideration is given to mineral resource conservation in the planning process (Pennington 1980). The concept of land use planning as a form (or rather part) of public policy making serves to place it within a larger environmental policy framework. As such, planning procedures and, on a more operational level, planning implementation, are determined by interaction between government (as representative for all consumers/the population at large) and a wide range of groups with special vested interests. The Surface Mining and Reclamation Act of the State of California (Beeby 1988) for example identifies and protects mineral resources in areas of high land use conflict and ensures reclamation of mined lands. This Act is not used as a coercive instrument but as a source of information to shed light on what is too often an emotional debate. The social and economic consequences of inadequate planning are extreme in terms of increased consumer cost and environmental damage.

Land use can take many forms, sequential land use and multiple land use are possible solutions allowing a socially optimal policy for aggregate production. Recycling of aggregates is often done in the context of multiple land use. Permanent recycling sites are on industrial vocation land and temporary sites can be in commercial areas because of specific demolishing activity. Conventional aggregates exploitation could also be seen as a multiple land use case, however because aggregate operation have a finite life, they eventually become a sequential land use. Mining is an episode in the occupation of the land by individuals. The absence of planning leads to sterilization and favors recycling by increasing the cost of the conventional aggregates.

2.1 *Multiple land use*

Multiple land use can exist if there exists a compatibility between activities such that overall benefits increase by having activities coexist. Multiple land use takes into account simultaneously all the uses of the land to maximize benefits to society, including uses that do not have market value such as environmental resources. Transactions involving the amenity services of natural resources do not take place in organized markets (Smith 1994) as environmental benefits are not valued in dollars but still need to be taken into account for land use purposes.

To evaluate the viability of multiple land use such as operating an aggregate recycling facility in a given area, four aspects have to be analyzed (van Kooten 1993). First is economic efficiency. Each use has its stream of benefit, *B*, but it also has conflicts or disbenefits between uses, *D*. This results in a trade-off situation. Based on economic results, uses can be classified as complementary, competitive, supplementary or incompatible. Maximization is achieved by identifying the mix of uses that give the greatest net benefit of the land. The following equation is a summary with *j* types of uses. Optimization is achieved by maximization of the net present value:
Where

$$\text{Max } PV = \sum_{j=1}^{n} \left\{ \int_{\tau}^{\infty} B_j e^{-i\tau} \ \mathbf{d}t - D_j e^{-i\tau} \right\}$$

PV = present value of a unit of land, B_j = relative benefit of type *j* use, D_j = elative disbenefit of type *j* = use, τ = time, i = interest rate.

The second aspect is income distribution. Who will benefit from the introduction of a new use or from its not being introduced? It is crucial to identify the parties involved, accurately evaluate the amount at stake involved in structural changes, and understand the motivation of the various groups. The third aspect is fairness. Existing property rights, especially for publicly owned resources, have been conceded over time to individual or groups. New uses could interfere with these rights raising moral and legal issues that could result in compensation costs. The fourth is political acceptability. A use that is economically efficient but vastly unpopular will be short lived.

3 AGGREGATE RECYCLING

Recycling does not pose technical problems for new products (Hansen 1986). Up to 50% of natural aggregate could be replaced by recycled aggregate without seriously affecting the properties of concrete, both in fresh and the hardened states (Bairagi et al. 1993). Equipment and techniques have been developed to process concrete rubble into recycled aggregate (Hillmann 1991). If characterized by lower quality, it may be acceptable for road base material.

Recycling accounts for a relatively small portion of aggregate production. However, its importance is expected to increase, based on trends associated with economics and conservation, while remaining modest. For example in France in 1991 recycling represented 3% of the total aggregate production (Anonymous 1992). Denmark and Holland recycle an even greater percentage of their rubble.

Attendants at an aggregate recycling seminar stated why they recycled aggregate (Prokopy 1993). Their reasons included: outside pressure, limited landfill access, increasing disposal and transportation cost, and being required to recycle. The range of reasons is vast and is related to perception, economics and legislation. Outside pressure relates to organized groups that use public platforms to advance conservation and are involved in the permitting process. Recycling is also market driven when specific conditions make recycled aggregate an economical cost option. Requirements to recycle are on a contractual basis but could become broader if policies are enacted as legislation.

The major problems in recycling encountered by the same seminar attendees were: permitting process, equipment for this new and different material, lack of knowledge of the capabilities of recycled material, and steady flow of quality material. These problems are expected with an emerging activity. Being a new activity, legislative and technical framework to control the recycling industry is lacking. The absence of standardization of material impose a marketing burden on producers to prove the value of their recycled aggregate. Classical tests used for conventional aggregate do not always apply to recycled aggregate.

The basic factors necessary to make industrial waste (and recycling in general) both advantageous for use as aggregate (OMNR 1992) and economically efficient are: the processed material must not be potentially harmful, the quantity of material available at a potential recycling location must be large enough to justify the development of handling, processing, stockpiling and transportation systems and, the transportation involved must be reasonable in terms of competition with conventional aggregates.

Normal disposal costs in urban areas may be substantial, particularly when both traffic congestion and high tipping fees are combined. As tipping fees increase so will the pressure to send all construction waste material to recycling points without discrimination of the material heterogeneity. If tipping fees are to be used as an incentive to recycle more, it needs to be coupled with quality control of source material or to establish minimum standards. New strategies to manage construction wastes have been introduced, such as 'selective demolition' (Lauritzen & Jakobsen 1991), a technique which is based on the separation of selected materials on the demolition site. The quality of rubble, as determined by the level of contamination (earth, wood and gypsum), affects the marketability of the recycled product. As well, industrial waste must not be chemically deleterious and be able to withstand long term leaching. Industrial waste as a source for aggregate to be competitive is limited to the geographic areas where it is produced because of transport cost (Miller & Collins 1976). Production of aggregate from that source is limited by market size and uses.

Recycling aggregate waste reduces demands on increasingly scarce landfills. Similar situations to Vancouver in Toronto and Los Angeles have resulted in the growth of recycled aggregate industries on a very large scale. Blue Diamond, one of several Los Angeles recycling companies, has an annual production capability of 2 million tonnes a year. Their portable recycling plants produce 300-400 tph (Zimmerman 1991). In 1990, Toronto recycled 968,150 tonnes of concrete and 533,600 tonnes of asphalt (OMNR 1992). European cities have used recycling for many years in response to diminished fresh aggregate reserves and fewer landfill sites. Holland contains over forty aggregate recycling plants and has attempted to import demolition waste from England. All new German structural designs must include plans for demolition and recycling. In Britain, 10 million tonnes of building demolition waste and 8 million tonnes of road asphalt are recycled annually (Zimmermann 1991).

The maximum supply of recyclable material is located in the same area as maximum demand, both being characteristic of urban centers. Construction materials such as gravel, sand, stone, cement, wood, metals and plastics are the most widely utilized solid materials in the anthroposphere (Brunner & Stämpfti 1993). The materials will become the largest waste category in the future and assure future supply. Critical mass in terms of quantity is achievable in urban settings.

Recycled resources and virgin resources are substitutes in society's economic activities. Programmes that lower the costs of virgin natural resources reduce the proportion of material recovered from the waste stream (Ezeala-Harrison 1995). Local prices for conventional aggregates, and the disposal costs associated with concrete rubble, are key parameters in determining the value of recycled aggregates. Prices received for recycled aggregates tends to be higher in direct relation to the distance to supplies of conventional material. Conversely, rates received for concrete rubble tend to be lower in direct relation to haul distance from the source. As with conventional aggregate, recycled aggregate economics is heavily dependant on transport cost.

The prime supporting factors for aggregate conservation are reuse and recycling. These will probably be advanced through government practices of sustainable development and ever increasing landfill constraints (OMNR 1992). To be economically viable, recycling must be performed in its urban surroundings. A multiple land use situation would permit the combined advantage of available supply of rubble and short transport (for dumping and for delivery) making recycled aggregate competitive. Multiple land use allowing recycling of aggregates would promote preservation while displaying impediments normal with municipal zoning. Demand on natural resource stocks and landfill dumps would decrease. Reduced transport would minimize infrastructure use. Because the cost of not recycling is growing rapidly and because it is socially desirable, zoning or rezoning costs are not an obstacle. Zoning would be used as a means of separating activities of adjacent but dissimilar firms and households that imposed externalities upon each other (van Kooten 1993).

Private income distribution change could be expected to be minimal. Recycling is performed mainly by aggregate producers and occasionally by wreckers. Examples of exhausted quarries being used as recycling centers are not uncommon. Those with an established marketing network see advantages to carry conventional and recycled aggregates simultaneously. Private operators of landfills would lose income as easily manageable materials would be diverted from dumps. However, since only a fraction of waste materials can be recycled, this effect would not be great. Recycling of aggregates could be classified as a complementary activity to construction. Public income could vary if a royalty or some form of mineral rent is collected on the exploitation of virgin resources. The replacement by recycled aggregate would reduce revenue proportionally.

4 THE EXAMPLE OF THE VANCOUVER AREA

The Lower Mainland of British Columbia (Figs 1 and 2), which encompasses Vancouver, is a region rich in aggregate deposits. Urban expansion has sterilized many high quality sand and gravel resources. Government policies with respect to taxation and regulation are outdated and have made conventional aggregate production inefficient and costly on Crown Land. The cost of operation on private land has increased also as a result of escalating land prices on the Lower Mainland particularly in areas enjoying rapid growth (the primary market for aggregate). As the population of the Lower Mainland continues to grow, enormous quantities of building materials are consumed. Of these materials aggregate is perhaps the most fundamental to the

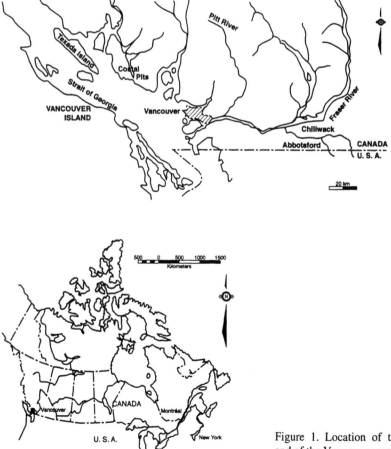

Figure 1. Location of the Vancouver and of the Vancouver region.

construction industry. A standard highway consumes aggregate at the rate of 12,000 tonnes/lane-km. A house with associated infrastructure uses 300 tonnes of aggregate.

There are 65,000 km of public roads within the province of BC, 45,000 km of which are provincial highways. Mountainous terrain is common in the province as are multiple freeze-thaw cycles and heavy precipitation (particularly on the Lower Mainland), each of which can cause damage to roads resulting in frequent maintenance. The BC Ministry of Transportation and Highways (MoTH) estimates that roughly 5 million tonnes of aggregate per year is required for maintenance of existing roads. During periods of highway building, such as during the Coquihalla project in the mid 1980s, MoTH annual consumption can reach 20 million tonnes of aggregate (Lee 1995). Between 1980 and 1988, British Columbians consumed an average of 14.8 tonnes of aggregate per capita, per annum. In 1994 the B.C. Ministry of Energy Mines and Petroleum Resources (MEMPR) estimated that 41,837,000 tonnes of aggregate were produced province wide of which roughly half (21,000,000 tonnes) was consumed in the Lower Mainland. Other estimates for Lower Mainland consumption range as high as 30,000,000 tonnes per year.

Over the next 30 years, the population in the core areas of the Lower Mainland is expected to double to 2 million. The Fraser Valley population is also expected to double to 346,000. Four billion dollars in private development is expected to be spent over the next 12 years. Transportation improvements over the same period will cost $5.7 billion. As aggregate materials are the fundamental commodity of development both in transportation and in the housing/industrial sectors, demand for aggregate will increase in proportion to the expansion of the Lower Mainland.

From industry information sources, the Lower Mainland is supplied with aggregate from the following locations (Fig. 1):
  – Coastal pits on tidewater and Texada Island Quarries, 9 million tonnes per year,
  – North side of the Fraser River, 4 million tonnes per year,
  – Matsqui/Abbotsford and Chilliwack, 4 million tonnes per year,
  – Fraser River (including dredged sand), 4 million tonnes per year,
  – Imports from the USA, 1 million tonnes per year.

The ownership of the land upon which these pits and quarries operate varies. Some are privately owned and operated, others operate on Crown Land, First Nations Land, or on private land owned by someone other than the pit operator. Production costs vary also. Pits on the North side of the Fraser River have high overburden removal costs. In some cases overburden to aggregate thickness exceeds four to one. Quarry operators bear the added costs of drilling blasting and crushing. Tidal operators must barge their product between 100 and 150 km to market as well as pay for on and off loading of the barge. The exact number of gravel pits in British Columbia is not known. The BC MEMPR estimates that 6000 pits exist in the province of which 2600 are active. The largest producer and consumer of aggregate in the province is the BC MoTH which estimates that they hold tenure (of various forms) over approximately 3000 gravel pits throughout the province (Lee 1995).

As aggregate is a bulk, low cost commodity transportation charges often play the greatest role in determining the final product cost. Truck haulage is particularly costly at approximately $55.00-85.00 per hour for a standard dump truck. Output from the 2600 commercial pits in BC was valued at $170 million in 1994. With transportation costs included this value rose to $370 million (Matysek & Bobrowsky

1995). Aggregate price has increased to the point at which alternatives or substitutes are considered. Recycled aggregate can in some circumstances be available at a lower cost.

4.1 *Recycled aggregate on the lower mainland*

Conditions on the Lower Mainland favour aggregate recycling. As the population expands up the Fraser Valley, aggregate producing operations are driven farther from municipalities such as Vancouver (Fig. 2). The nearest inland aggregate source to Vancouver is located 40 km away in the Coquitlam River Valley. Unmined aggregate deposits located nearer to Vancouver have been sterilized by urban development. Vancouver is depending more heavily on barged aggregate which is shipped 100-150 km. Recycled aggregate operations can be located much more conveniently for use by the westernmost municipalities on the Lower Mainland. Reduced transportation costs can allow recycled aggregate to be sold at an overall lower price than conventional material. In some cases transportation savings can be further enhanced by using aggregate transporting trucks to haul back recyclable waste.

Recycling operations avoid paying many of the taxes and royalties imposed upon the conventional aggregate producers. Mining tax and royalties, Crown Land fees and royalties as well as soil removal fees are all forms of taxation payable by aggregate operations but not applicable to aggregate recyclers. Environmental regulation of aggregate recycling is also reduced in comparison with the conventional industry. Sand and gravel and crushed stone operations must post environmental bonds and supply a detailed, comprehensive reclamation plan prior to operation. To extract aggregate resources from lands in the Agricultural Land Reserve (ALR) conventional aggregate producers must show extraction will maintain or enhance the effected land.

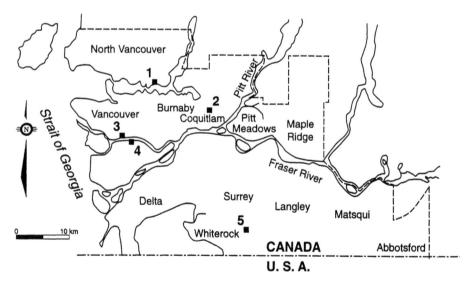

Figure 2. Limits of the Lower Mainland of British Columbia and location of aggregate recycling sites. Identification of numbered locations: 1. BA Blacktop, 2. Eco Tech, 3. Columbia Bitulithic, 4. Richvan, and 5. LM Recycling.

In comparison, aggregate recycling requires no environmental bond or reclamation plan. As the land upon which the recycling takes place is not harmed by the operation, it is not incompatible with agricultural land uses.

The Greater Vancouver Regional District (GVRD) estimates that in 1991, demolition, land clearing and construction accounted for 36% of the total waste generated in the district. Of this amount, 420,000 tonnes of waste concrete and asphalt (aggregate waste) were produced (GVRD 1993). Aggregate waste is produced during demolition or renovation of structures, or road construction or repair work. Demolition or renovation of structures can often produce 'unclean' waste which contains rebar or some other undesirable substance which may either damage the recycling equipment or somehow harm the quality of the recycled product. Although rebar can be removed with magnets during recycling it is preferred if it is removed at the demolition site rather than by the recycler. It is common for recyclers to charge an extra dumping fee (often substantial) if a large amount of rebar is contained in a load of waste. Other undesirable material such as wood, brick, or glass are also heavily penalized or refused outright at aggregate recycling facilities.

Road repair and demolition generates far fewer 'unclean' substances with respect to the recycling process. Concrete, asphalt and road based materials can all be easily recrushed, screened and reused. Of total demolition in 1991, 58% was residential in origin while 42% was commercial. Of these quantities both residential and commercial waste averaged 3.2% reinforced concrete (containing rebar) and 15% non-reinforced concrete. Residential waste contained 6.7% asphalt while commercial waste contained 3.8% asphalt (GVRD 1993). For 1993 and 1994 the GVRD estimates that 417,900 tonnes and 550,415 tonnes (respectively) of aggregate waste were produced (Marr 1995). No statistics were available for these years concerning quantities of recycled material produced.

4.2 *Aggregate recycling operations*

Aggregate recycling has been carried out on the Lower Mainland since 1983. Since this time four principle recycling stations have opened and become established with a fifth opening periodically (Fig. 2). Another two companies operate portable crushing and screening facilities which can be transported to demolition sites where concrete is recycled and either used on site or shipped elsewhere.

Lower Mainland Recycling (LMR) is located in South Surrey at the Surrey Municipal Pit, Location 5 on Figure 2. Until recently LMR also operated from another location in the same municipality where conventional aggregate material was also available. This location has now been reclaimed. The LMR circuit consists of a 22 × 36 single Toggle Cedarapids jaw crusher which feeds a multi-decked vibrating screen in closed circuit with a 42 × 46 roll crusher. Despite the high costs associated with roll crusher operation (rollers wear quickly), LMR feel that their system gives superior control over undersize and therefore is economic in the long term due to less over crushing. The tendency of the roll crusher to jam when fed long, thin, soft pieces of asphalt (which can pass uncrushed through the jaw crusher), has been overcome by a unique adaptation which allows the rolls to be quickly opened. This allows oversize to pass through the rolls and be re-circulated and slowly abraded to a more manageable size. The circuit is diesel powered and consumes 400 liters/day.

LMR charges a $2.00 tippage fee even for clean material. Additional fees are charged for disposal of unwanted materials including rebar which the LMR circuit is not equipped to easily remove.

The major purchasers of LMR material are private contractors doing road repair and construction work for local municipalities. The following example is from a suburb of Vancouver, the Municipality of White Rock (Fig. 2), where material from LMR was used in a road construction project. The material was applied by a private contractor at the direction of the city Public Works. The road shoulder, in which the recycled material was used, was unusually wide and served as both a parking surface as well as an additional traffic lane during congested periods. The road and shoulder is also used by heavy trucks and public busses. As a result of heavy use, previous road shoulder applications had required regrading three to four times per year. The Municipality had considered hot mix asphalt at this location but rejected it due to its high cost of placement of one dollar per foot. In comparison, the cost of replacement of the shoulder using recycled asphalt was ten cents per foot. The material was applied in June of 1993 over a length of approximately 1 km and at a depth of approximately 7 cm. Figure 3 is a view of the site in June 1993. The site was revisited in September of 1995 and Figure 4 is a view of the same area two years later. The recycled material showed good cohesion with only a limited loss of fines from the surface. No indication of potholes or rutting was apparent.

RichVan holdings operates a recycling facility on River Road in Richmond and is well located to serve customers in that municipality as well as those in Vancouver, Location 4 on Figure 2. RichVan provides a wide variety of material sizes and composition. The circuit is comprised of a single toggle Cedarapids crusher feeding an impactor. The circuit is equipped with magnets to remove rebar, chains and other ferrous objects. These materials can then be recycled at a neighboring metal recycling facility. Dumping of clean material is permitted free of charge and a penalty is imposed for undesirable waste. RichVan has a mobile rock breaker (Insley) that is capable of handling very large material. RichVan has been tested and approved for use

Figure 3. Road shoulder in White Rock made of recycled aggregate, June 1993.

Figure 4. Same site as Figure 3 but two years later, September 1995. Similar tire marks are visible because of the presence of a municipal bus stop.

by the Municipality of Richmond and, at the time of interview, was under consideration by the City of Vancouver.

Columbia Bitulithic operates an aggregate recycling facility on Mitchell Island in Richmond B.C., Location 3 on Figure 2. An additional waste material receiving site is located in Coquitlam. From there waste is transferred to the Mitchell Island facility for recycling. The Company itself has been in existence in the Lower Mainland since 1910 and has grown into one of the Lower Mainland's larger locally owned construction materials producers. Columbia Bitulithic has been recycling aggregate since 1983. The Columbia Bitulithic circuit is very similar to the RichVan circuit with a primary jaw crusher followed by vibrating screens in closed circuit with an impactor. The circuit is protected by a number of magnets to remove metal objects. In an effort to reduce the supply of waste, Columbia Bitulithic recently initiated tippage fees of between \$25-\$75 per load of asphalt waste. Concrete waste is dumped free of charge. Due to the diversity of its operations, Columbia Bitulithic is the largest consumer of its recycled product. The Mitchell Island site provides both a supply of aggregate for Columbia Bitulithic's road building and construction projects as well as a location for waste disposal from demolition projects. The facility is convenient to Richmond, Burnaby and Vancouver. Facilities exist at the Richmond location for loading and unloading barges.

BA Blacktop operates a recycling facility in North Vancouver, Location 1 on Figure 2. BA is primarily a paving contracting company although they have been involved in recycling for more than 15 years. Although asphalt is the primary recycled material, BA has recently begun handling waste concrete for both Lafarge and Ocean Cement, two of the largest building material producers on the Lower Mainland. This is expected to increase the quantity of waste concrete being recycled. The bulk of recycled material produced by BA is used in their own road construction and repair contracts. BA reports recycled material as being superior to conventional aggregate in applications such as road grade material. Recycled asphalt maintains its cohesive qualities after the recycling process which allows it to resist the deleterious actions of water when the material is applied in wet conditions. The binder material which is also present in recycled asphalt acts to repel water which improves drainage. The BA circuit consists of a 40 × 42 Traylor Bulldog jaw crusher that feeds material to a screen which in turn feeds oversize to a smaller Pioneer secondary jaw crusher, and undersize to a roll crusher. Each of these crushers is in closed circuit with the screen which allows material to be reduced over several passes until the desired product size is reached. The BA circuit is not equipped with a breaker or with protective magnets. As BA primarily recycles asphalt, neither of these pieces of equipment were required. With the increase in waste concrete however, BA will consider purchasing the appropriate equipment. The BA circuit is capable of producing 800-1000 tonnes per day. An inexperienced crusher operator however can reduce production to 300-500 tonnes per day. Like Columbia Bitulithic, BA has been faced with an oversupply of incoming waste resulting in periodic tippage fees being imposed.

Eco Tech recycles concrete and asphalt in Coquitlam, Location 2 on Figure 2. No operational details were made available by Eco Tech for this report. Although the GVRD has reported Eco Tech to have been in continuous operation, a previous aggregate recycling survey (Martin 1993) found that Eco Tech was not in operation and had opened and closed several times in the past. The Eco Tech recycling circuit is believed to be similar in production capabilities (by the GVRD) to other Lower Mainland aggregate recyclers.

Litchfield operates a portable recycling operation in the Lower Mainland. At the time of this report the portable crusher was in operation in East Vancouver at a warehouse demolition site. No operational details concerning the recycling operation were released by Litchfield. The circuit consists of a small primary jaw crusher which feeds an impactor. The material is then screened and sent to two stockpiles of oversize and undersize material. No recirculation of oversize material is present. The circuit has no apparent method of removing ferrous objects and no breaker is present indicating the waste material to be recycled must be carefully sorted. A brief survey of other demolition contractors within the Lower Mainland revealed no other companies that make use of portable recycling equipment.

5 WASTE ACCOUNTING AND POLICY MAKING

Although the recycled aggregate industry will never approach the size of the conventional aggregate industry, governmental estimates of the industry tend to underestimate the supply of waste generated. This fosters a belief that the recycling industry has grown to its maximum size and further encouragements to recycle would lead to diminishing returns.

In 1993 the GVRD along with the British Columbia Ministry of the Environment, Parks and Lands commissioned a study (GVRD 1993) that estimated that 85% of rubble was recycled. Based on this figure the GVRD reported that ...'significant further recycling may be difficult to achieve'. The installed recycling capacity at that time was three times greater than the figure given and recyclers were refusing material. The report underestimated not only the quantity recycled but the quantity available for recycling. As waste concrete is an inert, stable material it is desirable as fill, particularly in reclamation projects it is not always reported. In addition, the report failed to account for mobile recycling carried out by companies such as Lichfield. Without dependable data conclusions concerning recycling can only be qualitative.

There is no governmental incentives to promote recycling in a meaningful way in the Lower Mainland of British Columbia. There is no specific policy concerning recycled aggregate. This type of activity has grown only recently and the legislative body has not dealt with the matter. Recycling of aggregate in the Vancouver area is exclusively market driven, responding to a business opportunity. Specific conditions are met that make the venture profitable. It is in part fueled by the laissez-faire attitude of the provincial authorities in ignoring aggregate resources in its development planning. Sterilization of proximal virgin resources result in a geographic advantage to recyclers. The controlling parties relative to aggregate are the municipalities. They impose, through by-laws, indirect taxation in the form of a royalty for soil removal. Increased recycling would result in a decrease in revenue and control by this body. If standardization of recycled aggregate is to come about it will have to be regulated by a level of authority that already exercises control on construction standards.

6 CONCLUSIONS

Mineral aggregate represents one of the largest (in terms of volume) and most valuable segment of the mining industry in many industrialized countries. A prominent role in the economics of aggregate commodities is played by the transport cost element such that the location of resources in relation to areas of demand strongly influences the competitiveness of individual producers. At the same time the combination of urban sprawl and decreasing tolerance of environmental externalities has conspired to expose many production sites to environmental regulations which force either complete closure or significant cost increases. The producers of conventional aggregates are forced to internalize many of the externalities associated with the production of virgin resource. The full environmental costs are not accounted and mainly the difference between those associated with recycled aggregate in relation to conventional aggregate. This means that the environmental advantages of recycling is not visible because the savings are mainly conceptual from a private value point of view.

Recycled aggregate will become a partial substitute for conventional aggregate if the market recognizes the lower cost of recycled aggregate in specific applications. However, market failure to estimate all costs involved, particulary environmental, will create a gap between private and social value for using recycled aggregate, recycling less then the optimum quantity. A mandatory reuse/recycling policy requiring a minimum amount of recycled material in construction could act as a corrective measure. Demolishing will become an issue and buildings easily demolished will command a premium.

Multiple land use in the case of recycled aggregate requires a general framework policy. Technical expertise needs to be further developed as well as regulatory procedures. An equilibrium will exist between conventional and recycled aggregate. Public pressure, cost or legislation are factors that modify in a direction or another of this equilibrium.

REFERENCES

Anonymous 1992. Recyclage de materiaux de demolition. *Mines et carrières – Industries minérals* 74(11): 29-59.

Bairagi, N.K., Rovande, K. & Pareek, V.K. 1993. Behaviour of concrete with different proportions of natural and recycled aggregates. *Resources, conservation and recycling* 9(1&2): 109-126.

Beeby, D.J. 1988. Aggregate Resources, California effort under SMARA to ensure their continued availability. *Mining Engineering* 40(1): 42-45.

Brunner, P.H. & Stämpfli, D.M. 1993. Material balance of a construction waste sorting plant. *Waste Management and Research* 11: 27-48.

Carter, P.D. 1981. The Economics of mineral aggregate production and consumption in Newfoundland and Labrador. Newfoundland Department of Mines and Energy.

Carter, W.L. 1975. The Paradox of Quarrying in the Northeastern Megapolis. *Environmental Geology* 1(2): 67-68.

Ezeala-Harrison, F. 1995. Analysis of optimal recycling policy in a market economy. *Natural Resources Forum* 19(1): 31-37.

Goldman, H.B. 1961. Urbanization – impetus and detriment to the mineral industry. *Mining Engineering* 13(2): 717-718.

GVRD 1993. Demolition, Land clearing, Construction. Technical Memorandum 4 by CHM Hill Engineering for the Greater Vancouver Regional District.

Hansen, T.C. 1986. Recycled aggregates and recycled aggregate concrete, second state-of-the-art report development 1945-1985. *Material Structure* 111: 201-246

Hillmann, F.R. 1991. Recycling rubble: Does it have to be a 'no win' business proposition? *Canadian Aggregate* 5(1): 24-25

Hogberg, R.K. 1970. Urbanization, environmental quality and the industrial minerals industry. *Mining Congress Journal* 53(2): 108-114.

Lee, S. 1995. Ensuring ongoing economic sources of Highway construction aggregates through a gravel resource management program. Abstract In *Aggregate Forum Developing an Inventory that works for you,* B.C. Ministry of Energy, Mines and Petroleum Resources IC 1996-9.

Lauritzen, E.K. & Jakobsen, J.B. 1991. Nedriving af bygninger og anlaegskonstruktioner. SBI Anvishng 171, Horsholm, Denmark.

Libecap, G.D. 1989. *Contracting for property rights.* Cambridge: Cambridge University Press.

Martin, W.S. 1993. The Conventional and Recycled Aggregate Market of the Lower Mainland of British Columbia. BASc Thesis, University of British Columbia, Vancouver

Marr, A. 1995. Personal communication, Greater Vancouver Regional District.

Matysek, P. & Bobrowsky, P. 1995. Aggregate Forum – Developing an inventory that works for you! Abstract In *Aggregate Forum Developping an Iventory that works for you,* B.C. Ministry of Energy, Mines and Petroleum Resources IC 1996-9.

McDonald, G.T. 1989. Rural Land Use Planning Decision by Bargaining. *Journal of Rural Studies* 5(4): 325-335.

Miller, R.H. & Collins, R.J. 1976. Waste materials as potential replacement for highway aggregates. National Cooperative Highway Research Program Report 166. 94 pp.

OMNR 1992 Mineral aggregates conservation, reuse and recycling. John Emery Geotechnical Engineering Limited, prepared for the Ontario Ministry of Natural Resources.

Pennington, D. 1980. Minimizing regulatory problems of non coal surface mines. In *New York State Museum Bulletin 436*, pp. 6-12.

Prokopy, S. 1993. C & D debris recycler evaluate their industry, prospects. *Rock Products* 96(8): 25-26.

Smith, V.K. 1994. Natural resource damage assessments and the mineral sector: valuation in the courts. In W.E. Martin (ed.), *Environmental economics and the mining industry*: 15-52, Boston: Kluwer Academic Publishers.

Taylor, C.D. 1989. The Aggregate Industry in the Vancouver Area. MBA Thesis, Simon Fraser University, Burnaby.

Thomson, R.D. 1980. Mining of mineral aggregates in urban areas. Ph.D. thesis, University of Pittsburgh, Pennsylvania.

van Kooten, G.C. 1993. *Land resource economics and sustainable development: economic policies and the common good.* Vancouver: UBC Press.

Zimmerman, K. 1991. Riding the Recycling Wave. *Rock Product* 94(10): 50-56.

Modeling sand and gravel deposits and aggregate resource potential

PETER T. BOBROWSKY
BC Geological Survey Branch, Victoria, Canada

GAVIN K. MANSON
School of Earth and Ocean Sciences, University of Victoria, Canada

1 INTRODUCTION

British Columbia, Canada, supports a large, economically and socially important aggregate industry. Aggregate is used in the province of British Columbia primarily in road construction and maintenance, fill, concrete, asphalt and ice control (Hora 1988). As such, the major producer and consumer of aggregate is the provincial Ministry of Transportation and Highways (MoTH). For instance, BC statistics indicate MoTH consumed 28 million tonnes of aggregate compared to a commercial consumption of only 20 million tonnes during the same year (Hora 1988). It is estimated that there are some 2600 active aggregate operations, which directly and indirectly employ approximately 4000 individuals, resulting in a net industry value of over $370 million ($ Canadian) per annum (Thurber Engineering Ltd. 1990).

A growing number of regions in North America are currently experiencing an aggregate crisis. Existing reserves are slowly being depleted or sterilized as rapid urban expansion removes the resource from further use. In the absence of planned resource management prescriptions, maintaining the availability of this important and valuable, non-renewable commodity begins to exceed realistic costs which the public can be expected to support. In British Columbia, several areas, in particular the southwestern part of the province (Lower Mainland and Vancouver Island), are preparing for a resource crisis as local reserves of sand and gravel disappear. In 1994, production estimates totalled 41.8 million tonnes for the whole province, of which 8.4 million tonnes (20%) were produced on Vancouver Island (Manson 1995). This paper addresses the universal concern of aggregate depletion and sound resource management. The objective of the paper is to examine aspects surrounding one popular approach to modeling aggregate potential. A case study using data from Vancouver Island (Fig. 1), outlines the methods and results of a landform-based sand and gravel modeling exercise.

2 BACKGROUND

Aggregate is a low-cost, high-bulk commodity, where much of the cost to the consumer often arises from transportation (Langer & Glanzman 1993). In British Co-

Figure 1. Location map of Vancouver Island study area.

lumbia, transportation costs are estimated to be $200 million per year, somewhat higher than the $170 million per year for extraction (Thurber Engineering Ltd. 1990). High costs of sand and gravel throughout Canada dictate that the maximum distance of transportation not exceed approximately 100 km by truck and about twice this distance by barge and rail (Edwards 1989). The same rules apply today, and as a result it is necessary that most operations be located near high-demand areas, usually urban centres, which in turn creates problems for both the aggregate producer and land-use planner.

First, as population increases and urban centres expand on Vancouver Island, gravel operations tend to be engulfed by residential development, and the previously unnoticed externalities of noise, dust, truck traffic, and aesthetic disamenity associated with aggregate extraction suddenly impact on the new residents near the aggregate operation (Galbraith 1984). Often the aggregate producer may be forced to relocate to areas farther from the urban development as lands are rezoned or as higher costs of production are incurred because externalities are internalized. This could result in aggregates of lesser quality being utilized to offset the higher costs of using better quality aggregates from distant sources (Poulin et al. 1994).

Second, both known and unknown sand and gravel deposits located marginally to urban development frequently are sterilized as competing land-uses, such as building development, preclude aggregate extraction. Ironically, the process of urban development that is dependent upon the supply of aggregate is itself the major contributor to the reduction of that same supply.

Clearly the need for sound aggregate land-use planning and resource management on the periphery of urban centres is apparent. If deposits are identified, and subsequent gravel extraction is planned to occur prior to urban encroachment, the supply of high-quality, low-cost aggregate can be maintained. A method of accurately and easily predicting aggregate potential is, therefore, required. We propose that aggregate potential can be qualitatively estimated through the application and interpretation of pre-existing geological information, and then quantified through the use of deposit models. As a 'first approximation' the methodology provides a guideline for land-use planners, producers and developers to manage aggregate resources.

2.1 *Mineral deposit models*

Deposit models are tools for determining the probability that deposits of a certain volume, surface area, or other characteristic exist in a given region. They are defined as a systematic arrangement of information describing the essential attributes or properties of a class of mineral deposits (Bliss 1993). Models may be empirical (descriptive), in which case the various attributes are recognized as essential, even though their relationships are not known, or they may be theoretical, in which case the relationship of the deposit parameters are known (Singer & Cox 1988). Mineral deposit models essentially consist of a cumulative distribution of data for a given deposit parameter (e.g. volume, percent gravel, and so on) where the values of certain percentiles such as the 10th, 50th and 90th, are given.

Models have been extensively used in quantitatively assessing grades and tonnages of metal deposits (cf. Singer & Ovenshine 1979) and have recently been adapted to sand and gravel deposits (Bliss 1993). The adaptations to aggregate deposits have focused largely on volume and surface area, and only recently have they incorporated suitability for aggregate extraction by modeling the engineering and geotechnical characteristics of percent fines, grain size distribution, durability and reactivity (Bliss 1995). Empirical models are useful in estimating the range of deposits likely to be found in unknown areas, whereas theoretical models are useful in quantitatively predicting the parameters (e.g. volume) of a given deposit.

3 GEOLOGY OF SAND AND GRAVEL DEPOSITS ON VANCOUVER ISLAND

3.1 *Physiography and stratigraphy*

Located some 30 km off the southwest coast of mainland British Columbia, Vancouver Island is a narrow, 500 km long mountainous island, bordered to the east by the Georgia Basin, the Strait of Juan de Fuca to the south, Queen Charlotte Sound to the north and the Pacific Ocean to the west (Fig. 2). Most sand and gravel deposits are derived from Late Wisconsinan age drift, and are primarily confined to lowland areas on the east side and south end of the island. The Quaternary geological record on Vancouver Island is reasonably well known (Clague & Bobrowsky 1995, Table 1). Deposits containing suitable materials for economic extraction of sand and gravel include from oldest to youngest: Muir Point Formation (pre-Sangamon in age; Hicock & Armstrong 1983), Cowichan Head Formation (middle Wisconsinan in age;

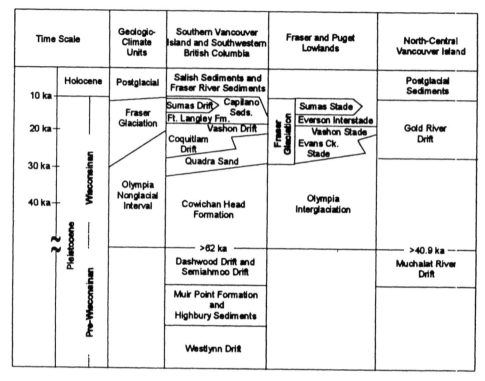

Figure 2. Correlation of surficial units between southern Vancouver Island, the Fraser Lowlands, and northern Vancouver Island (after Ryder & Clague 1989).

Table 1. Pleistocene and Holocene units which act as important hosts for sand and gravel on Vancouver Island according to population regions.

Region	Geological unit	Reference
Greater Victoria	Vashon drift recessional outwash; Capilano sediments	Leaming 1963
Saanich Peninsula	Quadra Sands – Saanichton Gravels; Vashon drift advance outwash	Leaming 1963
Shawnigan	Vashon drift recessional and advance outwash; Capilano sediment fluvial and marine deposits; Salish sediments	Leaming 1963 Halstead 1966
Duncan	Vashon drift advance outwash; Capilano sediment fluvial and marine sediments; Salish sediments	Leaming 1963 Halstead 1966
Nanaimo	Vashon drift recessional outwash; Capilano sediments; Salish sediments	Leaming 1963 Halstead 1966
Horne Lake/ Parksville	Vashon drift ice-contact deposits; Capilano sediment terraced deltas; Quadra Sands	Leaming 1963 Fyles 1960
Courtenay	Vashon drift glaciofluvial deposits; Capilano sediment fluvial terraces; Quadra Sands	Leaming 1963 Fyles 1960
Oyster River	Vashon drift glaciofluvial terraces and eskers; marine gravels	Leaming 1963 Fyles 1959

Armstrong & Clague, 1977), Quadra Sand (late Wisconsinan in age; Clague 1977), Capilano Sediments (recessional Fraser Glaciation; Fyles 1963) and Salish sediments (recent; Armstrong 1977).

3.2 *Geological units hosting sand and gravel deposits*

Little research has been done on Vancouver Island towards identifying geological deposits and landforms which may host sand and gravel. But, as the pressures on existing sand and gravel reserves become more acute, the need for geological aggregate data has increased. Early work by Fyles (1963), Leaming (1968) and McCammon (1977) has only recently been expanded to include local studies by Clague & Hicock (1976), Goff & Hicock (1992) and Blyth & Rutter (1993a). These compliment provincial syntheses by Hora & Basham (1980) and Hora (1988).

The majority of sand and gravel pits in British Columbia are related to various episodes of the last glaciation (= Fraser Glaciation; ca. < 29,000 years old). Only a small number of pits are hosted by pre-glacial sediments and post-glacial alluvial fans and terraces (Hora 1988). On the east side of Vancouver Island, Fyles (1963) and McCammon (1977) showed that most aggregate pits mine terraced deltaic deposits (Capilano Sediments) and glaciofluvial and other ice-proximal components of the Vashon Drift. A few pits are also known to occur in gravels correlative to the Quadra Sand. Near Victoria, local deposits called Saanichton Gravels separate Quadra Sands from the overlying Vashon till and these also provide a key source of aggregate (Alley 1979). In general, deposits correlative to Vashon Drift and Capilano Sediments are frequent hosts; Quadra Sands play a lesser role as deposits of this age are frequently sandy and buried too deeply to be economic, with the exception of the Saanichton Gravels in the south (Table 1).

3.3 *Quality of Vancouver Island sand and gravel*

Vancouver Island is fortunate to support a moderately abundant supply of high quality sand and gravel that can be used for a variety of purposes. The clast lithologies on Vancouver Island are variable, but most components are hard, inert, without coatings and consist of volcanic rocks, granodiorite, and sandstone. Although Vancouver Island gravels are generally free from deleterious clasts such as chert and volcanic glass that cause problems elsewhere in British Columbia (Hora 1988), shale, decomposed granite, and chert are locally present in small amounts. The impurities do not seem to affect the quality of concrete made from the aggregates (Fyles 1963).

The main problem with sand and gravel on Vancouver Island is the abundance of silt and sand that frequently necessitates screening prior to use. In extreme cases, the addition of pebbles from other sources has been required to produce usable aggregate (Leaming 1968).

3.4 *Geomorphic landforms as sand and gravel hosts*

Whereas previous aggregate work in the province has approached the importance of sand and gravel deposits from a stratigraphic framework, this study addressed suitability from a surficial mapping perspective. Terrain classification and surficial mapping in British Columbia follows the methodology of Howes & Kenk (1988). Terrain mapping at a scale of 1:50,000 was completed for most of Vancouver Island by the Ministry of Environment (MOE) during the period 1972 and 1979. In 1993, mapping of southern Vancouver Island was completed through the Ministry of Energy, Mines and Petroleum Resources (MEMPR).

The Howes & Kenk (1988) terrain classification system requires polygons of like-materials to be delimited through airphoto interpretation. The identified landforms are labeled using a string of letters and symbols which describe texture (e.g. silty, sandy, clayey), surficial materials (e.g. colluvial, marine, glaciofluvial, etc.), surface expression (e.g. blanket, plain, fan, veneer) and geological processes (e.g. inundation, channeled, etc.). Landforms can be composite elements in which case qualitative description of the relative proportions of each element in the polygon is given.

Few of the polygons and landforms on MOE maps have been ground-truthed and the accuracy of terrain polygon boundaries is largely dependent on the quality of the airphotos and the ability/expertise of the interpreter. Terrain maps are less reliable in areas of limited ground access, where dense forest cover limits air photo interpretation. The most difficult areas to map are forested valley-sides which support a mixture of till, colluvium and bedrock, whereas the easiest areas to map include distinctive landforms such as floodplains and fans, or landforms not masked by vegetation (Ryder & Howes 1984). On Vancouver Island, access is generally good, particularly in areas where gravel is being extracted, and the landforms hosting aggregate are thus easier to map. In general, the terrain maps are a good source of information on surficial geology, however, in most cases the information is confined to the uppermost stratigraphic layer. This represents a serious limitation of the map approach when, as in this study, data regarding the texture and genesis of subsurface deposits is also important.

4   METHODS

The methods followed in this study can be divided into three phases. In Phase 1, privately and publicly owned sand and gravel pits on Vancouver Island were inventoried. The assumption here being that an historical precedence of extraction, as indicated by the presence of a pit, provides key information on preferred types of landforms already in use. Phase 2 consisted of identifying the landform hosting the existing pit(s), estimating landform surface area and modeling distribution frequencies. Finally, in Phase 3, deposit volume data were collected, frequency distributions modeled and then tested against surface area-volume relationships.

4.1   *Phase 1 – Pit inventory*

In British Columbia, privately and publicly owned sand and gravel pits are treated differently and regulated by different agencies. Private pits are considered to be mines, and are regulated by the Ministry of Employment and Investment (MEI), which under the Mines Act and the Health, Safety and Reclamation Code, require operators to obtain reclamation permits prior to mine startup. The first step in this process involves completion of a Notice of Work (NOW) form with MEI (the form requires pit location data). Notice of Work data was consulted to develop an inventory of active and inactive privately owned sand and gravel pits on Vancouver Island, the Gulf Islands and parts of the mainland (MEI Mining Inspection District #8).

Public pit information in British Columbia is managed by MoTH. Data used in this study consisted of pit lists housed in two databases: Aggregate Deposit Inventory System (ADIS), and Aggregate Resource Management System (ARMS). The former database provided inventory identification, location and some volume estimates, whereas the latter included deposit identification, geotechnical data and some deposit volumes. Cross checking between the databases eliminated a number of errors, however, inaccuracies are expected as both database systems are currently being phased out and replaced by a new inclusive Gravel Management Survey System (GMSS).

All pits were hand plotted onto National Topographic Series 1:50,000 scale topographic maps. Locations were verified against form data. Pit locations were digitized as point data into Quikmap and a supporting dBase file on a GTCO Super L-Series digitizing table. The locations of digitized pits were checked against original data points. A total of 163 private pits and 177 public pits were plotted and digitized on the base maps (Fig. 3). Only public pits with volume estimates were digitized, and it is estimated that an additional 400 or so public pits were not plotted.

4.2 *Phase 2 – Landform identification and area modeling*

All digitized aggregate pits were plotted at 1:50,000 scale and overlaid onto corresponding terrain maps of the same scale. Landform types hosting individual pits were then determined. A total of 205 polygons on 34 maps sheets were identified to be hosting the 340 private and public pits (in several cases two or more pits occurred within a single terrain polygon). The 205 complex landform designations were simplified into 11 different types or families for statistical analysis. The families consist of the following: undifferentiated glaciofluvial (F^G), glaciofluvial fan (F^Gf), glaciofluvial terrace (F^Gt), glaciofluvial plain (F^Gp), undifferentiated fluvial (F), fluvial fan (Ff), fluvial terrace (Ft), marine (W), morainal (M), colluvial (C) and organic (O).

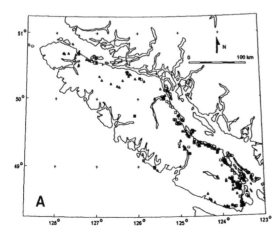

Figure 3. a) Locations of sand and gravel pits on Vancouver Island and the Gulf Islands.

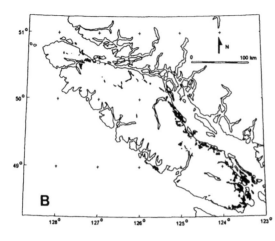

Figure 3. Continued. b) Locations of landforms hosting the sand and gravel deposits.

Letters in brackets correspond to the nomenclature of Howes & Kenk (1988) according to the BC Terrain Classification System.

Polygon shapes were digitized into Quikmap and a supporting dBase file, from which surface areas were calculated. An analysis of variance was conducted to determine if the landform types were significantly different in mean surface area. Cumulative plots of surface area vs. proportion of landforms was plotted following the method of Bliss (1993). The resulting frequency distributions were tested for lognormality by taking the natural log of the surface area and testing for normality using the mean and variance goodness-of-fit test described by Till (1974).

4.3   *Phase 3 – Volume determination and modeling*

Volume information compiled from MoTH databases was used to evaluate the statistical distribution of sand and gravel deposits. A total of 165 pits with volume information were used in this part of the study.

Deposit models require a definition of deposit extent to be established. Bliss & Page (1994) review several definitions and prefer to define deposit extent as pits that are within 1.5 km of each other. In this study, landform polygons defined deposit extent and pits located within the same polygon were considered to be the same deposit. Reported volumes of pits hosted by the same landforms were, therefore, totaled to determine deposit volume. The 165 pits with reported volumes were shown to occur within 146 landform polygons.

ANOVA was conducted to determine if there is a significant difference in mean volume of deposits hosted by different landform types, and the frequency distribution of volumes was further tested for lognormality. Finally, linear regression analysis of the deposit volumes vs. landform surface area was calculated. In this latter analysis, 114 deposit volumes were used.

5 RESULTS

The results of Phase 1 (Inventory) are shown in Figure 3. The 163 private and 177 public sand and gravel pits are primarily located on the east side of Vancouver Island and adjacent Gulf Islands, where most of the population is currently concentrated. For Phase 2 (Landforms and Areas) the locations of the 205 landform polygons are illustrated in Figure 3.

5.1 Pit/landform frequency

Rank-ordering the families according to frequency for hosting sand and gravel pits and deposits provided an indication of the relative importance of each family. Undifferentiated glaciofluvial landforms appear to be the most important in terms of hosting sand and gravel pits; with 70 of the total public and private occurrence falling in this category. Other landforms that are important hosts for pits include fluvial fans (46 pits in 32 landforms), marine landforms (43 pits in 31 landforms), glaciofluvial fans (36 pits in 32 landforms) and glaciofluvial terraces (35 pits in 18 landforms). The number of landforms in each family and the number of pits hosted is given in Figure 4.

5.2 Mean pit density

Of equal interest is the mean number of pits hosted per landform, which demonstrates the discrepancies in landform favourability. Glaciofluvial fans host 2.4 pits per landform, undifferentiated glaciofluvial landforms host 2.2 pits and glaciofluvial terraces host 1.9 pits. The marine and morainal families, containing large numbers of landforms, were found to host on average, only 1.4 and 1.0 pits per landform, respectively. The mean number of pits per landform is summarized in Figure 4.

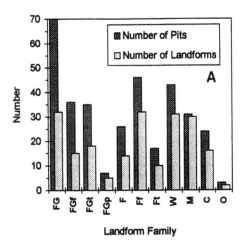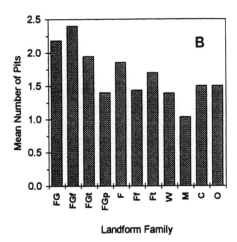

Figure 4. Histograms showing: a) Number of sand and gravel pits and landforms in each landform family, b) Mean number of pits per landform for each landform family.

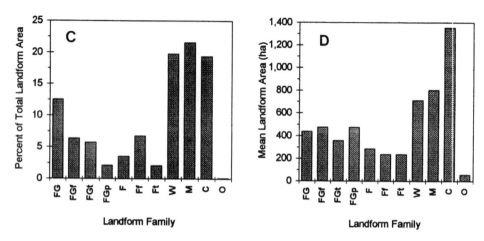

Figure 4. Continued. c) The percentage of each landform family of the total area of landforms identified as hosting sand and gravel pits and d) The mean landform area of each landform family.

5.3 *Landform areas*

Total and mean surface areas and standard deviations of each landform family are given in Table 2. The total surface area of all landforms identified is 111,958 ha, with a mean of 546 ha and a standard deviation of 702 ha. Morainal landforms have the largest total surface area of 24,113 ha, followed by marine (22,127 ha), and colluvial (21,676 ha) landforms. The percent of total landform area of each family is given in Figure 5. The computed mean surface area for the largest three is 1355 ha for colluvial, 804 ha for morainal and 714 ha for marine landforms. An analysis of variance between the mean surface areas indicates that there is a significant difference ($a = 0.05$) in mean surface area of all landform families. A Tukey test indicates that colluvial landforms are significantly larger than all other families except marine, morainal, organic and glaciofluvial plains. Given the small samples sizes for organic and glaciofluvial plains, their importance in this conclusion is likely minimal.

The surface area model for landforms hosting sand and gravel deposits is given in Figure 5. As colluvial landforms showed a significantly different mean area, they were excluded from the model following the precedence of Bliss (1993). The distribution is not significantly different from lognormal at the 0.05 level of significance, with a 10th percentile of 53 ha, 50th percentile of 281 ha and a 90th percentile of 1130 ha. Although excluded from the above analysis, the colluvial family was also found to be not significantly different from lognormal (0.05). The surface area model provides a 10th percentile of 239 ha, a 50th percentile of 767 ha and a 90th percentile of 2964 ha.

The surface area models of other landform families with sufficiently large samples sizes is given in Figure 6. They generally differ in curve shape and slope, reflecting differences in the distribution of surface areas.

Table 2. Summary landform data for BC aggregate research.

Landform family	Number of landforms	Number of pits	Mean no. pits/landform	Mean area (ha)	STD. dev.	Total area (ha)	% of total landform area
FG	32	70	2.19	4.38	487	14,001	12.51
FGf	15	36	2.40	478	768	7,164	6.40
FGt	18	35	1.94	358	367	6,446	5.76
FGp	5	7	1.40	477	287	2,383	2.13
F	14	26	1.86	286	210	4,004	3.58
Ff	32	46	1.44	237	193	7,569	6.76
Ft	10	17	1.70	236	136	2,362	2.11
W	31	43	1.39	714	725	22,127	19.76
M	30	31	1.03	804	545	24,113	21.54
C	16	24	1.50	1355	1475	21,676	19.36
O	2	3	1.50	58	29	115	0.10
Totals	205	338	1.65	546	656	111,958	100

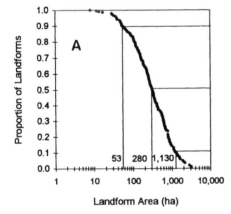
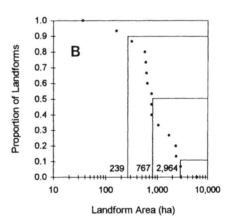

Figure 5. Landform surface area models for: a) All landform families except the colluvial family and b) The colluvial family.

5.4 *Volume determination and modeling*

Of the 165 pits originally grouped into families, 146 sand and gravel deposits were identified. Of these, 32 were either not located on published maps or were found to occur in suspect locations, leaving 114 deposits for use in the determination of the surface area to volume relationship. In the deposit volume modeling alone, all 146 deposits were used as location was not a critical factor. Summary deposit volume data for each landform family are given in Table 3.

The total volume of deposits determined in this study was found to be 38,882,000 m^3 with a mean of 270,000 m^3. Glaciofluvial terrace deposits were the largest family

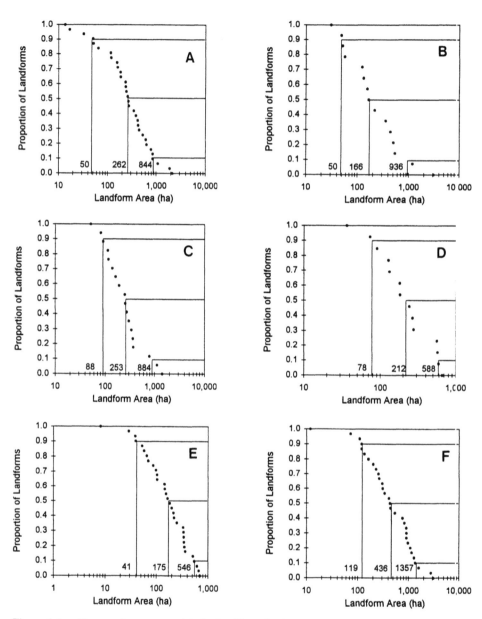

Figure 6. Landform surface area models for landform families important as sand and gravel deposit hosts including the: a) Undifferentiated glaciofluvial family, b) Glaciofluvial fan family, c) Glaciofluvial terrace family, d) Undifferentiated fluvial family, e) Fluvial fan family and f) Marine family.

totaling 11,338,500 m³. Undifferentiated glaciofluvial landforms hosted deposits to-taling 7,384,550 m³, fluvial fans hosted 4,576,200 m³ and glaciofluvial fans hosted 4,114,700 m³ of sand and gravel. The percentage of the total volume in each land-form family is shown in Figure 7. In terms of mean volume of deposits located in the

Table 3. Summary data of deposit volume statistics.

Landform family	Number of deposits	Percentage of deposits	Mean volume (m³)	Std. dev.	Total volume	% total deposit volume
FG	20	13.89	369,228	555,913	7,384,550	18.99
FGf	8	5.56	514,338	294,959	4,114,700	10.58
FGt	17	11.81	666,971	1,024,862	11,338,500	29.16
FGp	2	1.39	401,250	361,250	802,500	2.06
F	10	6.94	145,540	142,202	1,455,400	3.74
Ff	28	19.44	163,436	271,227	4,576,200	11.77
Ft	5	3.47	170,000	92,844	850,000	2.19
W	14	9.72	244,021	449,906	3,416,300	8.79
M	10	6.94	55,250	53,721	552,500	1.42
C	3	2.08	55,500	25,242	166,500	0.43
O	1	0.69	23,000	–	23,000	0.06
Unknown	26	18.06	161,608	302,005	4,201,800	10.81
Total	144	100.00	248,993	508,346	38,881,950	100.00

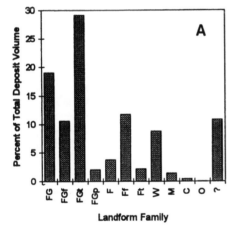 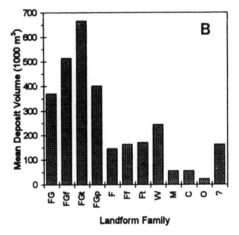

Figure 7. Histograms showing: a) The percentage of the total sand and gravel deposit volume hosted in each landform family and b) The mean deposit volume of each landform family.

different landform families, glaciofluvial terrace landforms were largest with a mean value of 667,000 m³, followed in turn by glaciofluvial fans, plains and undifferentiated glaciofluvial families with mean volumes of 514,300 m³, 401,300 m³ and 369,200 m³, respectively (Fig. 7). An analysis of variance and Tukey Test were applied to the mean volume data, indicating a significant (0.05) difference in deposit volumes by the landform families, with glaciofluvial terraces hosting deposits significantly larger than those of marine and fluvial fan landforms. ANOVA tests comparing mean deposit volume indicates no significant difference. Given the above, all landforms can be modeled collectively in the volume deposit model given in Figure 8. The distribution of deposit volumes is not significantly different from a lognormal

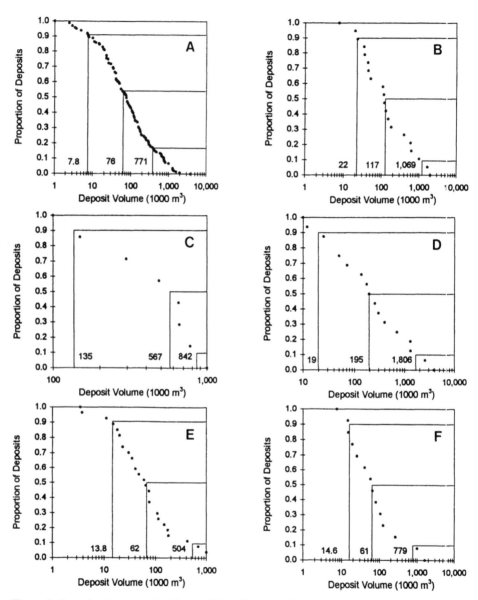

Figure 8. Deposit volume models for: a) All landform families, b) The undifferentiated glacioflu-vial family, c) The glaciofluvial fan family, d) The glaciofluvial terrace family, e) Fluvial fan family and f) The marine family.

distribution, providing a 10th percentile of 7755 m³, a 50th percentile of 76,500 m³ and a 90th percentile of 771,300 m³.

Deposit volume models for those landform families with sufficient data points are given in Figure 8. As with landform surface area models, these tend to vary in shape and slope for each family as a result of differences in the distributions of reported volumes.

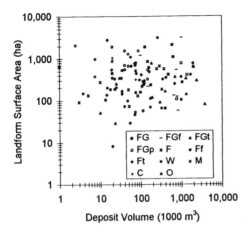

Figure 9. Graph of relationship of deposit volume to landform surface area. The correlation coefficient $R^2 = 0.0001$ indicates no correlation.

Figure 9 shows the deposit volume-landform surface area relationship determined for all landforms. The results show no correlation between landform surface area and the volume of the deposit hosted by that landform ($r^2 = 0.0001$). Treated individually, landform families still showed little or no correlation between the two variables.

6 DISCUSSION

6.1 *Distribution of deposits*

The location of sand and gravel pits is controlled by favourable geology, demand, zoning and regulations, and competing land-uses (Langer & Glanzman 1993). On Vancouver Island, where uninhabited land areas are common and competing land-use issues are of a lesser importance (in contrast to, for example, the Lower Mainland), the factors of geology and demand are most important.

Sand and gravel deposits identified in this study are primarily concentrated in a region called the Nanaimo Lowlands, a narrow coastal belt on the east side of Vancouver Island. Here, much of the population of the island is clustered in small urban centres such as Nanaimo, Courtenay and Campbell River. These and other communities are experiencing unprecedented population growth requiring ever increasing amounts of locally available aggregate. Several major roads and highways, including the Trans-Canada Highway and the Island Highway currently under construction, service these areas further increasing the demand for local aggregate.

In British Columbia as elsewhere, a significant portion of the cost of sand and gravel centres on transportation distance (cf. Langer & Glanzman 1993). In BC, the market cannot support truck transportation beyond 100 km and barge or rail transportation beyond 200 km (Edwards 1989). Thus, producing pits are all located within this transportation zone for local use.

In this area, it is fortuitous that the high demand is easily accommodated with a high supply. Sand and gravel deposits are located primarily in late Pleistocene to early Holocene sediments, which on eastern Vancouver Island, tend to occur in the coastal lowlands and broader inland valleys (Alley 1979). The majority of deposits are seen to lie below 160 m elevation and often within 25 km of the coast, although the highest observed deposit in this area is located at 310 m asl (map NTS 92L/1). The high supply also provides opportunity for considerable aggregate export to neighbouring centres such as Greater Vancouver. This contrasts with a region such as the Peace River of northeastern B.C., where availability and potential for local aggregate is extremely low.

   Because field verification of pit location could not be performed, misplot errors were common, but cross-referencing eliminated or corrected a number of misplots.

6.2   Geology of deposits

The results of this study indicate that sand and gravel deposits on Vancouver Island are hosted within a wide range of landforms. Several criteria presented earlier can be used to evaluate which of these landforms are geologically most important. Criteria include the total number of pits hosted, the mean number of pits per landform within each family and the mean deposit volume (volume per landform) of each family. Landform area is not considered to be a good indicator of aggregate potential as it has been shown that certain large landforms such as those within the colluvial family can host few deposits. The total number of landforms in each family is not considered a good indicator because a large number of landforms can host relatively few deposits in low concentrations. A simple system of scoring is presented to quantitatively measure the aggregate potential of each landform family, wherein for each criterion (e.g. total volume of deposits hosted) the eleven landform families are ranked in order of importance. Scores of 1 to 11 are assigned with a high score indicating high potential. Scores in each category are summed to give a total score indicating aggregate potential. The results of this scoring indicate that undifferentiated glaciofluvial landforms have the highest aggregate potential, followed closely by glaciofluvial terraces and fans, which in turn are followed by fluvial fans and marine landforms (Table 4).

Table 4. Categories and scores assigned in quantitive assessment of BC aggregate potential.

Landform family	Total volume	Score	Mean deposit volume (m^3)	Score	No. of pits	Score	No. of pits per landform	Score	Total Score
F^G	7,384,550	10	369,228	8	70	11	2.19	10	39
F^Gf	4,114,700	8	514,338	10	36	8	2.40	11	37
F^Gt	11,338,500	11	666,971	11	35	7	1.94	9	38
F^Gp	802,500	4	401,250	9	7	2	1.40	3	18
F	1,455,400	6	145,540	4	26	5	1.86	8	23
Ff	4,576,200	9	163,436	5	46	10	1.44	4	28
Ft	850,000	5	170,000	6	17	3	1.70	7	21
W	3,416,300	7	244,021	7	43	9	1.39	2	25
M	552,500	3	55,250	2	31	6	1.03	1	12
C	166,500	2	55,500	3	24	4	1.50	6	15
O	23,000	1	11,500	1	3	1	1.50	5	8

These results are in reasonably good agreement with common sources of aggregate potential previously identified on Vancouver Island (Section 2) and in other parts of Canada as well (cf. Edwards 1989). However, the importance of marine, and to a lesser extent, colluvial and morainal sources of aggregate on Vancouver Island is unique to both the Quaternary geology of this area and the use of landform maps.

The extent of marine transgression during the early Holocene is based on the upper limit of marine deposits. Evidence shows that most of the Nanaimo Lowland and portions of other coastal regions on Vancouver Island were under marine submergence during highest sea level. We conclude that surficial mapping from air photos in this area correctly identifies marine sediment cover, but that this cover is most likely a veneer which hides occasional sources of aggregate underneath. This was observed in several landforms where stratigraphic data indicated underlying landforms with a higher potential for aggregate. Sand and gravel is likely being extracted from the underlying glaciofluvial deposits, but since the landform descriptor frequently does not specify deeper deposit type, the landforms are classified in the marine family. Marine gravels also occur as raised relic beach deposits. Regardless of the context, gravely marine deposits are being mined for aggregate (cf. Table 1).

The same arguments apply to the morainal landforms which are generally recognized elsewhere as poor sources of aggregate due to the wide range of particles ranging from clay to boulder size (Edwards 1989). It has been demonstrated on Vancouver Island that till (Vashon drift) overlies potential sources of gravel. Fyles (1963) noted that sub-till gravels in Quadra Sand were being exploited in some areas. In these cases the overlying till was mapped as a morainal landform implying this was the source of gravel. Additionally, some maps such as those of Blyth & Rutter (1993a, b) tend to show large colluvial landform areas which may include smaller aggregate rich deposits. Pits located in these small hidden landforms are lumped within the surrounding larger colluvial map polygons.

An attempt was made to account for the limitation of maps while assigning specific landforms to families where polygon descriptors contain more than one type of deposit. Landforms were assigned to families on the assumption that sand and gravel was being extracted from the most probable sources based on the following rules. If any gravely terrain unit was indicated, the landform was assigned to that family, and if two or more gravely members were indicated, the most abundant was assigned. If a glaciofluvial or fluvial component was indicated, then that family was chosen. If both were described, the family was assigned to the most abundant. If marine, morainal or colluvial members were present, the most abundant was assigned. Surface expressions, if more than one was given, were assigned on the basis of relative abundance. For landforms with stratigraphic information, assignment into a family followed the same rules.

This method could not be applied to landforms with descriptors containing more than one deposit type. For those with only one deposit type, as was the case for many landforms, no subjective decisions were made in the grouping and landform family was assigned based only on the given terrain unit. This represents a source of error attributable to the landform maps as some pits are clearly misclassified. Examples include three different pits that occur in two different organic landforms. Correction could not be made and the organic landform was used in the absence of additional information and field verification.

6.3   *Deposit parameter models*

Maps were used to determine the area of landforms for application in generating an area model. The maps used in the model were produced by several different mappers which adds a problem of poor consistency between map sheets. Another source of error occurred during digitization, since the hardcopy versions of maps occasionally distort during reproduction as compared to the original mylar. The RMS error noted during map calibration reached as high as 800 m for some sheets, but as low as 100 m in others. Pit locations and landforms with an RMS greater than 200 m were not digitized. There were virtually no problems in digitizing landforms across map borders. Although landform boundaries were consistent across any two map sheets, descriptors were not, so the larger area of the two and its descriptor was always chosen.

It was found that there was no significant difference in mean deposit volumes for the different landform families, although for pit volumes glaciofluvial terraces were found to be significantly larger. This difference is strongly influenced by the fact that the largest gravel operation on Vancouver Island (3,825,000 m^3) is a glaciofluvial deposit and this large size distorts the mean pit volume. However, when pit volumes in the same landform are totaled to give a deposit volume, the influence of the single large pit is reduced and the glaciofluvial terraces are no longer significantly larger. Although the conclusions drawn from the volume data in this study are reasonable, there remains a significant inherent source of error. Specifically, deposit modeling is hampered when calculations are based on operator volume estimates rather than drilling and other subsurface data, especially in cases near urban centres where volumes are frequently underestimated as parts of a deposit may be excluded from production due to sterilization (cf. Bliss & Page 1994). Volume estimates used in this study were provided by MoTH which reports volumes of gravel remaining to be extracted. In many cases, multiple volume estimates were obtained from different databases for the same pit, in which case the largest value was used. It is believed that on average, most estimates are far short of the actual volume present.

Regardless of the sources of error in the data, both the landform surface area and deposit volume distributions were not found to be significantly different from lognormality. The defining characteristic is that the distribution of the logarithmically transformed data is normal whereas the distribution of the untransformed values is non-normal (cf. Till 1974).

The predictive ability of models is based on the percentiles calculated from the distribution which give the probability of there existing deposits of certain areas and volumes. For example, the 90th percentile of the deposit model derived here is 771,300 m^3, which means that 10% or 1 in 10 deposits have volumes of 771,300 m^3 or greater. The probability of finding deposits of this size or greater on Vancouver Island is therefore 0.1 or 10%. Likewise the probability of finding deposits > 76,500 m^3 is 0.5, and > 7750 m^3 it is 0.9. For the surface area model, the probability of finding landforms > 1130 ha that host sand and gravel is 0.1, whereas the probability of finding landforms > 281 ha is 0.5 and > 53 ha is 0.9. Since data were collected for deposits on Vancouver Island, the resulting models should only be applied to Vancouver Island. Bliss (1993) has received quite different results for deposits in California.

6.4 *Deposit parameter relationships*

Deposit volume and landform surface area are both lognormally distributed and can therefore be tested for autocorrelation. Bliss & Page (1994) showed that area and volume are strongly correlated for deposits in California and the UK, yet the results of this study demonstrate that there is no relationship between the two variables (Fig. 9). Several factors may collectively account for this lack of correlation.

One possibility is that the attempt to compare landform surface areas with deposit volumes is not meaningful and no correlation should be expected. As landforms are not necessarily complete aggregate deposits, but rather partial hosts, only a portion of the landform may be sand and gravel deposit. Thus, the surface area of the landform is therefore larger than that of sand and gravel deposits alone, and the degree to which it is larger likely depends on the landform family. Given this argument, it should be meaningful to compare surface areas of landforms in a certain family to the volumes of deposits they host. An attempt to determine the surface area to volume relationship of each landform family separately was unsuccessful possibly due to the small number of volumes available for most of the landform families. Two possibilities remain: 1. That for Vancouver Island, deposit volume is not correlated to landform surface area, potentially being controlled by some other factor such as deposit thickness or proximity to paleo-shoreline or paleo-stream channels, and 2. That the integrity of the data used in the correlation is suspect. Reiterating the possible error associated with MoTH volume data, the latter interpretation is considered most likely.

6.5 *Inferences based on models and distributions*

Of the 205 landforms identified in this study 179 lie on Vancouver Island, covering 2.9% of the total surface area. The landforms host 292 sand and gravel pits at an average of 1.63 pits per landform. The additional 400 public sand and gravel pits remaining to be inventoried can be expected to lie in some 245 landforms. As some of these additional pits likely lie in landforms already identified, this can be considered a maximum estimate. Given a mean landform area of 521 ha, the landforms are expected to cover a maximum of 127,726 ha, thus increasing the maximum percentage of Vancouver Island surface area covered by landforms hosting sand and gravel deposits to about 7.0%.

Estimates can be made regarding the volumes of deposits hosted by these landforms. If the surface area to volume ratio is known, the results of linear regression can be used to generate an estimate. If the relation is unknown, the derived models may be applied to return probabilities of finding deposits of certain sizes. Given the sources of error inherent in the volume data, these estimates are only considered preliminary and are presented as examples of how deposit models may be applied.

The volume model determined in this study for all deposits may be applied. The 10th percentile given is 7750 m^3, and the 90th percentile is 771,300 m^3. As there is a 0.1 probability of finding deposits less than 7750 m^3, and a 0.1 probability of finding deposits greater than 771,300 m^3, as a first approximation one can argue that there exists a 0.8 probability of finding deposits between these two volumes.

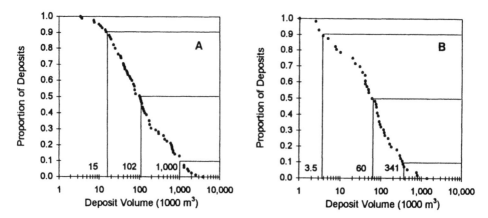

Figure 10. Deposit volume models of: a) The five highest ranking landform families including (in ranked order) the undifferentiated glaciofluvial family, glaciofluvial fan family, glaciofluvial terrace family, fluvial fan family and marine family and b) The remaining lower ranking landforms (in ranked order) including the undifferentiated fluvial family, fluvial terrace family, glaciofluvial plain family, colluvial family, morainal family and organic family.

This may be refined for exploration and land-use planning purposes by estimating according to different landform families. As stated earlier, for Vancouver Island the most important landform families for hosting sand and gravel deposits are undifferentiated glaciofluvial, glaciofluvial fans, glaciofluvial terraces, fluvial fans, and marine landforms. Undifferentiated glaciofluvial landforms host 13.9% of deposits. The volume model for this family gives a 10th percentile of 22,610 m^3 and a 90th percentile of 1,069,400 m^3. An estimated 13.9% of unknown deposits should, therefore, be located in undifferentiated glaciofluvial landforms with a probability of 0.8 of having volumes between 22,610 m^3 and 1,069,400 m^3. For glaciofluvial fans which host 5.6% of the deposits identified, the volume model indicates a probability of 0.8 of finding deposits between 135,000 m^3 and 842,500 m^3. Glaciofluvial terraces host 11.8% of the deposits and have a probability of 0.8 of finding deposits between 19,800 m^3 and 1,806,200 m^3. Fluvial fans host 19.4% of all deposits identified and with a probability of 0.8 are expected to have volumes between 13,800 m^3 and 504,500 m^3. Marine deposits host 9.7% of the deposits and at a probability of 0.8 should be between 14,650 m^3 and 779,500 m^3 in size.

The five families can be modeled together (Fig. 10) indicating that, as 60.4% of deposits are found in these landform families, 60.4% of unknown deposits will have a 0.8 probability of containing between 15,000 and 1,000,000 m^3 of sand and gravel. The remaining 29.6% of deposits located in glaciofluvial plain, undifferentiated fluvial, fluvial terrace, morainal, colluvial, or organic landforms should have volumes between 3480 and 341,000 m^3 of sand and gravel (Fig. 10). Land-use planning initiatives and sand and gravel exploration programs can incorporate quantitative analyses such as this, and concentrate their efforts on the better sources of gravel. These are identified in this preliminary analysis to be the undifferentiated glaciofluvial, glaciofluvial fans, glaciofluvial terraces, fluvial fans, and marine landforms. These landforms are seen to host more deposits of larger size.

7 SUMMARY AND CONCLUSIONS

The objectives of this study are to present the geological settings of sand and gravel deposits on Vancouver Island, British Columbia by identifying the geomorphic landform associated with each pit. Landform surface areas and deposit volumes were modeled, and the possibility for a covarying relationship between the two was tested. Preliminary models were presented and some applications tested. These applications are believed to be of use to land-use planning and sand and gravel exploration.

Sand and gravel deposits on Vancouver Island are located mainly on the east coast in the Nanaimo Lowlands and in some of the broader inland valleys. Known occurrences tend to be located near existing transportation corridors and near population centers, both for economic reasons and because these generally correspond to regions of favourable geology. A simple quantitative method for assessing aggregate potential is presented and it was found that in order of importance, undifferentiated glaciofluvial, glaciofluvial terraces, glaciofluvial fans, fluvial fans and marine landforms are the most favourable as hosts for sand and gravel deposits. Glaciofluvial sediments are thought to be associated with the advance and retreat of ice during the Fraser Glaciation which occurred some 29,000 to 10,000 years ago, whereas the fluvial and marine sediments were deposited during the last 10,000 years.

Landform surface areas and deposit volumes were modeled and the distributions were found to approximate lognormal functions, however, when the two parameters were compared to determine their covarying relationship they were found to be poorly related. The poor relationship is thought to be obscured by the manner in which volume data are reported. The model given here is purely empirical, and provides little opportunity to predict aggregate potential more than would a theoretical model. Ten percent of the landforms that host deposits on Vancouver Island are > 1130 ha, 50% are > 281 ha, and 90% are > 53 ha, whereas 10% of the deposits are > 771,300 m^3, 50% are > 76,500 m^3, and 90% are > 7750 m^3. Of the aggregate deposits yet remaining to be identified, it is estimated that 60.4% will be found in undifferentiated glaciofluvial, glaciofluvial terrace, glaciofluvial fan, fluvial fan, and marine landforms and will have a 0.8 probability of containing between 15,000 and 1,000,000 m^3 of sand and gravel. An additional 29.6% of deposits are expected to be located in glaciofluvial plain, undifferentiated fluvial, fluvial terrace, morainal, colluvial, or organic landforms and with a probability of 0.8 they are expected to have volumes of sand and gravel between 3480 and 341,000 m^3.

ACKNOWLEDGMENTS

We appreciate the financial and logistical support of the BC Geological Survey Branch and School of Earth and Ocean Sciences to complete this work. This paper benefited considerably from the editorial comments of Brian Grant and an anonymous reviewer.

REFERENCES

Alley, N.F. 1979. Middle Wisconsin stratigraphy and climate reconstruction, southern Vancouver Island, British Columbia. *Quaternary Research* 11:213-237.

Alley, N.F. & Hicock, S.R. 1986. The stratigraphy, palynology, and climatic significance of pre-Middle Wisconsinan Pleistocene sediments, southern Vancouver Island. *Canadian Journal of Earth Sciences* 23: 369-382.

Armstrong, J.E. 1977. Quaternary Stratigraphy of Fraser Lowland, Fieldtrip Guidebook, *Geological Association of Canada*, Annual Meeting. 20 pp.

Armstrong, J.E. & Clague, J.J. 1977. Two major Wisconsin lithostratigraphic units in southwest British Columbia. *Canadian Journal of Earth Sciences* 14: 1471-1480.

Bliss, J.D. 1993. Modeling sand and gravel deposits – initial strategy and preliminary examples. US Geological Survey Report 93-200. 31 pp.

Bliss, J.D. 1995. Three-part surficial aggregate assessment. In P.T. Bobrowsky, N.W.D. Massey and P.F. Matysek (eds), *Aggregate Forum: developing and inventory that works for you!*, Report of Proceedings, p. 34-35. March 30, 1995, Richmond. *BC Ministry of Energy, Mines and Petroleum Resources*, Information Circular 1995-6.

Bliss, J.D. & Page, N.J. 1994. Modeling surficial aggregate deposits. *Nonrenewable Resources* 3(3): 237-249.

Blyth, H.E. & Rutter, N.W. 1993a. Quaternary geology of southeastern Vancouver Island and Gulf Islands (92B/5, 6, 11, 12, 13, and 14). In B. Grant and J.M. Newell (eds), *Geological Fieldwork 1992, BC Ministry of Energy, Mines and Petroleum Resources*, Paper 1993-1: 407-413.

Blyth, H.E. & Rutter, N.W. 1993b. Surficial Geology of the Victoria Area, *BC Ministry of Energy, Mines and Petroleum Resources*, Open File 1993-23.

Clague, J.J. 1986. The Quaternary stratigraphic record of British Columbia – evidence for episodic sedimentation and erosion controlled by glaciation. *Canadian Journal of Earth Sciences* 23: 885-894.

Clague, J.J. 1977. Quadra Sand: a study of the late Pleistocene geology and geomorphic history of coastal southwest British Columbia. *Geological Survey of Canada*, Paper 77-17: 1-24.

Clague, J.J. & Bobrowsky, P.T. 1995. General Quaternary Geology of Southwestern BC. In Quaternary Geology of Southern Vancouver Island. B5: Field Trip Guidebook, p. 7-15. *GAC/MAC Annual Meeting*, Victoria, BC.

Clague, J.J. & Hicock, S.R. 1976. Sand and Gravel Resources of Kitimat, Terrace, and Prince Rupert, British Columbia. *Geological Survey of Canada*, Paper 76-1A: 273-276.

Edwards, W.A.D. 1989. Aggregate and non-metallic Quaternary mineral resources. In R.J. Fulton (ed.), Chapter 11 *Quaternary Geology of Canada and Greenland*, Geological Survey of Canada, Geology of Canada No. 1: 684-687.

Fyles, J.G. 1959. Surficial Geology, Oyster River, British Columbia. *Geological Survey of Canada*, Map 49-1959.

Fyles, J.G. 1960. Surficial Geology, Courtenay, British Columbia. *Geological Survey of Canada*, Map 32-1960.

Fyles, J.G. 1963. Surficial Geology of Horne Lake and Parksville Map Areas, Vancouver Island, British Columbia, 92F/7, 92F/8. *Geological Survey of Canada*, Memoir 318. 142 pp.

Galbraith, D.M. 1984. The Urban Problem of the Sand and Gravel Industry in British Columbia. Unpublished Masters Thesis, University of Victoria. 61 pp.

Goff, J. R. & Hicock, S.R. 1992. An evaluation of the potential aggregate resources for Sooke Land District, BC, 92B/5. In B. Grant and J.M. Newell (eds), *Geological Fieldwork 1991, BC Ministry of Energy, Mines and Petroleum Resources*, Paper 1992-1: 331-340.

Halstead, E.C. 1966. Surficial Geology of Duncan and Shawnigan Map-areas. *Geological Survey of Canada*, Paper 65-24.

Halstead, E.C. 1963. Surficial Geology, Nanaimo, BC. *Geological Survey of Canada*, Map 27-1963.

Hicock, S.R. & Armstrong, J.E. 1983. Four Pleistocene formations in southwest British Columbia: their implications for patterns of sedimentation of possible Sangamonian to early Wisconsinan age. *Canadian Journal of Earth Sciences* 20: 1232-1247.

Hora, Z.D. 1988. Sand and Gravel Study 1985 – Transportation Corridors and Populated Areas. *BC Ministry of Energy, Mines, and Petroleum Resources*, Paper 1988-27. 41 pp.

Hora, Z.D. & Basham, F.C. 1980. Sand and Gravel Study 1980 – British Columbia Lower Mainland. *BC Ministry of Energy, Mines, and Petroleum Resources*, Paper 1980-10. 74 pp.

Howes, D.E. & Kenk, E. 1988. Terrain Classification System for British Columbia. *BC Ministry of Environment*, MOE Manual 10.

Langer, W.H. & Glanzman, V.M. 1993. Natural Aggregate – Building Americas Future. *US Geological Survey* Circular 1110. 39 pp.

Leaming, S.F. 1968. Sand and Gravel in the Strait of Georgia Area. *Geological Survey of Canada*, Paper No. 66-60. 149 pp.

Manson, G.K. 1995. Modeling Sand and Gravel Deposits and Aggregate Resource Potential, Vancouver Island, British Columbia. Unpublished Honours Thesis, University of Victoria. 53 pp.

McCammon, J.W. 1977. Surficial Geology and Sand and Gravel Deposits of Sunshine Coast, Powell River, and Campbell River Areas. *BC Ministry of Energy, Mines, and Petroleum Resources*, Bulletin 65. 36 pp.

Poulin, R., Pakalnis, R.C. & Sinding, K. 1994. Aggregate resources: production and environmental constraints. *Environmental Geology* 23(3): 221-227.

Ryder, J.M. & Clague, J.J. 1989. British Columbia (Quaternary stratigraphy and history, Cordilleran Ice Sheet). In R.J. Fulton (ed.), Chapter 1, *Quaternary Geology of Canada and Greenland, Geological Survey of Canada*, Geology of Canada No. 1: 48-58.

Ryder, J.M. & Howes, D.E. 1984. Terrain Information: A User's Guide to Terrain Maps in British Columbia. *BC Ministry of Environment and Parks*. 16 pp.

Singer, D.A. & Cox, D.P. 1988. Application of mineral deposit models to resource assessments. US Geological Survey Yearbook, Fiscal Year 1987. pp. 55-57.

Singer, D.A. & Ovenshine, A.T. 1979. Assessing metallic mineral resources in Alaska. *American Scientist* 67(5): 582-589.

Thurber Engineering Ltd. 1990. Sand and Gravel Industry of British Columbia – Report to BC Ministry of Energy, Mines and Petroleum Resources. 72 pp.

Till, R. 1974. Statistical Methods for the Earth Scientist – An Introduction. Halstead Press: Toronto. 154 pp.

Aggregate modeling and assessment

JAMES D. BLISS
US *Geological Survey, Tucson, Arizona, USA*

1 INTRODUCTION

Trying to find ways to evaluate aggregate resources is the most recent development in the long history of mineral resource assessment. Government interest in what mineral resources may be present in a colony, region, or country is not a recent development. In fact mineral resource potential was one reason William E. Logan, the founder of the Geological Survey of Canada, made a general geologic survey of Upper Canada as early as 1846 (Gargill & Green 1986). The first quantitative mineral resource assessment was made by Allais (1957) who estimated the undiscovered mineral resources of the French Sahara. Government interest in mineral assessment is greatest in areas poorly explored or under dispute. This past interest by government and by others resulted in over 100 papers on regional assessment of nonfuel resources (Singer & Mosier 1981).

Mineral assessment techniques have generally advanced from qualitative to quantitative. In the USA assessment focus has broadened from deposits of metals, to industrial minerals, and recently to the consideration of aggregate. Assessment can be made in many different ways but the approach under consideration for use here is one described by Singer (1993). A full scale assessment of aggregate using this approach has not yet been completed. What follows are suggestions and ideas for those who are interested in modifying existing or developing new assessment methods. Also suggested are needed tools (including models) so aggregate can be better evaluated in the future. Other issues may need to be addressed which are not yet recognized.

2 DEFINITION OF AGGREGATE

Natural aggregate includes both crushed stone and sand and gravel. Processing is commonly limited to crushing, washing and sizing (Langer 1988). There is a fundamental division in aggregates between those produced by crushing stone and those produced from unconsolidated surface material. For most areas, aggregate is usually produced first from sand and gravel, then by crushed stone. Sand and gravel comes from surficial deposits of fluvial, lacustrine, marine, eolian or glacial sediments. Aggregate is also produced by ripping, quarrying, and blasting limestone, granite, traprock or other cohesive material from bedrock and crushing it.

Aggregates are described by grain size. Combined sand and gravel deposits should consist of at least 25% gravel-sized (4.76-76.2 mm) grains (Langer 1988). Coarse-grained aggregates include those with grains predominantly greater than sieve No. 4 (4.76 mm) while most fine-grained aggregates include those with grains predominantly less than sieve No. 4 (4.76 mm). Some grains are retained or passed on to the intervening sieves to the No. 200 sieve (0.074 mm). It is expected that most grains will not pass through the No. 200 sieve (0.074 mm). A few particles may be included between the 3/8-in sieve (9.52 mm) and No. 4 sieve (4.76 mm) in fine-grained aggregates (Huhta 1991). These rules describe aggregate at the level of a 'resource base' (Harris 1984) which includes all material suitable and unsuitable for extraction and without regard to economics.

Aggregate may be defined more precisely if the characteristics reflecting their suitability to a specific end use are included. Among other things, these geotechnical characteristics include sorting, impurities, durability, weathering susceptibility; these are discussed below. Cox et al. (1986, p.1) defines a mineral deposit as 'a mineral occurrence of sufficient size and grade that it might, under the most favorable of circumstances, be considered to have economic potential'. For aggregate deposits, the word 'grade' may be substituted with 'suitable geotechnical characteristics'. In some regional studies, data may limit assessment of aggregates at the level of the resource base.

Fill should not be confused with aggregate. It is another material used in construction but has no value beyond adding bulk. In fill, grain size distribution is usually unimportant but the material should be inert and clear of organic debris. Additional fill needed during road construction may be obtained away from the construction site in what is called borrow pits in the USA.

3   CLASSIC QUANTITATIVE ASSESSMENT AND MODELING

3.1 *Introduction*

The type of assessment under examination is one developed by Singer (1993). Over five million kilometers in 27 assessments have been evaluated since 1975, mainly in North America and primarily for metallic deposit types (Singer 1993). A similar type of assessment has been completed in British Columbia (Kilby 1996). These assessments attempt to provide answers concerning the possible size and grade of undiscovered mineral deposits, and how many undiscovered mineral deposits are likely. The format was developed so economic analysis was possible and the value of mineral resources could be compared to other land uses by land managers (Singer 1975). Three parts (Fig. 1) make up a quantitative mineral resource assessment: 1. Areas permissive for specific deposit type(s) are delineated, 2. The amount of metal is estimated using grade and tonnage models (see below for discussion on models), and 3. The number of undiscovered deposits for each deposit type is estimated (Singer 1993). Cox (1993) describes two ways to help in estimating the number of undiscovered deposits – the deposit density method, and the counting method. However, for

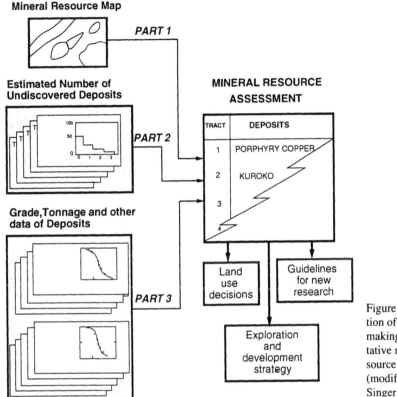

Figure 1. Interrelation of the three parts making up a quantitative mineral resource assessment (modified after Singer 1993).

most assessments, the estimate of numbers of undiscovered deposits has been done subjectively (Singer 1993).

Delineation is not possible if a geologic map is not available. Other kinds of information useful for delineation include mineral occurrences, geophysical, exploration, and geochemical maps. Delineation is guided by the descriptive model of deposit types (see below).

Because the estimated deposits are undiscovered, exact answers (e.g. the size of the deposit is 500,000 metric tons) are impossible – 'uncertainty is an integral part of the problem' (Singer 1993, p.70). The predicted size of an undiscovered deposit is not a single value, rather it is a distribution of values.

In order to answer questions assessments raise, two types of models have been developed – the descriptive model, and the grade and tonnage model.

3.2 *Descriptive models*

Descriptive models allow mineral deposits to be recognized as members of populations. Deposits should all share one or more essential features. Ideally, descriptive models provide sufficient detail about regional geologic settings (tectonic setting, regional features, age) so permissive terrains are recognizable for the deposit types on a geologic map of a scale of about 1:250,000 or less. Descriptive models also give local geologic attributes such as host rock, associated rocks, structural setting, ore controls and geometry, alteration (and typical dimension thereof), zoning, effect of weathering, ore and gangue mineralogy, isotopic signatures, fluid inclusions, and geochemical and geophysical signatures, all of which may help to identify member deposits and possible prospects which may indicate undiscovered deposits. Descriptive models may also contain a list of economic limitations important to many industrial mineral deposit types. This can include physical and chemical attributes significantly influencing production, and effecting end use. If applicable, other types of compositional or mechanical process restrictions are listed. Singer (1993, p.70) notes that 'deposit models represent the glue that bonds together diverse information on geology, mineral occurrence, geochemistry, and geophysics that is used in mineral exploration and assessments' as shown in Figure 2.

3.3 *Grade and tonnage models*

Grade and tonnage models provide distributions of grades and tonnages of mineral deposits grouped using descriptive models. Models are simple diagrams. Several examples are included in this report. Commodity grades, tonnages, and other variables important to commodity use are plotted on the horizontal axis. Cumulative proportion of deposits (the sample data) is plotted along the vertical axis. Construction is completed by plotting smoothed curves through the data points and the intercepts for the 90th, 50th, and 10th percentiles (Cox & Singer 1986). The curve commonly represents a lognormal distribution that has a mean and a standard deviation that fits the

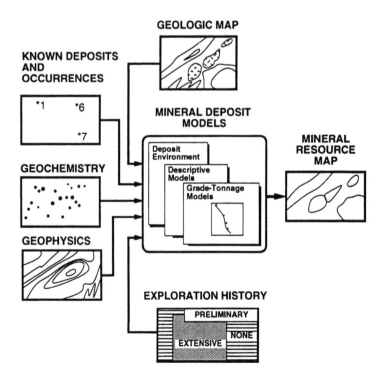

Figure 2. Network of how information is integrated by mineral deposit models used for mineral resource assessments (modified after Singer 1993).

data. These distribution of values are the sizes and grades of undiscovered deposits of the mineral deposit type under consideration. Also see Cox & Singer (1986) and Bliss (1992a) for about 100 descriptive and grade-tonnage deposit models. Examples of commonly known metallic deposit types with published descriptive and grade and tonnage models include kuroko massive sulfides (Singer 1986, Singer & Mosier 1986), sedimentary exhalative Zn-Pb (Briskey 1986, Menzie & Mosier 1986) and porphyry Cu (Cox 1986, Singer et al. 1986). Most models developed up to 1990 were for metal deposit types.

4 INDUSTRIAL MINERALS

Some problems faced in modeling and assessing aggregate have already been identified in attempts to model and assess other industrial minerals. Remember models are essential for successful assessments of the type under consideration here. Interest by federal land managers and others has shifted from metal to industrial mineral deposits. Work on industrial models has resulted in 22 descriptive models and 24 new or modified grade, tonnage and other types of related models compiled in two reports (Orris & Bliss 1991, 1992). Models of industrial minerals may use grade and tonnage data (as do metallic deposit types). However, many industrial minerals have other quantitative attributes related to end use that need to be considered (Orris & Bliss 1989). An example of this includes the content of impurities (e.g. iron in glass sands), or unique deposit-type specific characteristics (e.g. percentage of stones of gem quality in diamond placers, Bliss 1992b). Some industrial mineral deposit types can only be described by the amount of contained material (e.g. residual kaolin, Orris 1992).

Some industrial minerals and all aggregates are sensitive to proximity to markets, and availability of transportation. These and other factors mean that only some deposits in assessments are economic. Ultimately, assessment results would be greatly improved if they could undergo some type of economic filter to identify which deposits are economic and reasonable for use in forecasting possible future supply. These filters are currently absent. It would be highly desirable if they are developed.

5 AGGREGATE MODELS

5.1 *Descriptive models for aggregates*

Model development has just begun for aggregate deposits. The work by Bliss & Page (1994) was limited to sand and gravel deposits found in California and the United Kingdom. This accounts for the sand and gravel bias in the following discussion. Modeling of sand and gravel deposits near Vancouver, British Columbia is also underway (Manson 1996). 'Deposit' as defined by Bliss & Page (1994) was extremely general – it lumped sand and gravel bodies separated by 1.5 km or less as part of the same deposit. Many different types of sand and gravel bodies were lumped together using this rule. For example, sand and gravel in stream beds were lumped with occurrences in adjacent terraces. Sand and gravel deposits in glaciated areas involved joining bodies of sand and gravel in many different combinations including those from glacial terraces, outwash plains, eskers, kame terraces, and modern and raised beaches (Bliss & Page 1994).

Descriptive models for aggregate deposits have yet to be developed. This is due, in part, to the general way deposits have been initially defined. Development of descriptive models will be an important future activity in aggregate modeling and will likely be based on geomorphology, bedrock features, and sedimentary forms. Weathering is one important factor which will need to be considered.

5.2 *Size and other types of models for aggregates*

Quantitative models developed to describe aggregate deposits for use in assessment need to include deposit characteristics vital to end users. The primary consumers of aggregate are governments (for use in building roads and other infrastructure) and the construction industry. Suitable aggregate must behave in ways which meet minimum geotechnical criteria or standards (percent fines, grain-size distribution, durability, reactivity to cement) to insure roadways and structures constructed with these materials have acceptable longevity and are within acceptable safety limits. Expected difficulties in developing models for aggregate deposits using geotechnical criteria are: 1. Standards vary considerably reflecting local geology, climate and government attitudes concerning material suitability, and 2. Use of the material will vary as well. One way to measure aggregate quality is by tests using procedures developed by the American Society of Testing and Materials (ASTM), the American Association of State Highway and Transportation Officials (AASHTO), and by local and state governments. One approach (and one used below) is to model ASTM test results where the boundaries of acceptable aggregate behavior are also given.

A number of characteristics are important to suppliers and users and therefore need to be quantitatively modeled. Broadly, they include: 1. Deposit size and geometry, 2. Grain size distribution, and 3. Physical and chemical characteristics. A large number of models may be needed for each aggregate deposit type. The effort to compile a sufficiently large data base to develop models will be a major future challenge particularly if one considers the number of different deposit types that may be present for aggregates.

One major deficiency in the quantitative models for sand and gravel deposits developed by Bliss & Page (1994) is that the data are from deposits defined without regard to suitability for use in terms of geotechnical characteristics. These models describe sand and gravel deposits at the level of detail of a resource base, which is defined by Harris (1984) as including deposits for which extraction technology is both feasible and non-feasible, and economics are irrelevant.

5.2.1 *Deposit size and geometry models*

Unlike metallic deposits, sand and gravel deposit sizes are commonly reported by volume not weight. Nearly all sand and gravel deposit data used by Bliss & Page (1994) from California and the UK have volumes which can be described by a single volume model (Fig. 3). See the descriptive model section on how these deposits were defined. Manson (1996), using a more detailed classification of deposits types, found

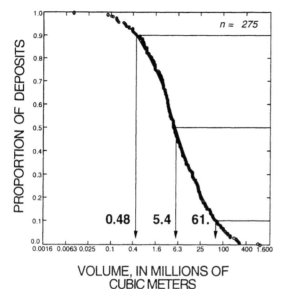

Figure 3. Volume model of sand and gravel deposits excluding those in alluvial fans (from Bliss & Page 1994).

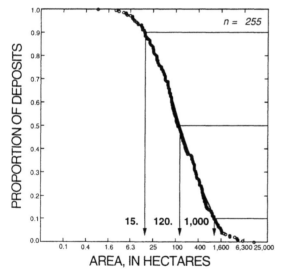

Figure 4. Area model of sand and gravel deposits excluding those in alluvial fans (from Bliss & Page 1994).

no significant difference among the volumes for deposits in different landforms on Vancouver Island, British Columbia.

Bliss & Page (1994) modeled deposit areas (Fig. 4) from the data from California and the UK. They also developed a regression model allowing volume to be forecast from area (Bliss & Page 1994, Fig. 7). However, Manson (1996) was not able to find a significant relation between area and volume for aggregate deposits on Vancouver Island, British Columbia.

At this time only a few deposit types with sufficient data for preliminary modeling are available. All need descriptive models to insure consistent definition. The first is for sand and gravel found in alluvial fans. To date, all data are from deposits in California as reported in Cole et al. (1987), Goldman (1961, 1964, 1968), Joseph et al. (1987) and Stinson et al. (1987). Models on geometry include area (Fig. 5) and thickness (Fig. 6). A volume model of deposit in alluvial fans was published previously in Bliss & Page (1994, Fig. 2) and has been updated here (Fig. 7).

Geology of alluvial fans may have quantitative links to sand and gravel volume, and to other characteristics. Does knowledge of fan geometry, geology, and hydrology allow quantitative forecasts to be made concerning the size and characteristics of fan-hosted sand and gravel deposits? In similar fashion, can the geology and hydrology of the stream basin on which the alluvial fan develops be useful in forecasting characteristics of the fan-hosted sand and gravel deposits? Relations like these may be valuable tools in future assessment.

Another preliminary model is the volume of sand and gravel deposits in eskers

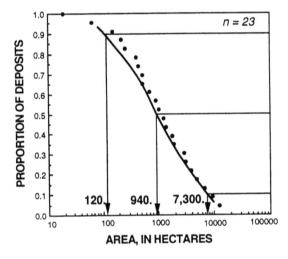

Figure 5. Area model of sand and gravel deposits in alluvial fans.

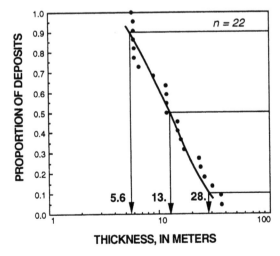

Figure 6. Thickness model of sand and gravel deposits in alluvial fans.

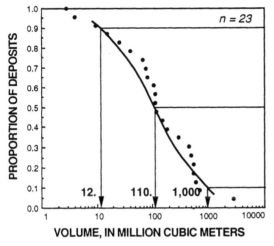

Figure 7. Volume model of sand and gravel deposits in alluvial fans.

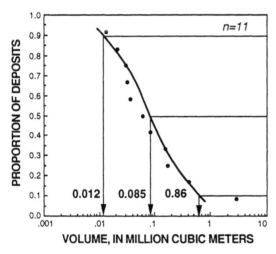

Figure 8. Volume model of sand and gravel deposits in eskers.

(Fig. 8). While deposits in alluvial fans are large, these are small. The data are for deposits in Manitoba, Canada (Groom 1986, 1987). The data set is also small. A descriptive model and more data for deposits in eskers will be needed before this model could be used in assessments.

Care will need to be taken in future size and deposit geometry modeling efforts. Deposits must be consistently defined spatially. Because geology of aggregate deposits can be complex, modelers will need to develop and report definitions of their 'deposits'. This will be help insure models are used correctly in assessment.

5.2.2 *Models of grain size distribution within deposits*
The distribution of grain sizes is a fundamental characteristic in the definition of aggregates. Models characterizing these distributions will be necessary. For aggregate produced from consolidated bedrock, grain size is a product of extraction. However,

the effect of bedrock geology on resulting grain sizes produced by blasting, ripping, and crushing will also need to be explored if assessments of bedrock as a source of aggregate are to be made. Information on joints, fractures, and other factors may be useful. For sand and gravel deposits, a grain-size distribution is already present.

Sand and gravel deposits contain not only sand and gravel, but cobbles and boulders as well as silt. In practice, those searching for surficial sand and gravel deposits find and describe other types of deposits. Thus six general types of deposits can be distinguished using grain sizes: 1. Sand deposits, 2. Sand deposits with gravel, 3. Sand deposits with gravel and boulders, 4. Sand and gravel deposits, 5. Sand and gravel deposits with boulders, and 6. Boulder deposits with sand and gravel (Bliss & Page 1994).

For aggregate producers and users, this classification is too simplistic. For them, the size distribution necessary to supply users is one that has a range of grains sizes which meets ASTM, AASHTO, local, or state government standards. Models should be able to do this and will need to be developed. However, those involved in qualitative mineral resource assessment prior to model development should consider this: well-sorted sand and gravel deposits will not be economic as they contain only a few grain-size classes. Most standards require a greater diversity of grains sizes than most well-sorted deposits can supply. What producers need are well-graded deposits which can supply a mix of grain sizes for users. Goldman (1994) suggested the minimum specifications for fine-grained aggregate deposits of no more than 45% of the material should pass between two consecutive standard sieve sizes as set by the American Society of Testing and Materials (ASTM), California Division of Highways, USA Army Corps of Engineers and the USA Bureau of Reclamation.

One size class of material is allowed only in small amounts in most standards. This is fines – usually considered to be minus No. 200-mesh (0.074 mm) material. As waste byproduct, the increased percentage of fines in deposits adds expense during extraction, dredging, hauling and disposal or stockpiling. The single (and often biggest) problem shared by nearly all aggregate facilities is the production of unusable fines.

Models giving the distribution of fines in sand and gravel deposits are not available. However, those involved in mineral resource assessment prior to model development should consider this: the aggregate industry usually processes deposits with no more than 10 to 15% fines (Drake 1995). Goldman (1994) suggested the minimum specifications for an aggregate deposit of no more than 5% less than the No. 200 (0.074 mm) sieve to be generally usable under requirements set by the organizations noted previously. Note that aggregate producers are working deposits with substantially more fines. As aggregate becomes more valuable, deposits with even more fines may be worked. Discovering a way to use fines is one of the biggest challenges facing the industry.

Large material, including boulder and cobble, may appear to be a problem to aggregate suppliers. However, these oversized materials can often be readily crushed to grain sizes found in short supply in the deposit.

5.2.3 *Models of physical and chemical characteristics*
The biggest challenge in modeling aggregate deposits is describing physical and chemical characteristics. Like other industrial minerals, physical and chemical char-

acteristics commonly determine if a particular deposit can be used. For example, the requirements for portland cement and asphalt concrete are the most stringent. Suitable sources of aggregate for this use will need to meet minimum standards concerning particle shape and surface texture, bulk and particle density, water absorption, mechanical properties, durability, susceptibility to leaching, and thermal incompatibility. Depending on use, other properties requiring evaluation include polish, and shrinkage. Limits are also set as to how much impurities can be tolerated including clays (silt, dust), carbonates, chlorides, shell, organic matter, alkalis, sulfates, chalk, mica, pyrite (and other metallic impurities), volcanic glass, zeolite, fissile shale, and opal (Goldman & Reining 1983, Smith & Collis 1993). Development of impurity models for aggregate deposits is quite plausible where allowable grades have values below some maximum 'economic' concentrations that can be tolerated or an 'intolerant-grade cutoff' (K.R. Long, unpubl. observ.). Some impurities occur as deleterious coatings on aggregate grains which disrupt concrete bonding. Aggregate requirements also are refined to reflect local conditions including climate. All these factors will need to be considered when using modeling in aggregate assessment.

Modeling chemical and physical characteristics is just beginning. Since a large number of characteristics are involved, perhaps the initial modeling effort needs to focus on a few. For example, Zdunczyk (1991) was interested in finding out which characteristics were most important for evaluating deposits in Caribbean countries. It was thought that these deposits would be sources for aggregate which can be imported into the Gulf and East Coast states of the USA. Goldman (1994) also identified some general minimum specifications for aggregate deposits which might meet requirements set by the American Society of Testing and Materials (ASTM), California Division of Highways, USA Army Corps of Engineers and the USA Bureau of Reclamation.

Examples of some of the tests identified by Zdunczyk (1991) and Goldman (1994) are listed below. They may need to be modeled first.

Test of soundness. Soundness measures aggregates' susceptibility to weathering; particularly by frost damage from expansion of water in fissures or by absorption of water under nonfreezing conditions.

Aggregate samples are immersed in solutions with sodium sulfate or magnesium sulfate, then dried. Crystallization of absorbed solutions during drying forces fissures to open causing grain fragments to break away. Five cycles of immersion and drying may be used as in ASTM Test C 88 (Marek 1991). Test results are determined by sieving the sample '...through specified sieves somewhat smaller than the original sieves on which a given size fraction was retained. The resulting weighted average loss for each size fraction is used' (Marek 1991, p.3-46). Soundness tests in Alabama, Florida, North and South Carolina, Maryland, Massachusetts, and Pennsylvania used sodium sulfate. Tests in Louisiana, Mississippi, Georgia, Virginia, and New York used magnesium sulfate. Only Texas required testing with both methods.

Test results for the targeted states (excluding New York) used a five-cycle procedure in which acceptable aggregate would have an average loss of 14% for both sodium sulfate and magnesium sulfate. Texas had the most tolerant standards, allowing a loss of 18%; Alabama, Massachusetts, and Pennsylvania standards were less tolerant at 10%. New York's standard required a loss of no more than 18% using a 10-

cycle procedure (Zdunczyk 1991, Table 1). If all states were designated as part of the targeted market, the New York test standards would be the standard to adopt because it is the most stringent; otherwise, aggregate in deposits used for export to these states should have a loss of no more than 10%. Goldman (1994) suggests that soundness for coarse aggregate as a general rule should exhibit a reduction of particle sizes of less than 10% using ASTM Test C 88.

Test of resistance to mechanical breakdown. The Los Angeles Degradation Test is used to measure the resistance of coarse aggregate to mechanical breakdown by abrasion and impact (Marek, 1991). This is important in the evaluation of aggregate for portland cement and asphalt concrete. Washed and dried samples are divided into several size fractions which are recombined into standard grading as defined in AASHTO T-96 (ASTM C 131) or required by end use. The material is placed into a drum with steel spheres and rotated a standard number of times, usually 500. The sample is removed and sieved dry over a No. 12 sieve (1.68 mm). 'The percent passing this sieve, sometimes termed the percent wear or percent loss, is the Los Angeles degradation value for the sample' (Marek 1991). The average maximum loss tolerated in the 12 Gulf and East Coast states with published standards is 46% for aggregate used in portland cement; the least tolerant loss is 35% which is the standard for New York. The average maximum loss tolerated by the nine Gulf and East Coast states with published standards is also 46% for aggregate used in asphalt cement; the least tolerant loss is 30% which is the standard for Massachusetts. Goldman (1994) suggest that acceptable coarse aggregate gives a loss of material passing the No. 12 sieve (1.68 mm) of less than 30% using ASTM Test C 131.

Two models of Los Angeles abrasion test data have been developed from data on 174 sand and gravel sites in New Mexico. Figure 9 shows how sand and gravel sites differ between two adjacent aggregate regions within New Mexico – the Basin and Range, and the Nonglaciated Central Region of the United States (Langer & Glanzman 1993). Nearly all the sites in the Basin and Range have test results which meet the loss criteria – less than 30% of material passing the No. 12 sieve. This is comparable to just 20% of those sites meeting the loss criteria in the Nonglaciated Central Region in New Mexico (Fig. 9).

Test of specific gravity and absorption. Specific gravity is the ratio of the mass of a given volume of aggregate to the mass of an equal volume of water. Specific gravity and absorption are important information for proportioning the ratio of binder to aggregate (Marek 1991). Testing is based on a procedure outlined by Archimedes' sometime between 200 and 300 BC. Different ways to measure specific gravel have been devised for fine- and coarse-grained aggregate. Because water is used, absorption is also addressed as described in AASHTO Test Methods T-84 and T-85 (ASTM Tests C 128 and C 127). As a general rule, material suitable for aggregate should have a specific gravity greater than 2.55 and absorption less than 3% using ASTM Test Method C 127 (Goldman 1994).

Test of fineness using modulus. The empirical fineness modulus is another way to express aggregate grading. It is particularly important for mix design as it helps determine the amount of coarse aggregate to be used (White 1991). It is expressed as an index giving the degree of fineness or coarseness of fine-grained aggregates. Calcu

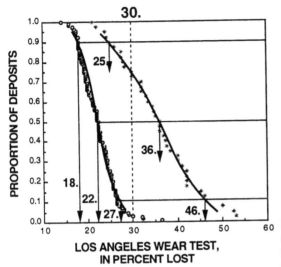

Figure 9. Two 'percent-lost' models as determined using the Los Angeles wear test. Results are from various types of sand and gravel sites found in the Basin and Range and the Nonglaciated Center Aggregate Regions as defined by Langer & Glanzman (1993).

lation of the fineness modulus involves adding the total percentage of sample retained on each of a specific series of sieves and dividing by 100 (Thrush 1968). As a general rule, suitable aggregate should have indices between 2.3 and 3.1 (Goldman 1994).

Test of plastic fines. A measure of plastic fines in aggregate is sand equivalent. It is the relative proportion of usually undesirable plastic fines and dust to all other grains passing the No. 4 sieve (4.75 mm) (Marek 1991). Measurement is made from a ratio of the height of sand to flocculated and suspended plastic fines in a cylinder and reported as a percent (Marek 1991). As a general rule, suitable aggregate should have a sand equivalent of not less than 75 using ASTM Test D 2419 (Goldman 1994).

Using these specifications will clearly restrict the definition of aggregate deposits in ways which will exclude a large part of the aggregate resource base. Developing and using models addressing chemical and physical characteristics will greatly improve mineral resource assessment. The study of Los Angeles abrasion test data in New Mexico is promising but needs to be refined with the addition of descriptive models and deposit definitions. Failure to consider geotechnical characteristics in past assessments has overly simplified the description of aggregate deposits and has created an illusion that there is likely more usable resource available than there is.

Developers of models of physical and chemical characteristics need to be aware that the discussion associated with ASTM test methods C 131 explicitly warns that the test results (particularly if just one type of test result is available) may not be sufficient to reject a aggregate source without careful evaluation. This is probably true for many other geotechnical tests as well.

6 AGGREGATE ASSESSMENT

As noted previously, the approach under consideration to systematically assess aggregate is one using three broad and interrelated activities similar to those developed by Singer (1993). To date, most assessments of this style have been for metallic deposit types. The activities are simple to state but may not always be easy to execute. The parts and order of events also have been modified slightly from Singer (1993) to fit the needs of aggregate assessment. They are:
  – Classify aggregate deposits by type, using characteristics important to end users and pit operators within the assessment area,
  – Consistently define boundaries of tracts known (or suspected) to contain aggregate deposits of specific types, using regional bedrock and surficial geology, and
  – Estimate number of known and possible undiscovered deposits.

Systematic classification and associated descriptive models of aggregate deposit types are needed before a better quantitative assessment can be made. Without this, deposit type identification and classification will be inconsistently done by those conducting assessments. Some of the characteristics important to end users are introduced here. Factors important to pit operators will need to be identified in the area being assessed.

Tracts are regions containing both known and possibly undiscovered aggregate deposits. All significant deposits, including past aggregate workings and occurrences, may also be shown. Those preparing assessments must classify aggregate deposits in the study area. Many pits have material that is only suitable as a source of construction fill material, not aggregate. Again, it is important to know geotechnical requirements. While it has not been shown yet, substantial simplification and reduction of effort may be possible if assessments and models address only deposits that meet, or can be readily upgraded to, specific geotechnical standards. This may only be true for aggregate needed for portland cement and asphalt concrete.

Tract boundaries are based on surficial or bedrock geology, not aggregate deposits. (A substantial part of the tracts include areas without aggregate deposits). For example, the outline of a glacial moraine may be used to define one tract; the adjacent outwash plain used for another; alluvium in a younger stream valley crossing both of these may be a third. The outcrop of a specific bedrock unit might be used as a tract boundary, but only those parts with sufficient close jointing might be considered suitable for aggregate mining and therefore defined as a deposit. These examples show that tracts do not define the boundaries of deposits but are areas which share one or more geologic characteristics. Removal of all known deposits during tract preparation should not change the tract boundaries. Areas within a tract that have been exhaustively explored for aggregate deposits should be shown. One can see that several types of tracts may be needed for aggregate deposits with different surficial or bedrock geology.

It has been suggested that assessment of aggregate in areas that aren't currently accessible or close to nearby markets is pointless. However, conditions change with time and land managers may need to know where resources are located. Changes in surface status may occur that might make an area available for aggregate extraction. Other markets may develop at new locations. And evaluation might have to be done before urbanization or other types of surface sterilization render some areas impossible to evaluate.

Langer et al. (1996) described a procedure used to estimate the resource base of sand and gravel in southwest New Mexico with some consideration of the possible suitability to end users. Three deposit types were used: modern stream alluvium, terrace, and alluvial fan. Materials were classified as satisfactory, unknown, or poor in probable performance in portland cement. Estimates of average deposit thickness for terraces and alluvial fans were taken from deposits in the study area and elsewhere. Area was estimated from outlines of alluvial fans and terraces. Area and thickness were used to calculate volume. No attempt to integrate the variability of deposit thickness, etc. was made and this would be a valuable addition to the exercise. The estimate provided is for the aggregate resource base and is a possible source of material for portland cement concrete. Only a portion of the estimated volume would be worked under the most reasonable foreseeable situations.

The surficial nature of most sand and gravel deposits may suggest the probable area containing resources. This can be based on data from water wells, geophysical surveys and other exposures. As a variation in assessment approach, the areas (not numbers) of suspected deposits might be empirically estimated. Bliss & Page (1994, Fig. 7) show that most of the variability in the volume of sand and gravel deposits is related to deposit area, rather than thickness. Thus, the volume of a deposit within a given area can be estimated using a simple regression of volume on area (Bliss & Page 1994, Eq. 2).

Assessment products can include maps that show tracts containing discovered and undiscovered aggregate resources. With a little additional effort, an estimate of the probable amount of undiscovered aggregate remaining in the assessed region can also be provided.

7 AGGREGATE PRODUCTION-CONSUMPTION DOMAIN (APCD)

Unlike high-unit value commodities, aggregate is marketed regionally and its significance is regional (Stinson et al. 1986). Therefore, aggregate production can not be viewed as similar to other commodities because aggregate is only economical within a restricted geographic area, the APCD (Fig. 10). This concept is based on the Production – Consumption (P-C) regions used by the California Division of Mines and Geology for mineral land classification of aggregate. Stinson et al. (1986, p.viii) define as 'the geographic area which includes the geologic deposits from which aggregate is produced and the market area which those deposits serve.'

Development of P-C regions in the San Francisco-Monterey Bay area of central California involved recognizing that aggregate operations, particularly those supplying aggregate for Portland cement concrete, are grouped into three areas of competing companies or production districts. This together with consideration of batch plant and hot plant locations, transportation costs, and marketing practices led to the definition of three P-C regions in the San Francisco-Monterey Bay area: South San Francisco Bay P-C Region, North San Francisco Bay P-C Region, and Monterey Bay P-C Region (Stinson et al. 1986).

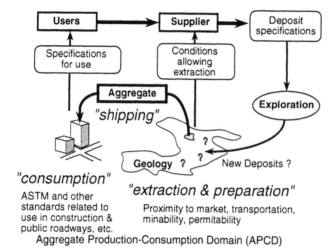

"consumption"
ASTM and other
standards related to
use in construction &
public roadways, etc.

"extraction & preparation"
Proximity to market, transportation,
minability, permitability

Aggregate Production-Consumption Domain (APCD)

Figure 10. Idealized aggregate production cycle.

Within each APCD, is a circular system driven by internal aggregate 'consumption' (Fig. 10). The cycle is driven by the demands of aggregate users, both from within and outside of government, including highway ministries, companies building private residences and office buildings, and others. These consumers seek aggregate with certain required characteristics or 'specifications for use' (Fig. 10) from suppliers. The 'suppliers' on the right of Figure 10, compete by marketing. Those selected by users extract, prepare, and ship the needed aggregate. Proximity to market, cost effective transportation, minablility of the deposits and permitability of the mining site are the essential factors.

On occasion, suppliers may need to locate new aggregate resources. These new deposits will need to meet the requirements given by 'deposit specifications' and include both conditions for use and extraction as identified in the box in the upper right hand corner of Figure 10. Exploration may discover new deposits which can supply additional aggregate. If none appear to be available within the APCD, adjacent areas outside may be considered. If deposits are found and shipped to markets within the APCD, the boundaries of the APCD will need changing.

The value of APCD can clearly be seen in aggregate assessments made by the California Division of Mines and Geology Maps showing APCD boundaries may be a useful addition in aggregate assessment. However, if APCD boundaries are too unstable, they may be valid only for a short time. Forces changing APCD boundaries include changes in population and infrastructure requirements. Changing (and perhaps more restrictive) requirements set by users may eliminate deposits. Extraction technology will also improve – some unsuitable aggregate occurrences will become suitable in the future. As the price of aggregate goes up and the number of readily available suitable deposits decrease within an APCD, many producers will become more tolerant of deposits which have, or produce, more silt; of deposits that have other deficiencies (like being too well sorted). They might seek and work deposits beyond the old boundaries of a given APCD. The APCD concept is an interesting one and may be important as long as aggregate resources are regional ones. But whether development of APCDs is essential to aggregate assessment is still not clear.

8 CONCLUSIONS

Assessment involves the integration of region-specific data with quantitative mineral deposit models. Uncertainty assigned to part of a regional aggregate assessment cannot be eliminated, only minimized. The total amount of remaining aggregate forecast to exist must, therefore, be reported as a distribution. The level of certainty increases as the map scale of the assessment region increases. At best, resource assessment will become more like exploration at the level of detail described by Timmons (1995). Perhaps the California Division of Mines and Geology comes closest in their evaluation of 83,000 km^2 (32,000 miles^2) adjacent to urban areas of coastal and central California (Beeby 1996). The Alberta Geological Survey (Edwards 1996) has prepared detailed deposit descriptions perhaps comparable to ones a commercial producers would want.

Highway and other infrastructure-related agencies conduct studies with adequate geotechnical detail to insure their own aggregate supply. They may also provide information for agency contractors, although most of these efforts are not regional in scope nor readily available outside their organization. Practically speaking, most government agencies lack sufficient funding to do detailed evaluation. Therefore reasonable estimates of aggregate quantity and quality will come from regional assessments and will have to be used in lieu of detailed exploration.

Quantitative assessment of aggregate requires new and different types of models to better characterize deposits. This is necessary so assessments will be more useful to end users and producers. Assessment methods may need to be changed as well. An important challenge is to find quantitative links between regional geology and deposit characteristics. Many apparent sources of aggregate are not usable because of deficiencies in quality. Some of the types of models suggested here may help to better define deposits. Models, and the knowledge to apply them in assessment can provide better estimations of available and suitable aggregate resources.

ACKNOWLEDGMENTS

Dr Peter T. Bobrowsky and the British Columbia Ministry of Employment and Investment provided funds so that I might attend the Aggregate Forum held in Richmond, BC, March 30-31, 1995. I wish to thank Peter Bobrowsky for encouraging me to complete this work during a very difficult time. I would also like to thank all reviewers, known and unknown, who asked many of the right questions which resulted in what I hope is a significant improvement to this paper.

REFERENCES

Allais, M. 1957. Method of appraising economic prospects of mining exploration over large territories. *Management Sci.* 3(4): 235-247.
Beeby, D.J. 1996. Successful integration of aggregate data in land-use planning; a California case study. In P.T. Bobrowsky, N.W.D. Massey & P.F. Matysek (eds), *Aggregate forum – developing an inventory that works for you!: Richmond, British Columbia. Mar. 30, 1995:* 25-26. Victoria: British Columbia Ministry Energy Mines & Petroleum Resources, Inform. Circ. 1996-6.

Bliss, J.D. & Page, N.J. 1994. Modeling surficial sand and gravel deposits. *Nonrenewable Res.* 3(3): 237-249.

Bliss, J.D. (ed.) 1992a. *Developments in mineral deposit modeling.* Washington: US Geol. Surv. Bull. 2004.

Bliss, J.D. 1992b. Grade, volume, and deposit specific models of diamond placers. In G.J. Orris & J.D. Bliss (eds), *Industrial minerals deposit models; grade and tonnage models:* 73-77. Washington: US Geol. Surv. Open-File Rep. 92-437.

Briskey, J.A. 1986. Descriptive model of sedimentary exhalative Zn-Pb. In D.P. Cox & D.A. Singer (eds), *Mineral deposit models:* 211-213. Washington: US Geol. Surv. Bull. 1693.

Cargill, S.M. & Green, S.B. (eds) 1986. Prospects for mineral resource assessments on public lands. *Proceedings of the Leesburg Workshop.* US Geological Survey Circ 980.

Cole, J.W, Miller, R.V. & Kohler, S. 1987. *Mineral land classification of the greater Los Angeles area, Part VI, Classification of sand and gravel resource areas, Claremont-Upland Production-Consumption Region.* Sacramento: Calif. Div. Mines & Geol. Spec. Rep. 143.

Cox, D.P. 1986. Descriptive model of porphyry Cu. In D.P. Cox & D.A. Singer (eds), *Mineral deposit models:* 76. Washington: US Geol. Surv. Bull. 1693.

Cox, D.P. 1993. Estimation of undiscovered deposits in quantitative mineral resource assessments – examples from Venezuela and Puerto Rico: *Nonrenewable Res.* 2(3): 82-91.

Cox, D.P. & Singer, D.A. (eds) 1986. *Mineral deposit models.* Washington: US Geol. Surv. Bull. 1693.

Cox, D.P., Barton, P.B. & Singer, D.A., 1986. Introduction. In D.P. Cox & D.A. Singer (eds), *Mineral deposit models:* 1-10. Washington: US Geol. Surv. Bull. 1693.

Drake, B. 1995. Comment – finding answers to the fines problem. *Rock Prod.* 98(9): 15.

Edwards, W.A.D. 1996. Mineral aggregate map case study, Alberta. In P.T. Bobrowsky, N.W.D. Massey & P.F. Matysek (eds), *Aggregate forum – developing an inventory that works for you!*: *Richmond, British Columbia. 30 Mar. 1995:* 33. Victoria: British Columbia Ministry Energy Mines & Petroleum Resources. Inform. Circ. 1996-6.

Goldman, H.B. 1961. *Sand and gravel in California, Part A – northern California*: Sacramento: Calif. Div. Mines & Geol. Bull. 180-A.

Goldman, H.B. 1964. *Sand and gravel in California, Part B – central California*: Sacramento: Calif. Div. Mines & Geol. Bull. 180-B.

Goldman, H.B. 1968. *Sand and gravel in California, Part C – southern California*: Sacramento: Calif. Div. Mines & Geol. Bull. 180-C.

Goldman, H.B. 1994. Sand and gravel. In D.D. Carr et al. (eds), *Industrial minerals and rocks:* 869-877. Littleton, Colo.: Society for Mining Metallurgy, and Exploration, Inc.

Goldman, H.B. & Reinig, D. 1983. Sand and gravel. In S.F. Lefond (ed), *Industrial minerals and rocks*: 1151-1166. Littleton, Colo.: Soc.Min.Eng.

Groom, H.D. 1986. *Aggregate resource in the rural municipality of Miniota.* Edmonton: Manitoba Energy & Mines Aggreg. Rep. AR85-3.

Groom, H.D. 1987. *Aggregate resource in the rural municipality of Shellmouth.* Edmonton: Manitoba Energy & Mines Aggreg. Rep. AR86-2.

Harris, D.P. 1984. *Mineral resource appraisal – mineral endowment, resources, and potential supply; concepts, methods, and cases.* Oxford: Clarendon Press.

Huhta, R.S. 1991. Introduction to the aggregate industry. In R.D. Barksdale (ed.), *The aggregate handbook*: 1-1 to 1-14. Washington: National Stone Association.

Joseph, S.E., Miller, R.V., Tan, S.S. & Goodman, R.W. 1987. *Mineral land classification of the greater Los Angeles area, Part V, Classification of sand and gravel resource areas, Saugus-Newhall Production-Consumption Region, and Palmdale Production-Consumption Region.* Sacramento: Calif. Div. Mines & Geol. Spec. Rep. 143.

Kilby, W.E. 1996. Mineral potential project – overview. In B. Grant & J.M. Newell (eds), *Geological Fieldwork 1995*: 411-416. Victoria: British Columbia Ministry Energy Mines & Petroleum Resources. Paper 1996-1.

Langer, W. 1988. *Natural aggregates of the conterminous United States.* Washington: US Geol. Surv. Bull. 1594.

Langer, W. & Glanzman, V.M. 1993. Natural aggregate – building America's future: US Geol. Surv. Circ. 1110: 39.

Langer, W.H., Knepper, D.H., Green, G.N. & Bliss, J.D. 1996. *An objective system for interpretation of geologic maps in regional assessments of natural aggregate.* Annual Meeting. Phoenix, Arizona. *11-14 March 1996.* Littleton, Colo.: Soc. Min. Metall. & Expl. Inc. Preprint.

Marek, C.R. 1991. Basic properties of aggregate. In R.D. Barksdale (ed.), *The aggregate handbook*: 3-1 to 3-81. Washington: National Stone Association.

Manson, G.K. 1996. Modeling aggregate resource potential, Vancouver, Island, British Columbia. In P.T. Bobrowsky, N.W.D. Massey & P.F. Matysek (eds), *Aggregate forum – developing an inventory that works for you!. Richmond, British Columbia. Mar. 30, 1995*: 36-37. Victoria: British Columbia Ministry Energy Mines & Petroleum Resources, Inform. Circ. 1996-6.

Menzie, W.D. & Mosier, D.L. 1986. Grade and tonnage model of sedimentary exhalative Zn-Pb. In D.P. Cox & D.A. Singer (eds), *Mineral deposit models*: 212-215. Washington: US Geol. Surv. Bull. 1693.

Orris, G.J. 1992. Preliminary contained material model of residual kaolin. In G.J. Orris & J.D. Bliss (eds), *Industrial mineral deposit models; grade and tonnage models*: 71-72. Washington: US Geol. Surv. Open-File Rep. 92-437.

Orris, G.J. & Bliss, J.D. 1989. Industrial-rock and mineral-resource-occurrence models. In E.W. Tooker (ed.), *Arizona's industrial rock and mineral resources of Arizona – workshop proceedings*: 39-44. Washington: US Geol. Surv. Bull. 1905.

Orris, G.J. & Bliss, J.D. (eds) 1991. *Industrial mineral deposit models – Descriptive deposit models.* Washington: US Geol. Surv. Open-File Rep. 91-11A.

Orris, G.J. & Bliss, J.D. (eds) 1992. *Industrial mineral deposit models; grade and tonnage models.* Washington: US Geol. Surv. Open-File Rep. 92-437.

Singer, D.A. 1975. Mineral resource models and the Alaskan Mineral Resource Assessment Program. In W.A. Vogely (ed), *Mineral materials modelling – a state-of-the-art review*: 370-383. John Hopkins Univ.Press.

Singer, D.A. 1986. Descriptive model of kuroko massive sulfide. In D.P. Cox & D.A. Singer (eds), *Mineral deposit models*: 189-190, 194. Washington: US Geol. Surv. Bull. 1693.

Singer, D.A. 1993. Basic concepts in three-part quantitative assessment of undiscovered mineral resources. *Nonrenewable Res.* 2(2): 69-91.

Singer, D.A. & Mosier, D.L. 1981. A review of regional mineral resource assessment methods, *Econ. Geol.* (76): 1006-1015.

Singer, D.A. & Mosier, D.L. 1986. Grade and tonnage model of kuroko massive sulfide. In D.P. Cox & D.A. Singer (eds), *Mineral deposit models*: 190-197. Washington: US Geol. Surv. Bull. 1693.

Singer, D.A., Mosier, D.L. & Cox, D.P. 1986. Grade and tonnage model of porphyry Cu. In D.P. Cox & D.A. Singer (eds), *Mineral deposit models:* 77-81. Washington: US Geol. Surv. Bull. 1693.

Smith, M.R. & Collins, L. (eds), 1993. *Aggregates; sand, gravel and crushed rock aggregates for construction purposes.* Bath, UK: Geol.Soc.Spec.Publ. 9.

Stinson, M.C., Manson, M.W. & Plappert, J.J. 1986. *Mineral land classification; aggregate materials in the San Francisco – Monterey Bay area, Part I, project description, Mineral land classification for construction aggregate in the San Francisco – Monterey Bay Area.* Sacramento: Calif. Div. Mines & Geol. Spec. Rep. 146.

Stinson, M.C., Manson, M.W., Plappert, J.J., Bortugno, E.J., Levis, E., Miller, R.V., Loyd, R.V. & Silva, M.A. 1987. *Mineral land classification; aggregate materials in the San Francisco – Monterey Bay area, Part V, Classification of aggregate resource areas, South San Francisco Bay Production-Consumption Region.* Sacramento: Calif. Div. Mines & Geol. Spec. Rep. 146.

Thrush, P.W. (ed.) 1968. *A dictionary of mining, mineral, and related terms.* Washington: US Bureau of Mines.

Timmons, B.J. 1995. Prospecting for natural aggregates; an update. *Rock Prod.* 98(1): 31-37.

White, T.D. 1991. Aggregate as a component of portland cement and asphalt concrete. In R.D. Barksdale (ed.), *The aggregate handbook.* 13-1 to 13-69. Washington: National Stone Association.

Zdunczyk, M.J. 1991. Importing construction aggregate to the continental United States. In Y. Drakapoulos (ed.), *Caribbean industrial minerals; new frontiers for investment:* 99-103. Kingston: Jour. Geol. Soc. Jamaica Spec. Issue No. 11.

Geologic characterization of natural aggregate: A field geologist's guide to natural aggregate resource assessment

W.H. LANGER & D.H. KNEPPER, JR.
USGS, Denver Federal Center, Denver, Colorado, USA

1 INTRODUCTION

This report is intended to help the field geologist determine the aggregate potential of an area. For purposes of this report aggregate is defined as crushed stone, and sand and gravel. The report bridges the gap between the geologist and the aggregate specialist by acquainting the geologist with some of the technical aspects of the aggregate industry. This report also provides the geologist with suggested field observations that will aid in identifying relationships between the geologic properties of rocks and sand and gravel and their likely performance as an aggregate.

The physical and chemical properties of aggregate result from the geologic origin and mineralogy of the potential source and its subsequent weathering or alteration. The properties of aggregate include: Texture, mineralogy, pore space, and weathering products. These properties can be observed and described by traditional geologic methods and terminology. Knowledge of which physical or chemical properties that determine the suitability of aggregate, will enable geologists to properly assess the potential of aggregate sources.

The largest single use of natural aggregates is in construction, and much of that aggregate is used in portland-cement concrete or bituminous mixes. The specifications for natural aggregate in portland-cement concrete or bituminous mixes are generally more rigorous and specific than for other construction-related uses. If aggregate can meet the specifications for these uses, it will satisfy almost any other use. Therefore, the specifications required for portland cement and bituminous mixes, and the geologic observations that help describe them, are emphasized in this report.

Aggregate characteristics important in defining its use in portland cement concrete or bituminous mixes include: 1. Physical properties, 2. Chemical properties, and 3. Contaminants. The major physical properties include: gradation of particle sizes; particle shape; particle-surface texture; porosity; pore structure; specific gravity; thermal properties; and susceptibility to volume changes. The major chemical properties include: Oxidation of certain minerals; short-term reactions; and longer-term reactions. The presence of certain contaminants can prevent cement from hydrating or bitumen from adhering to the aggregate.

Common tests and measurements (geologic observations) used to characterize aggregate are made in the field or laboratory by an experienced geologist. These include: 1. Resonance when struck with a hammer, 2. Friability or pulverulence when squeezed between the fingers, 3. Ease of fracturing, 4. Nature of fracture surfaces and fracture fillings, 5. Odor on fresh fracture, 6. Color and variations in color, 7. Internal structure such as porosity, granularity, seams, and veinlets, 8. Reaction to water, such as absorption of droplets on fresh fractures, expulsion of air or slaking, softening, or swelling when immersed, and capillary suction against the tongue, and 9. Reaction to acid (Mielenz 1994).

More extensive engineering tests are conducted to determine compliance with specification requirements (commonly set by purchaser), to ensure thorough quality controls; to assure that the customer is receiving the same material that is being produced at the plant site; and to obtain measurements of the physical properties used by the engineer in design of pavements, foundations, portland cement concrete, bituminous mixes, etc. (Marek 1991). These tests commonly expose aggregate to conditions that simulate the conditions under which the aggregate will be used. The tests are expensive and time consuming and are generally only used for detailed resource appraisals.

The most common guidelines that outline testing procedures and specifications for natural aggregates are those described by the American Society for Testing and Materials (ASTM). ASTM standards are based on exhaustive material testing, and the service records of those materials in actual use to estimate the quality of similar materials. Selected ASTM procedures for testing aggregate are shown in the Appendix. The approach to assessing aggregate described by this relates ASTM Standard 295-90, *Standard Guide for Petrographic Examination of Aggregate for Concrete* to geologic parameters, practices and terminology.

2   TYPES OF POTENTIAL AGGREGATE DEPOSITS

Naturally-occurring aggregate deposits, whether sand and gravel or source rock for crushed stone, are formed by a variety of geologic processes. Volcanoes, earthquakes, glaciers, rivers and streams, and marine processes have each contributed to the formation of the materials we use as aggregate. Consequently, the key to locating suitable deposits and assessing the potential for new aggregate sources is understanding the geologic processes that form them.

2.1   *Crushed stone*

Crushed stone is 'The product resulting from the artificial crushing of rock, boulders, or large cobblestones, substantially all faces of which have resulted from the crushing operation' (Langer 1988). More crushed stone is produced in the United States than any other mineral, mainly for uses in the construction industry. As noted by Tepordei (1993), 'Despite the relative low value of its basic products, the crushed stone industry is a major contributor to and an indicator of the economic well-being of the Nation'.

Crushed stone is classified according to the type of rock from which it was produced: Sedimentary, igneous, and metamorphic. Table 1 shows the most common types of rocks and which of those rock types often are good sources for crushed

Table 1. Common rocks and their potential for use as crushed stone (modified after Smith & Collis 1993).

Sedimentary rocks	Igneous rocks**	Metamorphic rocks
Argillite	Andesite†	Amphibolite
Breccia*	Basalt	Calcareous schist
Chalk	Dacite†	Gneiss*
Chert*†	Dunite	Granite-gneiss*
Claystone	Diabase	Graywacke
Conglomerate*	Diorite	Green (chlorite) schist
Diatomite	Gabbro	Hornblende gneiss*
Dolomite*	Granite	Hornfels*
Dolomitic limestone*	Granodiorite	Marble*
Limestone*	Granophyre	Phyllite
Oolitic limestone	Obsidian†	Quartz-mica schist
Quartzite*	Peridotite	Quartzite*
Sandstone, with varieties	Porhyrite	Schist
according to cement	Porphyry	Slate
(siliceous*†,	Pumice	Slaty marble
calcareous, clayey,	Quartz diorite	
ferruginous)	Quartz monzonite	
Sandstone with constituents	Rhyolite†	
other than quartz	Scoria	
(arkose*, graywacke*)	Syenite	
Shale	Trachyte	
Siltstone	Tuff	

* Sedimentary and metamorphic rocks commonly used as aggregate. ** Most igneous rocks are hard, tough, and dense, and make good sources of aggregate. Tuffs and certain lavas which are extremely porous may be exceptions. † May be reactive with alkali in portland-cement concrete.

stone. Detailed discussions of the rock types are contained in Tepordei (1993), Dunn (1991), and ASTM Standard C294 (American Society for Testing and Materials 1994).

The aggregate industry does not strictly adhere to a common petrological classification to describe the source rocks for crushed stone. For example, 'limestone' may refer to a limestone, a dolomite, or a marble that does not take a polish. On the other hand, the term 'marble' may refer to a true metamorphic marble or to a limestone or dolomite that takes a polish. Similarly, the coarse-grained, dark-colored mafic intrusive rock called gabbro may be identified as 'traprock' or 'black granite' by the aggregate industry; however, the term 'traprock' is also used to describe any fine-grained, dark-colored extrusive igneous rock such as basalt and andesite. These differences in nomenclature need to be considered when the geologist describes source rocks to the aggregate industry.

2.1.1 *Sedimentary rocks*
Of the chemically or biochemically deposited sedimentary rocks, hard dense limestones and dolomites, composed of carbonates, generally make good sources of crushed stone and make up approximately 71% of crushed stone production in the

US (Tepordei 1993). However, some limestone and dolomite may be soft, absorptive, and friable, which results in poor quality aggregate. Chert may be used as crushed stone; however, it is hard to crush and may cause adverse chemical reactions with alkali when used as concrete aggregate.

Clastic sedimentary rocks, including conglomerate, breccia, and sandstone, are locally used for crushing. Of these rocks, hard and dense sandstone is most commonly used for crushed stone. Even so, sandstone makes up less than 3% of the total US production (Tepordei 1993).

2.1.2 *Igneous rocks*

The aggregate industry identifies igneous rocks based primarily on grain-size and color: Light-colored, coarse-grained (intrusive) rocks are termed granite (this term, however, may also include light-colored, coarse-grained metamorphic gneisses as well); dark-colored, fine-grained (extrusive) rocks (called traprock); light-colored, fine-grained volcanic rocks like rhyolite and trachyte (termed either traprock or granite); and dark-colored, coarse-grained igneous rocks like gabbro (which may be referred to as black granite or traprock). Volcanic cinders and scoria are also used as sources for crushed stone, but together, they make up only about 0.2% of the total US crushed stone production (Tepordei 1993).

Igneous rocks commonly are hard, tough, and dense, and make an excellent source of crushed stone. However, certain extrusive rocks are too porous to make good aggregate and other highly siliceous igneous rocks tend to chemically react with alkali when used as aggregate in cement concrete. Fractures along cleavage in some coarse grained igneous rocks can result in crushing strengths too low for aggregate use. Some extrusive rocks may be unsuitable for aggregate if they are flow-banded, strongly jointed, vesicular, or highly fractured. Pyroclastic volcanic materials such as ash and tuff may be unsuitable unless they have become indurated by heating (welding) or compacted and cemented during burial.

2.1.3 *Metamorphic rocks*

Except for rocks formed by predominantly thermal activity (skarn), most regionally metamorphosed rocks show planar foliation caused by crystallization of parallel phyllosilicate minerals (micas). This foliation produces planes of weakness that are undesirable in aggregate, but only highly-foliated schist appears to have little or no use as aggregate. Marble as defined by producers accounts for 0.34% of the total crushed stone production and slate for 0.14% in the US (Tepordei 1993). Metamorphic quartzite physically resembles sedimentary silica-cemented quartz sandstone and the two are not separated in reported production figures. Similarly, to aggregate producers, gneiss resembles coarse-grained igneous rocks. Consequently, reported production figures include a substantial amount of gneiss under the granite category.

2.1.4 *Sand and gravel*

Sand and gravel deposits are accumulations of durable rock fragments and mineral particles. Such deposits result from the weathering of bedrock, and the subsequent transport, abrasion, and deposition of the rock detritus. Ice and water are the principal geologic agents that weather rock, and transport and deposit sand and gravel. Consequently, most gravel in the United States is found in glaciated regions in the northern parts of the country, in alluvial basins in the desert southwest, or as fluvial deposits near rivers and streams throughout much of the country. Aeolian deposits are too fine grained to be an important source of natural aggregate. Unlike sources of crushed stone, which may be any age, sand and gravel was primarily deposited during the Quaternary, although some Tertiary sand and gravel is mined as aggregate.

3 AGGREGATE ASSESSMENTS

Regional assessments (smaller than 1:50,000 map scale) and preliminary site investigations (1:50,000 map scale or larger) require the same basic information, however, the level of detail between the two commonly differs, especially in the amount and kind of field observations. For example, regional assessments rely heavily on published data and remote sensing interpretations, supplemented by reconnaissance field observations to define the characteristics of the regional deposits, and on geometric models of typical deposits to estimate volume.

Site investigations use detailed field studies to determine physical and chemical characteristics of potential aggregate, and spatial variability. Deposit thickness and geometry are determined from measured sections, well logs, and geophysical data. Regardless of the level of detail, an assessment of natural aggregate potential involves determining the location, quality, and the volume of the potential aggregate.

3.1 *Locating potential aggregate sources*

The preparation of a map showing the distribution of potential aggregate sources is essential for resource assessments. The key to mapping potential sources of aggregate is an understanding of the geology of the region: Focusing on the surficial geology (primarily Quaternary geology and geologic history) for deposits of sand and gravel, and a general study of the regional bedrock stratigraphy, origin, and structural history of the region for sources of crushed stone. Existing geologic maps serve as an excellent source of information for determining the location of potential sources of aggregate (Dunn 1991). Varnes (1974) described techniques to translate information on geologic maps into interpretive products. The mapped distribution of surficial deposits restricts the area in which sand and gravel deposits are likely to occur. The distribution of bedrock units suitable for crushed stone, (such as granite, limestone, dolomite, and basalt) as shown on geologic maps is sufficient for identifying the general areas where the target bedrock type is at or near the surface (Bottge et al. 1965, Timmons 1994), and for identifying the source rocks of sand and gravel particles. Commonly, available geologic maps are too general to allow the user to confidently select potential resources, and additional field work is necessary.

Aerial photos and other types of remote sensing data provide an excellent means for preparing reconnaissance geologic maps when none exist or existing maps are too generalized (Ray 1960, Schwochow et al. 1974). Aerial photographs are especially useful for identifying landforms associated with sand and gravel deposits and possible areas of bedrock outcrop. Detailed topographic maps can also be used effectively as they often show the location of existing pits and quarries, which provide clues to where minable aggregate is known to exist.

3.2 *Quality of potential aggregate sources*

National specifications exist for aggregate (ASTM 1994), however, specific requirements are determined by the users of the material, which include federal, state,
county, and city governments. Aggregate used in road building and concrete construction are subject to very rigorous specifications, but these specifications, as well
as the specifications for other applications, can vary from area to area; US Department of Interior (1981), Mielenz (1994), Barksdale (1991), and American Society for
Testing and Materials (1994) describe many of the tests and factors that must be
considered for concrete aggregate (see Appendix).

Many of the ASTM quality tests (see Appendix) used by aggregate producers require considerable time and/or specialized equipment to conduct. For assessment
purposes, however, good estimates of the quality of potential aggregate sources can
be obtained by relatively routine geologic field observations (or published descriptions) of the physical properties, potential contaminants, mineralogy, and weathering
characteristics of the materials that make up the deposit. Each potential aggregate
source should be characterized as thoroughly as possible according to the factors
summarized in Tables 2-5.

Table 2. Commonly measured physical properties of aggregate, and their relationships to performance in portland cement or asphaltic concrete.

Definition	General effects on use in portland cement concrete or bituminous mixes
Particle size – Particle grain-size distribution as determined by mechanical screening or, in the field by screening with portable sieves or by visual estimates	Concrete aggregate should contain a broad range of grain sizes throughout the sand-and-gravel range. Gap grading (aggregates with certain grain sizes missing) can be used and may be necessary in some applications Grading of aggregates for bituminous mixes varies depending on pavement design In most cases grading can be improved by processing
Particle shape: *Round* – Fully water-worn or completely shaped by attrition *Irregular* – Is naturally irregular, or partly shaped by attrition and having rounded edges *Angular* – Has well defined edges formed at the intersection of roughly planar faces *Flaky* – Has one dimension significantly smaller than the other two dimensions *Elongated* – Has one dimension significantly larger than the other two *Flaky and elongate* – Has three significantly different dimensions, i.e. length significantly larger than width and width significantly larger than thickness. (Smith & Collis 1993)	The shape of both coarse aggregate and sand can have marked effects on the workability of fresh concrete and on the strength properties of hardened concrete. These effects tend to be beneficial when the predominant particle shape is equidimensional, and are detrimental when the predominant shape is flaky and/or elongated (Smith & Collis 1993, US Department of Interior 1981) Intergranular contact provides the strength in bituminous mixes, making angular particles generally desirable for these mixes. Aggregate for bituminous mixes should be reasonably free of flaky or elongate particles. Smooth surfaces on aggregates may be easy to coat with bitumen, but offer little assistance to hold the aggregate in place. ASTM specifications call for a specified minimum amount of particles with fractured faces to be used in bituminous mixes

Table 2. Continued.

Definition	General effects on use in portland cement concrete or bituminous mixes
Particle surface texture – degree of roughness or irregularity of the surface of an aggregate particle: *Glassy* – Conchoidal fracture *Smooth* – Water-worn or smooth from fracture of laminated or very finely-grained rock *Granular* – Surface fractures show more or less uniform size rounded grains *Rough* – Surface fractures show fine- or medium-grained rock containing no easily visible crystalline constituents *Crystalline* – Contain easily visible crystalline constituents *Honeycombed* – Visible pores and cavities (Smith & Collis 1993)	Particle surface texture primarily affects the bond between the aggregate and cement paste in hardened concrete. Concrete flexural strengths and compressive strengths decrease with increasing particle smoothness. Recent advances in cement concrete technology have made it possible to produce very high strength concrete with aggregate having a relatively smooth surface texture (Smith & Collis 1993)
Porosity – Porosity is the percentage of the total volume of aggregate occupied by pore spaces	An approximate inverse correlation exists between aggregate quality and rock porosity. Porosity affects the strength and elastic characteristics of aggregate, and may affect permeability, absorption, and durability. Rock with a water absorption of 2% or less will usually produce good aggregate, whereas otherwise suitable rocks with a water absorption that exceeds 4% may not (Smith & Collis 1993) For some applications surface pores contribute to a rough surface texture. When used in portland cement concrete, aggregate with surface pores can absorb water that can be released at a later time thus improving the curing conditions (Barksdale 1991)
Pore structure – Pore structure is the size, shape, and volume of the spaces within an aggregate particle. Pores can be impermeable (isolated, enclosed cavities) or permeable (interconnected and connecting to the surface of the particle)	Large volumes of permeable pores are not desirable in aggregate for most applications. Large pore volumes allow aggregate to absorb large volumes of water or salt solutions, thus reducing soundness. For bituminous mixtures a large volume of pores also increases the absorption of binder, thus increasing the cost of paving mixture (Barksdale 1991)
Grades of fracturing: *Massive* – Fracture spacing > 3 ft *Moderately fractured* – Fracture spacing 8 in. to 3 ft *Very fractured* – Fracture spacing 4 in. to 8 in. *Extremely fractured* – Fracture spacing 2 in. to 4 in. *Crushed* – Fracture spacing < 2 in. (Dunn 1991)	Fractures are natural pathways for groundwater movement and subsequent weathering or alteration Rocks with considerable fracturing tend to make poor aggregate (Dunn 1991) Fracturing in bedrock can affect blasting and mining operations

Table 2. Continued.

Definition	General effects on use in portland cement concrete or bituminous mixes
Strength: *Strong* – Makes metallic sound, and breaks with difficulty, when struck with hammer *Moderately strong* – Makes dull sound, and breaks with moderate hammer blow *Weak* – Cuts easily with knife *Very weak* – Breaks with finger pressure (Dunn 1991).	Weak or very weak particles commonly perform poorly in use and break down during handling
Specific gravity – Ratio of the mass of a given volume of aggregate to the mass of an equal volume of water	Specific gravity may be a useful general indicator of the suitability of an aggregate. Very low specific gravity frequently indicates aggregate that is porous, weak, or absorptive; high specific gravity generally indicates high-quality aggregate Specific gravity of aggregate is of significance when design or structural considerations require that concrete have a maximum or minimum weight
Volume change – wetting and drying – Change in the volume of aggregate as the moisture content of the aggregate changes over time	Aggregate should exhibit little or no volume change with wetting and drying. Swelling or shrinkage produces disruptive forces that can crack concrete or cause popouts in the mixture (Barksdale 1991)
Coefficient of thermal expansion – Change in the volume of aggregate produced by a variation in temperature	Aggregate should have a coefficient of thermal expansion that is approximately equal in all directions, and all minerals in the aggregate should have the same coefficients of thermal expansion (Barksdale 1991). If concrete contains ingredients that expand at different rates and to different degrees of severity, internal stresses may crack the concrete (Rexford 1950)
Thermal conductivity – Ability of an aggregate to conduct heat	Aggregate with low thermal conductivities are desirable to prevent the penetration of frost through pavement (Barksdale 1991)

A variety of properties are used to characterize aggregate, many of which can be measured using standardized tests. Table 2 summarizes commonly measured physical properties of aggregate, and their relationships to performance in portland cement or asphaltic concrete. Contaminants within aggregates, when used in portland cement concrete, may: Cause decreased strength and durability of the concrete; affect the quality of the bond between the cement and the aggregate; cause an unsightly appearance of the concrete; and inhibit the hydration of the cement. Contaminants within aggregate commonly can be reduced to acceptable levels during processing by washing and screening. Specific sources of contamination are shown in Table 3.

Chemical properties of aggregate are important in the manufacture of concrete or bituminous mixes. Ideally, aggregate is an inert filler and should not change chemically within the concrete or bituminous mixes. However, some aggregates contain minerals that chemically react with, or otherwise affect, the concrete or bituminous

Table 3. Contaminants that may affect portland cement concrete or bituminous mixes.

Contaminant	Affects on use in portland cement concrete or bituminous mixes
Structurally soft or weak particles	Soft or weak particles, in significant amounts, can affect the integrity of portland-cement or bituminous concrete. In small proportions they probably are not detrimental except that they may result in 'pop-outs' when exposed to freeze-thaw action (Smith & Collis 1993) Micaceous minerals are soft, have a perfect cleavage parallel to one plane, and have low compressive and flexural strength Aggregates with minerals with a fibrous structure such as satin-spar, and fibrous varieties of amphiboles and serpentines, may be soft
Fine-grained materials that occur as surface coatings, lumps, or disseminated throughout the aggregate	Fine-grained materials increase water content of cement, may be misrepresented in grain-size when conducting sieve analyses, and decrease aggregate-matrix bond (Smith & Collis 1993). In addition, where present as lumps, fine-grained materials can affect soundness
Organic materials	Some organic matter, including sugar, fuel oil, and humus can retard or prevent the hydration of cement and hardening of concrete when present even in trace amounts. Other forms of organic matter, such as coal and lignite are regarded as undesirable, mainly because they are weak and unsound, and because they cause unsightly stains on the surface of concrete (Smith & Collis 1993). Some organic impurities have been shown to entrain large amounts of air in concrete which may reduce the unit weight and compressive strength of concrete, and may interfere with proper air-entraining agents (Swenson & Chaly 1956)
Chlorides	Chlorides, usually sodium chloride, occur in marine and some coastal sources of aggregate, and in some inland sedimentary sources. Presence of chloride in cement concrete may cause corrosion of imbedded steel reinforcing bar or mesh in reinforced concrete
Chemical contaminants	Any of the materials listed Table 6 showing chemical properties are considered deleterious material
Soluble particles	Solution of soluble materials is seldom a serious problem in aggregates. However, some rock or sand and gravel contain sufficient quantities of water soluble substances (such as gypsum), as coatings or seam fillings, to cause difficulties when used as concrete aggregate (McLaughlin et al. 1960)

mixes. In concrete, these chemical processes are reactions between the aggregate and cement or oxidation of minerals. In bituminous mixes, chemical factors may influence oxidation of asphalt or the stripping of bituminous film from aggregates. Minerals that can cause adverse chemical reactions in aggregate used in bituminous and concrete mixes are shown in Tables 4a and 4b.

Weathering of bedrock or gravel clasts decreases the strength of aggregate, increases the overall cost of separating the sound from the unsound rock, and affects the blasting and extractive techniques (Fookes 1991). The suitability of a natural aggregate source based on the degree of weathering of the bedrock is shown in Table 5.

Table 4a. Chemical properties of aggregate important in the manufacture of portland cement concrete or bituminous mixes.

Reaction	Affects on use in portland cement concrete or bituminous mixes
Alkali-reactive silica	See Table 6b – Alkali from cement in concrete pore-water solution can react with aggregate that contains certain silica minerals, forming a gel around the aggregate. The gel imbibes water, causes expansion of the aggregate, and subsequently of the concrete (Dolar-Mantuani 1983, Mather & Mather 1991)
Alkali-reactive carbonate	Reaction is similar to the alkali-silica reaction although no visible gel is formed. Rocks potentially susceptible to alkali-carbonate reaction are dolomitic limestones in which the dolomite constitutes 40-60% of the total carbonate fraction of the rock, in which there is a 10-20% clay fraction, and in which small dolomite crystals is scattered throughout a matrix of extremely fine grained calcite and clay (Hadley 1961, Mather & Mather 1991, Ozol 1994)
Electrochemical properties	Adhesion of bitumen to aggregate is related to the type of binder and type of stone. If aggregate and bitumen are improperly matched, the bitumen will strip from the aggregate. Bitumen is slightly negatively charged and adheres better to positively charged rocks such as basic igneous and metamorphic rocks, limestone, and dolomite. Cationic agents are added to bitumen when negatively-charged aggregates such as siliceous rocks are used. Many aggregates such as basalt, porphyries, and siliceous limestones have mixed charges (Hoiberg 1965)
Metallics	Some metal compounds, such as lead or zinc oxides, can seriously affect the setting rate of concrete. Some pyrite is able to oxidize, and together with creating an unsightly appearance, may cause expansion problems (Smith & Collis 1993)
Periclase	Periclase (magnesium oxide) hydrates in portland cement paste that causes increased volume. Periclase may not hydrate until long after the concrete has hardened and can no longer accommodate the stress induced by volume increase (Dolar-Mantuani 1983)
Sulfides	Pyrite, marcasite, and pyrrhotite are frequent accessory constituents of many potential sources of aggregate. If sufficient oxygen is available, all three minerals may oxidize and cause stains and loss of concrete strength. In addition, oxidation may generate soluble sulfate compounds that react with the cement matrix and cause volume increases and associated popouts or cracks (US Department of Interior 1981)
Sulfates	When present in sufficient quantities, or when in wet or damp locations, sulfates can react with cement compounds resulting in excessive expansion and ultimately disruption of hardened concrete. Gypsum ($CaSO_4\ 2H_2O$) is deleterious because it can affect the setting time of concrete. Magnesium and sodium sulfates are readily soluble, and therefore aggressively react with concrete. Calcium sulfate is less soluble, and is capable of slow, although progressive reactions
Zeolites	Natrolite and heulandite, two zeolites rich in sodium, can exchange the sodium for calcium from the cement paste, and increase the alkalis in the cement paste. Laumontite and leonhardite may undergo volume change during wetting and drying

Table 4b. The occurrence of alkali-silica reactive minerals and rocks.

Reactive material	Occurrence
Opal	Deposited at low temperatures from silica-bearing waters. Formed as deposits of thermal springs, in cracks and cavities of igneous rocks, in mineral veins, in hydrothermally altered rocks, and in some shales, sandstones, and carbonate rocks. May form entire rock mass, but more commonly occurs in accumulations in voids, fractures, or incrustations. Chert may contain opal, or any combinations of opal and other silica minerals. May form as coatings on gravels and sand, as a cement in sands, and as secondary weathered products of some rocks such as granite
Chalcedony	Cryptocrystalline to microcrystalline variety of silica with a fibrous structure that can be viewed under polarizing microscope. Chalcedony most commonly occurs in chert but also occurs in cherty limestone and dolostone, shale, phyllite, slate, and volcanic rock (Mielenz 1978)
Crypto crystaline to micro-crystalline quartz	Principal constituent of most varieties of chert. Chert most often occurs as nodules, lenses, or beds in limestone and dolomite, and less often as layers in a bedded deposit. It may also contain inclusions of disseminated clayey material, and calcite or dolomite remnants of the host rock Cryptocrystalline to microcrystalline quartz also fills veins and vugs in a variety of rock types, and may be a cement in some sedimentary rocks. Thus normally nonreactive rocks such as sandstone, basalt, granite, and other rock types may be reactive if coated or impregnated with opal, chalcedony, or cryptocrystalline to microcrystalline quartz
Tridymite and cristobalite	Minor constituents in shallow intrusive volcanic rocks with glassy or partially glassy groundmass. Rare in aggregate except where volcanic rocks are abundant. Rocks include cryptocrystalline rhyolite, latite, dacite, some andesite, and rocks of similar composition but with microcrystalline structure. Reactivity is affected by composition and texture of groundmass and is enhanced by large internal specific surface that develops when volatile constituents of rocks expand during eruption (Dolar-Mantuani 1983)
Volcanic glasses	Occur as rocks such as obsidian, perlite, or pumice, or as a portion of the groundmass of some volcanic rocks. Microstructure is very similar to opal. Very unstable and potentially reactive
Macro-crystalline quartz	Some macrocrystalline quartz is alkali-reactive, especially quartz that has a deformed crystal lattice from having been intensely fractured, strained, or shocked, quartz with surface irregularities, and quartz with pores and inclusions. These minerals tend to be slowly to very slowly reactive. The poorly ordered silica at the grain boundaries may be responsible for the reactive nature of strained quartz (Smith & Collis 1993, Dolar-Mantuani 1983). Most common in metamorphic rocks, but also occurs in some igneous rocks that have been subjected to high stresses. May occur as detrital material in clastic sediments. Undeformed macrocrystalline quartz generally is not expansively alkali-reactive (Dolar-Mantuani 1983)

Table 5. Weathering of bedrock or gravel clasts and its relevance to the performance of aggregate (modified after Fookes 1980).

Fresh	No visible sign of rock weathering	Aggregate properties not influenced by weathering. Mineral constituents are fresh and sound
Faintly weathered	Discoloration on major discontinuity surfaces	Aggregate properties not significantly influenced by weathered minerals. Mineral constituents sound
Slightly weathered	Discoloration indicates weathering of rock and discontinuity surfaces. All rock material may be discolored by weathering and may be somewhat weaker than fresh rock	Aggregate properties may be significantly influenced by weathered minerals. Strength and abrasion characteristics may be weakened. Some altered mineral constituents and microcracks
Moderately weathered	Less than half the rock is decomposed and/or disintegrated to a soil. Fresh or discolored rock is present as a continuous framework or as corestones	Aggregate properties significantly influenced by weathered minerals. Soundness characteristics markedly affected. Altered mineral constituents common; many microcracks
Highly weathered	More than half the rock is decomposed and/or disintegrated to a soil. Fresh or discolored rock is present as a continuous framework or as corestones	Generally not suitable for aggregate
Completely weathered	All rock is decomposed and/or disintegrated to a soil. The original mass structure is still largely in tact	Not suitable for aggregate
Residual soil	All rock material is converted to soil. The mass structure and material fabric are destroyed. There is a large change in volume, but the soil still has not be significantly transported	Not suitable for aggregate

3.2.1 *Sand and gravel deposits*

Regional studies that provide general information regarding the quality of aggregate should include:
- Predominant lithologies of clasts,
- Particle-size distribution (texture),
- Color,
- Weathering,
- Presence of deleterious constituents, including chemical reactants.

Particle-size distribution of sand and gravel deposits can be quite variable, both laterally and in the third dimension. To assist characterizing particle-size distribution, it is important to base the description on how a deposit might be mined. For example, to reduce processing costs, a relatively thick layer of gravel overlying a thick sand deposit would probably be excavated separately and, consequently, the gravel and sand units should be described separately. A coarse-grained esker in a sandy outwash plain would also be mined separately, and therefore should be described separately from the outwash plain. In contrast if a deposit was characterized by interfingering layers of sand and gravel or pockets or lenses of contrasting grain sizes, either laterally or in the third dimension, the operator would mine the materials collectively and process them later. This type of deposit can be described as a single unit, although the characteristics of the particles in the pockets and lenses should also be characterized. Detailed information about the quality of the gravel can be included on a map or in a report for preliminary site studies. Most quality information can be determined by careful visual examination, scratch and acid tests, and hitting the sample with a hammer.

Descriptions for gravels at individual sites should include:
– Predominant lithologies,
– Particle-size distribution,
– Particle shape,
– Particle surface texture,
– Internal structure of clasts, including observations of pore space and fractures,
– Grain packing, degree of cementation, and type of cement,
– Mineral composition,
– Hydrothermal alteration,
– Weathering,
– Coatings or incrustations,
– Presence of deleterious constituents, including chemical reactants.

Identification and percentages of gravel clasts should be made. Sample size of each size fraction should comprise at least 150 clasts (Mielenz 1994). Each lithologic type should be listed as a percentage of the total pebble count. Using criteria described in Table 6, the physical quality of aggregate can be characterized as satisfac-

Table 6. Terms used to describe relative degrees of physical and chemical quality of natural aggregate (from Woods et al. 1960).

Term	Definition
Physical quality	
Satisfactory	Particles are hard to firm, relatively free from fractures, and not chip like; capillary absorption is very small or absent; and the surface texture is relatively rough
Fair	Particles exhibit one or two of the following qualities: Firm to friable; moderately fractured; capillary absorption small to moderate; flat or chip like; surface relatively smooth and impermeable; very low compressibility; coefficient of thermal expansion approaching zero or being negative in one or more directions
Poor	Particles exhibit one or more of the following qualities: Friable to pulverant; slake when wetted and dried; highly fractured; capillary absorption moderate to high; marked volume change with wetting and drying; combine three or more qualities under 'fair'
Chemical quality	
Innocuous	Particles contain no constituents which dissolve or react chemically to a significant extent with constituents of the atmosphere, water, or hydrating portland cement while enclosed in concrete or mortar under ordinary conditions
Deleterious	Particles contain one or more constituents in significant proportion which are known to react chemically under conditions ordinarily prevailing in portland cement concrete or mortar in such a manner as to produce significant volume change, interfere with the normal course of hydration of portland cement, or supply substances which might produce harmful effects upon concrete or mortar

Table 7. Results of hypothetical pebble count and evaluation of a potential gravel aggregate source.

Constituents	Amount as percent of particles in fractions indicates		Degrees of quality	
	1 1/2-3/8 in.	3/8-3/16 in	Physical	Chemical
Granite and granitic gneiss	65.7	78.0	Satisfactory	Innocuous
Weathered granite and granitic gneiss	11.8	10.8	Fair	Innocuous
Deeply weathered granite and granitic gneiss	5.4	3.3	Poor	Innocuous
Rhyolite	0.7	0.6	Satisfactory	Deleterious*
Dacite porphyry	0.6	0.2	Satisfactory	Innocuous
Basalt	0.7	0.3	Satisfactory	Innocuous
Pumicite	0.3	–	Fair	Deleterious*
Biotite and sillimanite schist	0.8	–	Satisfactory	Innocuous
Quartz and quartzite	8.9	4.7	Satisfacatory	Innocuous
Sandstone	2.8	1.5	Satisfactory	Innocuous
Weathered sandstone	0.3	–	Fair	Innocuous
Hard siltstone	1.3	0.2	Satisfactory	Innocuous
Porous ferruginous siltstone	–	0.1	Poor	Innocuous
Chalcedonic chert	0.6	0.3	Satisfactory	Deleterious*
Fissile shale	0.1	–	Poor	Innocuous

*Deleterious with high-alkali cements.

tory, fair, or poor; chemical quality can be characterized as innocuous or deleterious. An example of a hypothetical pebble count is shown in Table 7. The results of geologically characterizing a potential aggregate source can then be used to estimate or predict general engineering properties which include physical soundness, hardness, strength, and toughness (Table 8).

3.2.2 *Potential sources of crushed stone*
Usually the performance of a potential source of crushed stone can be judged by considering a few elementary features. Descriptions should include:
- Predominant lithologies,
- Color,
- Layering characteristics (schistosity, bedding, or banding),
- Location and spacing of fractures and parting planes,
- Hydrothermal alteration,
- Weathering,
- Presence of deleterious constituents, including chemical reactants.

Descriptions of rock and core samples for site studies should contain information from examination of hand specimens, polished sections, thin sections, or core samples, and should include:
- Predominant lithologies, and their percentage of the total rock mass,
- Grain (crystal) size, texture, and variation,
- Layering characteristics (schistosity, bedding, or banding),
- Location and spacing of fractures and parting planes,
- Alteration,

Table 8. Terms used by aggregate industry to describe the engineering properties of natural aggregate.

Physical soundness: The ability of an aggregate to resist weathering, particularly freezing-thawing and wetting-drying cycles	Generally aggregates that contain weak, cleavable, absorptive, or swelling particles are not suitably sound. Examples are shales, sandstones, limestones, clayey rocks, some very coarse crystalline rocks, and porous cherts (Gillott 1980, Neville 1973). Weathered rock types such as weathered igneous rocks where secondary clay minerals are produced, can also be unsound (Fookes 1980). However, because the physical properties of the rocks, not their composition, controls frost susceptibility, not all these types of rocks have durability problems. The most important physical property of rock particles affecting weathering resistance (particularly freezing-thawing) is the size, abundance, and continuity of pores, channels, and fractures (McLaughlin et al. 1960). These provide conduits for the passage of water, which in turn accelerate the weathering process. It is generally accepted that there is an approximate correlation between quality (soundness) and rock porosity. Specifications for soundness are similar for aggregates to be used in concrete or bituminous mixes
Hardness, strength and toughness: Hardness (resistance to load), strength (resistance to abrasion), and toughness (resistance to impact) of aggregates determine their ability to resist mechanical breakdown	Hardness, strength, and toughness are generally controlled by the individual mineral constituents of rock particles, the strength with which these minerals are locked or cemented together, and the abundance of fractures. Particles consisting of minerals with a low degree of hardness are considered to be soft; those which are easily broken down, due to weak bonding or cementation or to fracturing, are considered to be weak (McLaughlin et al. 1960). Soft or weak particles are deleterious in aggregates because they perform poorly in use and because they break down during handling, thus affecting the grading of the aggregates. Mechanical breakdown of aggregates due to the action of mixers, mechanical equipment, and (or) traffic, or breakdown due to weathering is referred to as aggregate degradation. Degradation can occur due to compressive failure of grains at points of contact, as well as to abrasive action of grains on each other. Degradation generally is of greater significance in bituminous pavements than in concrete pavements. A good average aggregate has a crushing strength several times greater than that of the concrete (Fookes 1980)

– Weathering,
– Porosity,
– Presence of deleterious constituents, including chemical reactants.

Descriptions of core samples should also include length of core recovered; amount of core loss and location; and type or types of breakage. Descriptions should include the relative degrees of physical and chemical quality (Table 6), which requires subjective estimates of what the rock will look like if crushed to gravel-size clasts.

3.3 *Quantity of potential aggregate sources*

Calculating the reserves of a sand and gravel deposit or a source of crushed stone involves determining the three-dimensional extent (volume) and variability (yield of usable material) of the deposit. For most regional assessments, the quantity estimates of aggregate resources should be considered inferred reserves. Inferred reserves are quantitative estimates based largely on a generalized knowledge of the geologic character of the deposits with few, if any, samples (Blondel & Lasky 1956). The estimates commonly are based on assumed continuity of materials made by comparing the geologic characteristics of the deposit against other similar, but better understood, deposits.

Maps of potential sources of granular aggregate should delineate the areal extent of the sand and gravel deposit, and the general distribution of material textures, both laterally and in the third dimension. Texture of surface materials can be shown as map units, overprints or colors, numbered codes, or as descriptive text. Texture distribution of subsurface materials can be shown as stack-map units (Kempton & Cartwright 1984), descriptive text, or as logs accompanying the map.

Regional studies which compile data in short periods of time and with limited field reconnaissance may not be useful in estimating the amount of potential sand and gravel aggregate resources present. Aggregate thickness may be included as part of descriptive texts or logs accompanying geologic maps, although detailed descriptions of potential deposits may be lacking from the text or, at best, limited to a few specific sites. When such limited data exist locally, better understood 'typical' deposits from other areas can be used to predict overall geometry and the lateral and third dimensional variability of texture of the local deposit (Bliss 1993).

Maps of potential sources of crushed stone aggregate should delineate the areal extent of lithologic units and, where possible, unit thickness. The structural attitude of layered units should be noted so that the volume of the units can be calculated from the map data. Regional studies should include general information regarding the quality of potential aggregate.

For site evaluations, estimating the volume of a sand and gravel or bedrock deposit is a complex process that involves determining the configuration of the deposit. Outcrop observations, drilling, coring, augering, and trenching are routine methods for determining the thickness of the deposit and overburden (cover), as well as the variation in the distribution of the deposit laterally and in the third dimension. These values are required for accurate reserve estimates. Visual estimates of particle-size classes separated by on-site sieving of auger and outcrop samples is a convenient method (Moore 1995). Seismic, ground-based resistivity, ground-penetrating radar, and electromagnetic measurements are especially useful supplements for determining the lateral and vertical distribution of the deposit between drill holes, pits, and trenches (Odum & Miller 1988, Dunn 1991). Detailed geologic cross sections and isopach maps derived from the various data provide the primary information necessary for calculating reserves. For site maps aggregate thickness commonly is shown with isopachs or as logs accompanying the map. Dunn (1991) describes the geologic information required to calculate reserves, the methods for calculating various types of reserve information, and the critical nature of the reserve information in evaluating the economic risk of developing the aggregate.

APPENDIX

ASTM Tests – Common laboratory tests to determine physical, chemical, and engineering properties of aggregate.

ASTM designation	Description
C 294 Standard descriptive nomenclature for constituents of natural mineral aggregates	This nomenclature briefly describes some of the more common, or more important, natural materials of which natural aggregate are composed
C 295 Petrographic examination of aggregates for concrete	This guide outlines procedures for petrographic examination of gravel, crushed stone, rock core, and bedrock outcrops
C 136 Sieve analysis of fine and coarse aggregates	A weighed sample of aggregate is separated through a series of sieves of progressively smaller openings for determination of particle-size distribution
C 123 Lightweight pieces in aggregate	A heavy liquid is used to separate light-weight particles from the aggregate
C 127 Specific gravity and adsorption of coarse aggregate	A sample of aggregate is immersed in water to fill the pores, the water dried from the surface of the particles, and the sample weighed. The sample is then weighted while submerged in water. Finally the sample is oven dried and weighed
C 289 Potential alkali-silica reactivity of aggregates (chemical method)	Chemical determination of the potential reaction of an aggregate with alkalis in portland cement concrete
C 227 Potential alkali reactivity of cement-aggregate combinations (mortar-bar method)	Aggregate is mixed with cement and water to form a bar. The bar is stored under specific moisture and temperature conditions and is measured at specific intervals over a period of months and years
C 1260 Potential alkali reactivity of aggregates (mortar-bar method)	Aggregate is mixed with cement and water to form a bar. The bar is stored immersed in a solution of sodium hydroxide (NaOH) at a controlled temperature. The bar is measured after 14 days, and specific intervals thereafter
C586 Potential alkali reactivity of carbonate rocks for concrete aggregates (rock cylinder method)	Determines the expansive characteristics of carbonate rocks while immersed in a solution of sodium hydroxide (NaOH)
C 131 Resistance to degradation of small-size coarse aggregate by abrasion and impact in the Los Angeles Machine C 535 Resistance to degradation of large-size coarse aggregate by abrasion and impact in the Los Angeles Machine	The Los Angeles test is a measure of degradation of mineral aggregates of standard gradings resulting from a combination of actions including abrasion or attrition, impact, and grinding. The sample is rotated in a steel drum containing a specified number of steel spheres. As the drum rotates, a shelf picks up the aggregate and the spheres and drops them to the opposite side of the drum, creating an impact-crushing effect. After the prescribed number of revolutions, the contents are sieved to measure degradation as a percent loss
C 88 Soundness of aggregates by use of sodium sulfate or magnesium sulfate	Test involves alternate cycles of immersion in saturated solutions of sodium or magnesium sulfate and drying to precipitate salt in permeable pore spaces. The internal force derived from hydration of salt simulates the expansion of water on freezing

REFERENCES

American Society for Testing and Materials 1994. Annual book of ASTM standards, Section 4, Volume 04.02, Concrete and mineral aggregates: Philadelphia, Pa., 878 pp.

Barksdale, R.D. (ed.) 1991. *The aggregate handbook*. National Stone Association, 16 chapters numbered separately.

Bliss, J.D. 1993. Modeling sand and gravel deposits – initial strategy and preliminary examples: US Geological Survey Open-File Report 93-200, 31p.

Blondel, F. & Lasky, S.G. 1956. Mineral reserves and mineral resources. *Economic Geology*: 686-697.

Bottge, R.G., McDivitt, J.F. & McCarl, H.N. 1965. Prospecting for natural aggregates, Part 2. *Rock Products* 68(6): 93-95.

Dolar-Mantuani, L. 1983. *Handbook of concrete aggregates, a petrographic and technological evaluation*. Park Ridge, N.J., Noyes Publications, 345 pp.

Dunn, J.R. 1991. Chapter 4 to Geology and Exploration. In R.D. Barksdale (ed.), *The Aggregate Handbook*: 4-2 to 4-45. Washington, D.C., National Stone Association.

Fookes, P.G. 1980. An introduction to the influence of natural aggregates on the performance and durability of concrete. *The Quarterly Journal of Engineering Geology* 13(4): 207-229.

Fookes, P.G. 1991. Geomaterials. *The Quarterly Journal of Engineering Geology* 24(1) 3-16.

Gillott, J.E. 1980. Properties of aggregates affecting concrete in North America. *The Quarterly Journal of Engineering Geology* 13(4): 289-303.

Hadley, D.W. 1961. *Alkali reactivity of carbonate rocks*. Skokie, Ill., Portland Cement Association, Resource Department Bulletin 139, 14 pp.

Hoiberg, A.J. (ed.) 1965. *Bituminous Materials: Asphalts, Tars, and Pitches*, Volume II: Asphalts, John Wiley & Sons, New York, New York.

Kempton, J.P. & Cartwright, K. 1984. Three-dimensional geologic mapping: A basis for hydrogeologic and land-use evaluations. In J.R. Keaton (ed.), Engineering geology mapping symposium. *Bulletin of the Association of Engineering Geologists* 21(3): 317-336.

Langer, W.H. 1988. *Natural aggregates of the conterminous United States*. US Geological Survey Bulletin 1594, 33 pp.

Marek, C.R. 1991. Chapter 3 – Basic Properties of Aggregate. In R.D. Barksdale (ed.), *The Aggregate Handbook*: 3-1 to 3-81. Washington, D.C., National Stone Association.

Mather, K. & Mather, B. 1991. Aggregates. In G.A. Kiersch (ed.), *The Heritage of Engineering Geology: The first hundred years*. 323-332. Centennial Special Volume 3. Geological Society of America, Boulder, Colorado.

McLaughlin, J.P., Woods, K.B., Mielenz, R.C. & Rockwood, N.C. 1960. Distribution, production, and engineering characteristics of aggregates, section 16, In K.B. Woods, D.S. Berry & W.H. Goetz (eds), *Highway engineering handbook*. 16-1 to 16-53. New York, McGraw-Hill Book Company, Inc.

Mielenz, R.C. 1978. Petrographic examination (Mineral aggregates): ASTM STP 169B-1978, pp. 539-572.

Mielenz, R.C. 1994. Petrographic evaluation of concrete aggregates. In *American Society of Testing and Materials*, Concrete and concrete-making materials: ASTM STP 169C, pp. 341-364.

Moore, D.M. 1995. Sand and gravel assessment, Nambe Pueblo, northern New Mexico, US Geological Survey administrative report, 39 pp.

Neville, A.M. 1973. *Properties of concrete*. New York, John Wiley and Sons, 686 pp.

Odum, J.K. & Miller, C.H. 1988. Geomorphic, seismic, and geotechnical evaluation of sand and gravel deposits in the Sheridan, Wyoming, area: US Geological Survey Bulletin 1845, 32 pp.

Ozol, M.A. 1994. Alkali-carbonate rock reaction, ASTM STP 169C, pp. 372-387.

Ray, R.G. 1960. Aerial photographs in geologic interpretation and mapping. US Geological Survey Professional Paper 373, 230 pp.

Rexford, E.P. 1950. Concrete aggregates for large structures. *Transactions, American Institute of Mining Engineering*. 187: 395-402.

Schwochow, S.D., Shroba, R.R. & Wicklein, P.C. 1974. Sand, gravel, and quarry aggregate resources, Colorado Front Range counties: Colorado Geological Survey, Department of Natural Resources Special Publication 5-A, 43 pp.

Smith, M.R. & Collis, L. (eds) 1993. *Aggregates – Sand, gravel and crushed rock aggregates for construction purposes*: London, Geological Society Engineering Geology Special Publication No. 9, The Geological Society, 339 pp.

Swenson, E.G. & Chaly, V. 1956. Basis for classifying deleterious characteristics of concrete aggregate materials. *Journal of the American Concrete Institute* 27(9): 987-1002.

Tepordei, V.V. 1993. Crushed stone – 1991: US Bureau of Mines Annual Report, 43 pp.

Timmons, B.J. 1994. Prospecting for natural aggregates; An update. *Rock Products* 97(8): 43-45.

US Department of Interior. 1981. Concrete manual. Water and Power Resources Service Technical Publication, 627 pp.

Varnes, D.J. 1974. The logic of geologic maps, with reference to their interpretation and use for engineering purposes. US Geological Survey Professional Paper 837, 48 pp.

Woods, K.B. & Lovell, C.W., Jr. 1960. Distribution of soils in North America, section 9. In K.B. Woods, D.S. Berry & W.H. Goetz (eds), *Highway engineering handbook*: New York, McGraw-Hill Book Company, Inc.

Ground Penetrating Radar: Applications in sand and gravel aggregate exploration

H.M. JOL
Department of Geography, University of Wisconsin-Eau Claire, Wisconsin, USA

D. PARRY
Associated Mining Consultants Ltd., Calgary, Alberta, Canada

D.G. SMITH
Department of Geography, University of Calgary, Calgary, Alberta, Canada

1 INTRODUCTION

Aggregate – sand, gravel, crushed stone, and other materials – is an essential ingredient to a nation's infrastructure. It is used in nearly all construction projects including residential, commercial and public works (e.g. transportation networks). With the demand for aggregate likely to increase in the foreseeable future, two keys to the aggregate industry's ability to maintain its strength involve long-range planning for replacement quarries, and education of the public (Nelson 1995).

Natural aggregate consists of rock fragments that are used in their natural state or are used after mechanical processing such as crushing, washing, and sizing. Sources of aggregate are: Products of erosion of bedrock and surficial material, and subsequent transport, abrasion, and deposition of the particles. The principal geologic agent that affects the distribution of sand and gravel deposits is water. Consequently, sand and gravel is widely distributed and abundant in glaciated areas, in alluvial basins, and in, adjacent to, or near rivers and streams (Edwards 1989, Langer & Glanzman 1993).

Each region has a demand for aggregate, but because it is a high-bulk, low-cost commodity, the transportation cost to the site of use is a significant part of the total aggregate cost. Therefore, natural aggregate is commonly used within 50 to 100 km of the place of extraction. Ultimately, the parameters defining the supply area are geology, ownership of the land, zoning or other land-use restrictions, socio-economic considerations, and transportation routes. Unless the distribution, availability, and quality of aggregate are known, it is difficult to formulate a reasonable plan to set aside or develop aggregate. Therefore, in planning for future aggregate needs, it is necessary to determine the reserves for the planning area. After preliminary investigations, detailed exploration of an identified source of sand and gravel should occur.

One geophysical method that is being increasingly applied to sand and gravel exploration is ground penetrating radar (GPR). Developments in GPR technology have made a near continuous subsurface profiling system commercially available; it features high resolution (decimeter-meter scale) at shallow depths (0-50 m), combined with the practical features of being digital, highly portable, robust, easy to operate and maintain, and economical. Interpretations derived from the radar data, coupled with stratigraphic logs from exposures, can aid in understanding the complex nature of depositional environments associated with aggregate deposits.

In general, GPR has been used in a wide variety of environmental and earth-science applications. There is, however, little previous research on the application of GPR to characterize aggregate deposits. Coarser sediments (sands and gravels) have been investigated by Ulriksen (1982), Davis & Annan (1989), Jol & Smith (1991, 1992), Fitzgerald et al. (1992), Smith & Jol (1992a, b), Huggenberger (1993) and Jol (1993). From this research of geological features with coarser sediments it became apparent that GPR has good potential for use in high energy depositional systems (such as river/fluvial, deltaic and coastal locations) as well as thicker sequences of coarse-grained sediments; environments which are typically the location of aggregate deposits. The good potential for successful use is due to the low electrical conductivity (resistive nature) of the sediments being examined which, as a consequence, reduces the attenuation rates of the GPR electromagnetic signal.

The main objective of this project is to demonstrate, through examples, that GPR can be used to map the depth and lateral extent of sand and gravel deposits. This will allow the aggregate industry to determine the vertical and horizontal limit of an aggregate deposit, thus accurately assessing the volume of such deposits.

2 METHOD

GPR is an electromagnetic (EM) measuring device used to examine sediment of low electrical conductivity (resistive environments). This EM sounding technique uses a transmitting antenna to introduce a high frequency EM pulse to the earth. In a fashion analogous to seismic reflection, a portion of the signal is reflected back from subsurface sediment boundaries. The amplitude of the reflected signal is largely a function of the magnitude of contrasts in dielectric properties which are in turn dependent on lithology or grain size. The returned energy is received by an antennae and is recorded as a function of time. The signal attenuation is proportional to electrical conductivity, and so the technique is of little value in regions of high electrical conductivity – particularly ones containing substantial amounts of mineralogical clays. Further GPR background can be acquired from Ulriksen (1982), Davis & Annan (1989) and Jol (1993).

The GPR acquisition system used was a Sensors & Software pulseEKKO™ IV from the University of Calgary. This system allows the operator to obtain digital recordings of time series traces for radar amplitude at discrete locations within the survey area. The approach taken is usually to obtain a sequence of such traces at some standard increment and observe the juxtaposition of these traces as a cross section of the dielectric structure of the earth in time. The data collected used the 12.5, 25, 50, 100 or 200 MHz antennae with a constant source-receiver offset. Frequency is an important design consideration for any GPR survey. Higher frequency antennae will provide finer resolution but will not allow penetration of the signal to depths as great as with lower frequencies. As well, continuity of reflections and depth of penetration can be aided by increasing the transmitter voltage (Jol 1995, Smith & Jol 1995). Profiles were collected with either a 400 V or 1000 V transmitter and each trace was vertically stacked with a sampling rate of 800 ps. To provide detailed horizontal resolution of the sedimentologic structures one meter station spacings were used for most surveys. The profiles were processed and plotted (wiggle trace/variable area format) using pulseEKKO™ IV software (Version 3.1). It is possible to evaluate the velocity of the radar pulse within the earth by calculating the near surface velocity measurements from common mid-point (CMP) surveys. This allows a conversion of the time section to a depth section.

It should be noted that all profiles have some degree of vertical exaggeration. The two uppermost continuous reflections in all profiles represent air and ground wave arrivals respectively, and should not be considered part of the data for stratigraphic interpretation as the lower and latter arriving reflections represent subsurface interfaces.

3 GPR EXAMPLES

Since the major geologic agent that affects the distribution of sand and gravel is water, many of the resulting deposits occur in river (fluvial) and deltaic environments. This section will look at a variety of former high energy depositional settings which are active or potentially active sand and gravel aggregate mines.

3.1 *Braid river/delta deposits*

The term braid delta can be described as laterally extensive, sheet-like sand and gravel bodies dominated by trough and planar, tabular cross-bedding underlain by la-

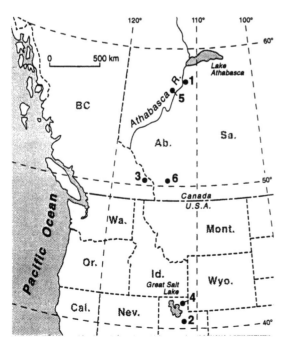

Figure 1. Location map of the study sites in western North America where GPR field experiments were carried out. 1. Embarras Airfield, Alberta, 2. American Fork, Utah, 3. Golden, B.C., 4. Brigham delta, Utah, 5. Ft. McMurray region, Alberta, and 6. Calgary, Alberta.

custrine mud or bedrock (Smith 1991). These features are a result of a braided river depositing sediments into a shallow body of water.

A GPR survey was carried out on the late Pleistocene, Athabasca River delta plain at Embarras airfield in northeastern Alberta (Fig. 1, Site 1). The sandy braided river deposits in this area are part of a vast 4000 km^2 deltaic braid plain deposited into glacial Lake McConnell at about 9900 years BP (Rhine & Smith 1988). At Embarras airfield, there is up to 14 m of trough cross-stratified sand. Between the sand and mud facies is a transitional zone of interbedded sand and mud.

The GPR survey at Embarras airfield extended for 1280 m in a west-northwest direction along the airfield runway; a portion of this profile is shown in Figure 2. The survey used 50 MHz antennae with a spacing of 2 m and a step of one meter. The depth of penetration is approximately 16 m (based on a near-surface velocity of 0.13 m/ns). Figure 2 shows an upper, 11-13 m thick unit with distinct, continuous and semi-continuous, wavy reflections interpreted as braided river deposition. The prominent continuous reflection at 11 m deep, below which little signal energy is returned, is interpreted as the sand-mud interface and water table. Below the lower-most continuous reflection little signal energy was returned due to energy attenuation caused by the lacustrine mud. The fact that fluvial sand and gravel, in such fluvial/deltaic settings, overlie lacustrine mud or bedrock aids in assessing the depth of sand and gravel deposits.

A second example is from the late Pleistocene American Fork River delta which deposited into former Lake Bonneville (Bonneville level) near Provo, Utah State (Gilbert 1890, Fig. 1, Site 2). The internal structure of the delta is exposed in an active sand and gravel pit on the north side of the river approximately 1 km west of the

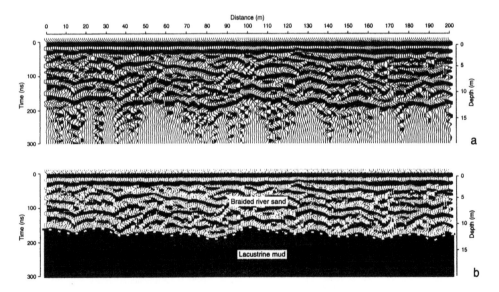

Figure 2. a) Radar profile along the Embarras airfield, northeastern Alberta. The profile shows the radar stratigraphy of a late Pleistocene braid delta. The lowest horizontal, continuous reflection delineates the contact between fluvial sand and lacustrine mud, b) Interpreted line. (after Jol & Smith 1991).

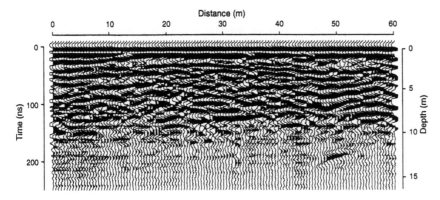

Figure 3. GPR profile of the late Pleistocene American Fork braid delta at the Bonneville level of Lake Bonneville, near Provo, Utah (after Jol & Smith 1992a).

Wasatch mountain front. A 60 m thick gravel facies occurs beneath an excavated bench 12 m below the surface. Horizontally-stratified, moderately sorted gravel is exposed in the pit highwall.

A GPR profile was run 60 m south along an excavated bench. The survey used 100 MHz antennae with one meter spacing and a one half meter step. The maximum depth of penetration is approximately 13 m based on a near surface velocity of 0.13 m/ns. Figure 3 shows distinct, continuous-to-semi-continuous, wavy reflections above 11 m with signal loss below 13 m. These reflections above 11 m are interpreted as braided river deposits in a deltaic environment. The thickness and style of deposition suggests a braided river deltaic plain extended over a relatively large area. Signal loss below 11-13 m is attributed to increased silt content, including intercalated lacustrine silt beds within the gravel. This signal loss would indicate that sand and gravel deposits would not be very clean and thus may need further aggregate processing.

3.2 *Fan-foreset delta deposits*

Fan-foreset deltas are dominated by coarse-grained, steeply-inclined layers. The prominence of the steeply inclined strata, often at 25°, implies a high energy depositional environment. Most of the sand and gravel sediment is stored in the dipping foreset facies (Smith 1991).

A GPR survey was carried out at a gravel pit excavated into the late Pleistocene Kicking Horse River delta at Golden, British Columbia (Fig. 1, Site 3). The deltaic exposure is located on the north bank of the Kicking Horse River, adjacent to the Canadian Pacific railroad tracks. The delta prograded into former glacial Lake Invermere, a short-lived proglacial lake in the Rocky Mountain Trench (Sawicki & Smith 1992).

A lithostratigraphic survey of the delta reveals three classical deltaic facies: bottomsets, foresets and topsets. The bottomset facies consists of clean stratified sand resting on lake-bed mud. Thick foreset facies contain inclined gravel beds dipping at 25°. The topset facies consists of horizontally stratified boulder-to-cobble gravel which is partially draped by a silt package (Sawicki & Smith 1992). Figure 4 shows a topographically corrected GPR profile of the late Pleistocene Kicking Horse River delta. The survey used 50 MHz antennae with a separation (source to receiver) of 1½ m and a step (station spacing) of 1 m and 400 V transmitter. The profile extends 200 m eastward along a gravel pit road cut into the dip slope of the delta. Depth of radar penetration is approximately 30 m, based on a near surface velocity of 0.10 m/ns.

Nearly parallel, horizontal reflections are present in the lower left portion of the profile. These are overlain by up to 22 m of sediments characterized by continuous, well defined, steeply-inclined reflections, dipping approximately 25° towards the west. The basal horizontal reflections are interpreted as sandy bottomset facies. The steeply dipping reflections represent the gravely foreset facies. The topset facies is not present between 0-150 m (horizontal scale) on the profile due to gravel excavation, but does occur between 150-200 m at a depth of 3-5 m. The mud drape (overburden) is recognizable by the increased separation between the air and ground waves which indicates a decrease in near-surface velocity caused by the increase in clay content of the sediments. The mud drape also caused energy attenuation which reduced penetration at the east end of the profile. Two results from this study that aided the gravel pit operator were: 1. An indication of the overburden thickness, and 2. The location of coarser aggregate.

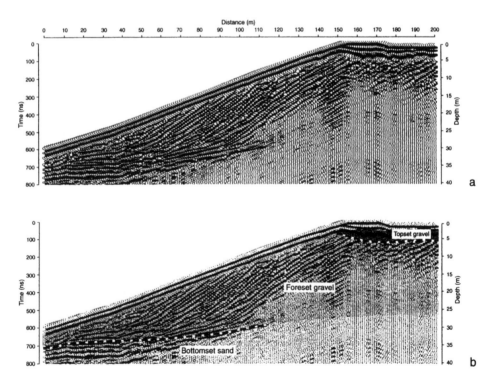

Figure 4. a) GPR profile from a gravel pit northwest of Kicking Horse River and the Canadian Pacific railroad tracks, Golden, British Columbia. The profile is oriented in the dip direction of foreset beds of the late Pleistocene Kicking Horse fan-delta, b) Interpreted line (after Jol & Smith 1991).

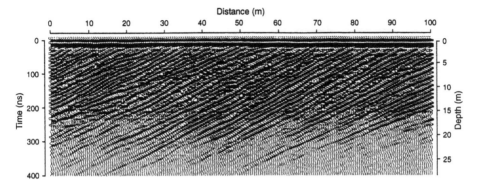

Figure 5. GPR profile of the late Pleistocene fan-foreset Brigham delta (Provo level) at Brigham city, Utah, USA (after Jol & Smith 1992).

A second example is from the Pleistocene Box Elder Creek delta (Brigham delta; Provo level) at Brigham City, Utah, which prograded into former Lake Bonneville (Fig. 1, Site 4). Part of the delta is currently exposed in a 70 m gravel pit highwall on the east side of Brigham City. Well records indicate an additional 50 m of silica-rich gravel lying beneath the base of the pit (G.A. Wilkes, pers. comm.).

A GPR profile, using the 100 MHz antennae with a 1 m separation and a one half meter step provided a compromise between resolution and depth of penetration of the dipping strata (Fig. 5). The survey was conducted across the pit floor approximately parallel to the dip of the exposed strata in the highwall. The profile shows continuous, inclined reflections having dip angles of 20-25° to a depth of 28 m. The profile across the pit floor indicates at least another 28 m of gravel deposit. The depth of the deposit and inclined strata indicates a possible continuation of the foreset succession observed in the highwall. The expected topset facies near the upper portion of the delta has been excavated. As well, the basal portion of the deposit was not recorded by this GPR survey. Subsequent surveys using lower frequency antennae (25 MHz) have penetrated to bottomsets, identified on the basis of reduced angle of reflections, at 45-50 m beneath the pit floor (Smith & Jol 1995). This information will aid in the further development of the sand and gravel pit.

3.3 *Catastrophic flood deposits*

As sand and gravel becomes harder to find in some localities, other types of deposits have to be explored. In the Fort McMurray region, Alberta, Fisher et al. (1995) collected and described numerous GPR profiles from catastrophic flood deposits (Fig. 1, Site 5). By keeping the antennae step and separation to a meter or less, detailed sedimentological data was gathered. A comparison between the radar and lithostratigraphic data acquired parallel to highwall exposures was made by Fisher et al. (1995). The change in grain size, often from cobble/boulder to pebbles, was sufficient to affect the radar wave. Sedimentary structures represented within the lithostratigraphic logs, and/or photographs were directly compared to wiggle traces on the GPR transect; thus it was possible to locate, map and assess the deposits.

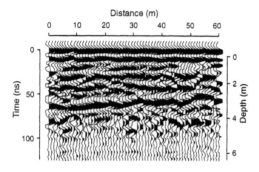

Figure 6. GPR profile of the Poplar Lake pit, Ft. McMurray, Alberta (after Fisher et al. 1995).

The GPR profiles were interpreted to represent high energy sand and gravel that was deposited rapidly (flood deposits). This interpretation was based upon both the sedimentology and radar reflection patterns. Energy attenuation within all transects showed a signal loss usually occurring within 5 m of the surface (Fig. 6). At some locations a strong continuous reflection was found at the base (Fig. 6). Where exposures were available, McMurray formation bedrock, silt, and clay-rich till, or glacial Lake McMurray sediment, was underlying the gravel. The juxtaposition of relatively transparent (resistive) sand and gravel sediment overlying an energy attenuating (conductive) material is an ideal environment in which to determine aggregate thickness. A grid network of GPR profiles would provide a 3-D evaluation of the lateral and vertical extent of a deposit and thus its volume.

3.4   *Calgary gravel pit – Case study*

A series of geophysical surveys were undertaken at a site south of the city of Calgary, Alberta (Fig. 1, Site 6). The purpose of these surveys was to map the thickness of a known aggregate resource and, if possible, the presence and thickness of overburden. Gravel removal is presently underway at the site, and operations have been ongoing for some time. The gravel deposit mined to date is on the order of 5 m thick, with overburden thickness of about 1 m. The operators of the pit were interested in determining the extent of mineable reserve left in situ with special consideration given to overburden thickness. The present operations and preliminary test pit analysis suggest that overburden is thickening appreciably as the pit expands into adjacent lease holdings.

The survey site lies in the floodplain where gravel deposits blanket the shales and sandstones of the local bedrock and are in turn overlain by glacial tills. There is a strong electrical conductivity contrast between the coarser grained deposits here and the overlying tills and underlying (predominant) shales. Given the extremely shallow section of interest, seismic methods were not considered. Gravity methods were dismissed as expensive and unlikely to discriminate well among the materials at this locale.

A variety of geophysical techniques measure the variation of electrical properties of the earth. It was not immediately clear which of the methods would be most suitable at the survey site, so a series of tests were undertaken. Three geophysical methods which measure variations in electrical properties of the earth were tested at the site: fixed frequency electromagnetics (FEM), direct current resistivity (DC) sound-ings, and ground penetrating radar (GPR). The tests were conducted along three lines. Both the top and the base of aggregate were observed on the GPR data sec-tions. A GPR velocity estimation was obtained locally, enabling depth conversion of the time sections. It was determined that GPR would be the most time efficient, cost effective method that could provide the necessary vertical and horizontal resolution to fulfill the survey objectives. Given its excellent lateral resolution, GPR was cho-sen for application to the whole survey area.

Seventeen lines of GPR data were collected using 50 MHz antennae spaced at 2 m apart. The resulting time sections were interpreted for top and base of gravel, incor-porating test pit data correlations where possible. The test pits were used as a further check against the time to depth conversions. The top and base of aggregate inter-preted from GPR sections were digitized and used to generate isopach maps of over-burden and of aggregate. Interpreted overburden varies in thickness from 0.3-3.2 m and thickens slightly to the southeast (Fig. 7). Interpreted aggregate thickness ranges from 0.5-4.4 m with a north-south trending thin just to the west of the present ex-traction location (Fig. 8). Thickness increases again to the west of survey station 1400 east.

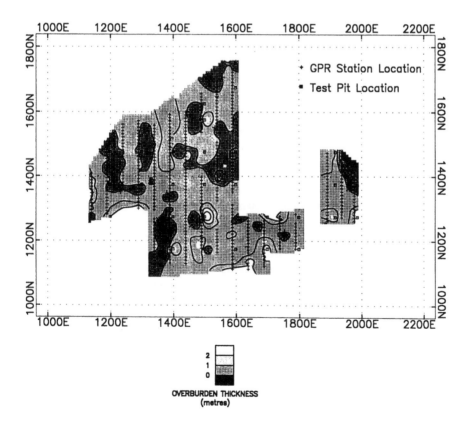

Figure 7. Isopach map showing overburden thickness from GPR surveys and test pit results.

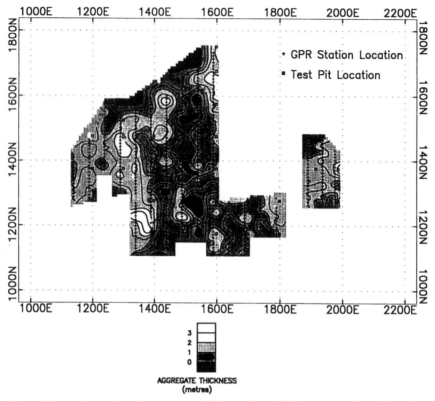

Figure 8. Isopach map showing aggregate thickness from GPR surveys and test pit results.

4 CONCLUSIONS

With the demand for aggregate likely to increase, future planning for aggregate needs is important. GPR can aid in preliminary investigations as well as in detailed exploration for sand and gravel aggregate. The deposits investigated in this study represent former high energy fluvial/deltaic environments. These environments are ideal for reconnaissance and detailed exploration involving the GPR method because there is little fine grained material interbedded within the coarser sand and gravel which would attenuate the GPR signal. As a result: 1. Slow, more labour intensive and costly drilling programs are not necessary, 2. Fewer test excavations are required to determine aggregate volume and quality, and 3. Aggregate data can be gathered in a non-invasive manner.

The study demonstrated that GPR can assess depth and lateral extent of a deposit. As well, it can aid the gravel pit operator by locating coarser aggregate, helping in the direction of further pit development, and, in some locations, indicating thickness of overburden. If attenuation of the GPR signal occurs, this could indicate the lower extent of the sands and gravels or that the deposit has some silt content which may require further aggregate processing.

At the Calgary site, although each geophysical method tested revealed information regarding the aggregate deposit, it was determined that GPR was the most suitable method to meet the survey objectives. However, it is strongly suggested that the approach of testing complementary and supplementary geophysical methods to determine their suitability at specific aggregate deposit sites be adopted as a matter of course. Each site is lithologically and geophysically unique. The ultimate goal of a geophysical investigation (determination of in situ volume) at such a location is best served by an approach which allows for the collection of the most appropriate suite of data possible given time and budgetary constraints.

ACKNOWLEDGMENTS

The following organizations are thanked for their support: Natural Sciences and Engineering Research Council of Canada (NSERC), and the University of Calgary (Department of Geography; Department of Geology and Geophysics). We appreciated the access by many gravel pit operators to their pits. Drs B. Broster and L. Jackson are thanked for their helpful reviews.

REFERENCES

Davis, J.L. & Annan, A.P. 1989. Ground-penetrating radar for high-resolution mapping of soil and rock stratigraphy. *Geophysical Prospecting* 37: 531-551.

Edwards, W.A.D. 1989. Aggregate and nonmetallic quaternary mineral resources. In R.J. Fulton (ed.), Chapter 11 of *Quaternary Geology of Canada and Greenland*: 684-687. Geological Society of America, The Geology of North America, K-1.

Fisher, T.G., Jol, H.M. & Smith, D.G. 1995. Ground penetrating radar used to assess aggregate in catastrophic flood deposits, N.E. Alberta, Canada. *Canadian Geotechnical Journal* 37: 817-879.

Fitzgerald, D.M., Baldwin, C.T., Ibrahim, N.A. & Humphries, S.M. 1992. Sedimentologic and morphologic evolution of a beach-ridge barrier along an indented coast: Buzzards Bay, Massachusetts. In C.H. Fletcher III & J.F. Wehmiller (eds), *Quaternary Coasts of the United States: Marine and Lacustrine Systems*: 65-75. Society of Sedimentary Geology (SEPM), Special Publication No. 48.

Gilbert, G.K. 1890. Lake Bonneville. *United States Geological Survey*, Monograph 1.

Huggenberger, P. 1993. Radar facies: recognition of facies patterns and heterogeneities within Pleistocene Rhine gravels, NE Switzerland. In J.L. Best & C.S. Bristow (eds), *Braided Rivers*: 163-176. Geological Society of London, Special Publication No. 75.

Jol, H.M. 1993. Ground penetrating radar (GPR): a new geophysical methodology used to investigate the internal structure of sedimentary deposits (field experiments on lacustrine deltas). Ph.D. Dissertation, Univ. Calgary, Calgary, Canada.

Jol, H.M. 1995. Ground penetrating radar antennae frequencies and transmitter powers compared for penetration depth, resolution and reflection continuity. *Geophysical Prospecting* 34: 693-709.

Jol, H.M. & Smith, D.G. 1991. Ground penetrating radar of northern lacustrine deltas. *Canadian Journal of Earth Sciences* 28: 1939-1947.

Jol, H.M. & Smith, D.G. 1992. Geometry and structure of deltas in large lakes: a ground penetrating radar overview. In Hänninen, P. & S. Autio (eds), *Proceedings, Fourth International Conference on Ground Penetrating Radar, Rovaniemi, Finland*: 159-168. Geological Survey of Finland, Espoo, Finland, Special Paper 16.

Langer, W.H. & Glanzman, V.M. 1993. Natural aggregate building America's future. *United States Geological Survey*, Public Issues in Earth Science, Circular 1110.

Nelson, K. 1995. Reconstruction realities. *Geotimes* July: 21-23.

Rhine, J.L. & Smith, D.G. 1988. Late Pleistocene Athabasca braid delta of northeastern Alberta, Canada: a paraglacial drainage system affected by aeolian sand supply. In W. Nemec & R. Steel (eds), *Fan deltas: sedimentology and tectonic settings:* 158-169. Glasgow: Blackie & Son.

Sawicki, O. & Smith, D.G. 1992. Glacial Lake Invermere, upper Columbia River Valley, British Columbia: a paleogeographic reconstruction. *Canadian Journal of Earth Sciences* 29: 687-692.

Smith, D.G. 1991. Lacustrine deltas. *Canadian Geographer* 35: 311-316.

Smith, D.G. & Jol, H.M. 1992a. GPR results used to infer depositional processes of coastal spits in large lakes. In Hänninen, P. & S. Autio (eds), *Proceedings, Fourth international conference on ground penetrating radar, Rovaniemi, Finland:* 169-172. Geological Survey of Finland, Espoo, Finland, Special Paper 16.

Smith, D.G. & Jol, H.M. 1992b. Ground-penetrating investigation of a Lake Bonneville delta, Provo level, Brigham City, Utah. *Geology* 20: 1083-1086.

Smith, D.G. & Jol, H.M. 1995. Ground penetrating radar: antenna frequencies and maximum probable depths of penetration in Quaternary sediments. *Journal of Applied Geophysics* 33: 93-100.

Ulriksen, C.P.F. 1982. Application of impulse radar to civil engineering. Ph.D. Dissertation, Lund Univ. Technology, Lund, Sweden.

Application of petrography in durability assessment of rock construction materials

PETER N.W. VERHOEF & ALEXANDER R.G. VAN DE WALL
Delft University of Technology, Faculty of Applied Earth Sciences, Delft, Netherlands

1 INTRODUCTION

Geological construction materials form the major part of all construction materials used in civil engineering projects. Despite their low unit value these 'Geomaterials' are of prime economic importance because of the huge tonnages used each year (Table 1). Like all construction materials, geomaterials must comply with certain specifications to be suitable for an application. These specifications concern the physical, chemical and mechanical durability characteristics. By meeting these durability specifications, the geomaterials should maintain their physical, mechanical and chemical integrity throughout the lifetime of the construction in which they are applied.

Because of the compositional variability of geomaterials, which is inherent to their natural origin, the assessment of their durability properties is not obvious. Many attempts have been made to relate the durability characteristics measured in the laboratory to their behaviour in engineering structures, often without much success. This paper examines the different aspects that relate to these problems. A general introduction, describing the special nature of geomaterials with their varying mineralogical composition and texture, is presented. This variability distinguishes them from engineered construction materials such as steel. Most rocks near the earth surface have been affected by natural weathering processes. In every rock, even if classified as 'fresh', minor amounts of secondary minerals (such as swelling clays) may be present. Their presence can have detrimental effects and lead to accelerated degradation when the material is 'in-service' in the engineering structure. The concept of durability is discussed in this context. Using this information laboratory testing methods of geomaterials are reviewed and the importance of petrography is explained. The paper provides several examples that illustrate the role of petrography and the need for a new direction in aggregate research.

2 GEOLOGICAL CONSTRUCTION MATERIALS

The geological construction materials that are commonly used in civil engineering construction are mainly natural soil aggregates (sand, gravel, boulders) and crushed

Table 1. Production figures and value of some materials in 1987 (after Lüttig 1994).

Product	Production ($\times 10^6$ tons)	Total value ($\times 10^9$ DM)	Unit value (DM tons)
Sand and gravel	8500	68	8
Hard rock and dimension stone	3285	304	89
Coal	3405	32	10
Oil	2907	640	220
Iron ore	940	48	51

rock or rock blocks. Rocks and soils are composed of minerals. These can be single minerals, mineral aggregates, or minerals cemented together to form rock.

2.1 *Distinction between soil and rock*

In this paper, the distinction between soil and rock is based on their level of cohesion. Soil is an assemblage of loose mineral and/or rock particles, while rock is an assemblage of bound minerals, either cemented or crystalline. Rocks have, due to the cementation or crystallinity, a certain tensile strength. The magnitude of the tensile strength depends on the nature of the bond between the mineral particles. Generally, the tensile strength of rocks is measured in MPa. Cohesive fine grained soils tend to have lower tensile strengths, which are measured in kPa.

2.2 *Microstructure of soil and rock materials*

Rocks and soils often have pore spaces between the mineral grains, or very fine hairline cracks along the grain boundaries. Minerals in these materials may also exhibit a preferred shape and/or orientation. The microscopic structure, just like the internal structure of other materials such as steel or concrete, determines to a large extent the mechanical behaviour of the geological materials.

2.3 *Mineral composition and weatherability*

A mineral is a crystalline structure with a specific chemical composition. Minerals are in fact chemical compounds. As with all chemical compounds, minerals that are not in chemical equilibrium may react with other minerals or with chemical substances carried in water or air. The reaction results in a compound with a chemical composition that is stable in a particular environment. The weathering of rock and soil is nothing more than the process just described.

Rock and soil are not generally thought of as chemically active. Most earth scientists have the idea that, due to the slow kinetics of most geochemical reactions, weathering reactions are so slow that they take 'geological time' to take effect. Often this is indeed the case, but sometimes such reactions occur within 'engineering time' (the lifetime of an engineering construction). For example, alkali-aggregate reactions in concrete normally show their effect in a time span of 10 years. The dissolution of gypsum rock can take place in short time spans; depending on the specific conditions, gypsum rock surfaces in contact with flowing water are known to retreat with rates in the order of meters per year (James & Lupton 1978).

Due to the very slow geochemical weathering reactions that take place near the earth surface, nearly every rock, even those considered 'fresh' or unweathered in engineering rock classifications, could contain secondary minerals. This may have a significant impact on the engineering behaviour of the rock. The products of weathering reactions could be clay minerals or iron hydroxides. Such secondary minerals could have detrimental effects, even if present in minute amounts (less than 1% by volume of the rock). When rocks, sands or gravel are extracted from a certain location, we want to know whether reactions could take place in the engineering environment, whether secondary minerals may have detrimental effects or whether the extracted material already contains secondary minerals that have a negative influence on the behaviour of the material.

2.4 *Durability of geomaterials*

The durability of geomaterials is defined as the ability of these materials to keep their mechanical and physico-chemical properties during engineering time in the engineering environment and in relation to the intended usage.

The mechanical and physico-chemical properties of geomaterials have relevance for the application in which they are used. For example, at least 75% of the volume of concrete is taken by aggregate. The most pronounced effect of the aggregate type chosen is on the density and thermal properties of the concrete. The strength and deformability of the concrete are largely affected by the stiffness ratio between the aggregate and the cement matrix. The workability of a concrete mix is influenced by the shape and surface texture of the aggregate. The bond between aggregate and cement matrix can be related to shape and surface texture, but is also dependent on the mineralogical composition of the aggregate. Chemical reactions between some types of aggregates and the alkalis in the cement may cause preliminary degradation of the hardened concrete (Verhoef et al. 1994).

Alternatively, the shape, size and material properties of rock blocks determine to a large extent the strength of a breakwater structure. The dynamic and static loading of structures by chemical, physical or mechanical processes may have detrimental effects on the properties of geomaterials. If this leads to a significant decrease in the performance of the engineering structure, they are considered to be 'not durable' in the engineering environment and the engineering structure in which they are applied.

Even if mechanical and physical laboratory tests indicate that a geomaterial has suitable properties for the engineering structure considered, the presence of weathering minerals or unfavourable micro texture could be problematic. For example, minor amounts of swelling clay minerals could lead to the breakdown of rocks used for coastal shoreline protection. Due to the repeated wetting and drying of the rock and resulting swelling and shrinkage of the clay particles, cracks grow in the rock and degradation occurs. This time-dependent degradation is an important aspect of durability.

The durability of geomaterials is related to their intended use and function; durability is only required in relation to those properties that are relevant for the functionality of the engineering structure. It should be noted that durability is not an overall constant. It relates to different properties in a variety of ways. For example, the abrasion resistance of rock blocks may deteriorate at a different rate than the strength of the same rock block. Also, certain properties may change without influencing the functionality of the engineering structure.

In summary, the durability of geomaterials is to be considered in relation to the engineering environment to which they are exposed, the expected life time of the engineering structure and the function they have to perform in this structure.

3 LABORATORY TESTING OF ROCK MATERIALS

For the design of a structure, the civil engineer commonly wants to obtain the following information from the engineering geologist or geoscientist:
   – Which rock or soil properties are relevant for the engineering structure considered?
   – How are these properties measured?
   – Is the material durable in relation to these properties?
The present state-of-the-art of aggregate investigation is not very sophisticated, much debate is going on concerning all above questions. Over the years a variety of so-called standard tests have been developed. These tests should assess the different properties that are relevant for the functionality of the engineering structure. Commonly such tests, especially those that should assess the durability, were developed to simulate as accurately as possible the conditions that would occur in practice. This has proven to be a very difficult undertaking and most tests are known to relate to practice in only a remote, qualitative, way. Reasons to explain these difficulties include:
   – Many different loads and processes act on the engineering structure, not all of which are clearly understood,
   – The interactions between these loads and processes are often very complex and vary from one location to another,
   – In the laboratory the material, the loads and processes involved and the duration of the processes have to be scaled down. This introduces additional unwanted effects in the simulation,
   – It is often not clearly understood what rock properties are the most relevant and how they influence the behaviour of the engineering structure,
   – Many tests have an empirical basis with only a vague underlying model, even though they are used to model real behaviour.
Much effort has been undertaken in the past two decades to select those standard tests that have relevance to applications in road, hydraulic or construction engineering. In Europe this exercise is currently repeated to make European standards and norms (CEN: Comité Européen de Normalisation). However, there is still a large gap between the parameters that a design engineer would like to use for the design and the rock or soil properties expressed as index values from standard tests received from the laboratory.

3.1 *Variability of geotechnical properties of geomaterials*

An additional problem with geomaterials is the large variability of their properties (see Table 2). This variability occurs on all scales and is the main difference between these materials and wholly engineered materials such as steel or, to a large extent, concrete. Variability is inherent to natural materials. Whilst manufactured materials are produced under controlled circumstances, natural materials have been formed under constantly changing circumstances over prolonged periods. Also after their formation they are still subject to alteration by geological events such as consolidation, interaction with ground water and cementation. They may, for example, be altered due to weathering. As weathering is not a homogeneous process, variation in the degree of weathering is common, even within a single lithological unit. Weathering can lead to considerable variation of the geotechnical properties within a rock unit. Without knowledge of the variability, relating the measurements to a certain level of performance is impossible.

A related aspect is the concept that the result of a measurement or test can only be an approximation of the 'true' value of the property under study. Every measurement will give a result that may, and probably will, be different. Obtaining a 'true' characteristic value for the material properties or the loads imposed in the engineering environment is thus impossible. The ideal approach to the evaluation of geomaterials would make use of this aspect and incorporate probabilities, based on the variability of observations and test results, rather than fixed values. Such an approach, which is common practice in many fields of engineering, seems warranted as it allows the estimation of the probability of failure with the different materials applied on a quantitative basis, rather than a simple statement concerning 'failure/no failure'. The probability of failure may be used to decide whether the risk is acceptable or not, depending on the specific circumstances.

Table 2. Variability in the strength of rocks.

	Number of tests	Coeff. of variation[4] (%)
Variability in compressive strength (UCS test)		
Sandstones[1]	40	19.8
Bentheim sandstone[2]	24	4.4
Calcareous sandstone[2]	172	12.4
Limestones[1]	16	22
Limestones[3]	16	27.5
Limestones[3]	12	33.5
Mudstones[3]	11	29.1
Variability in tensile strength (Brazilian test)		
Sandstones[1]	17	26.8
Bentheim sandstone[2]	36	17.9
Calcareous sandstone[2]	121	15.1
Sandstones[3]	15	18.7
Limestones[3]	76	29.3
Limestones[3]	66	23.1
Mudstones[3]	22	24.4

[1] Roxborough (1987), [2] Piepers (1995), [3] Van der Schrier (1988), [4] Coeff. of variation = Standard deviation expressed as percentage of the mean.

3.2   *Current methods of durability assessment*

Current practice in construction requires that material testing is performed, as much as possible, using test methods described in official standards (e.g. ISO, BS, ASTM). New European standards (CEN) are developed that will be used in the near future. One would like to know the relationship of the standard test results with the behaviour in practice. Case histories and monitoring studies (or generally 'practical experience') have given some idea of boundary values for test results. For specific purposes lists of requirements usually exist, such as that recently developed for rock armouring stone (Table 3). By using a combination of different tests, it is hoped that all the relevant properties are covered and that the results can be related to the engineering structure in its particular engineering environment.

Discussion may occur concerning the applicability of the test results, especially with marginal materials. Sometimes materials, though suitable in practice, fail certain test requirements and are therefore not allowed to be used. The opposite of this scenario may be worse. The possibility that materials may pass the required standard tests but do not fulfill their intended function shows the weakness of the present methods of suitability assessment. This is one of the reasons why research is now concentrating on the study of processes that act upon and within the engineering structure, as well as on the mechanisms involved in laboratory tests. An example of

Table 3. Guide to rock durability from test results (after CIRIA/CUR 1991).

Index test	Excellent	Good	Marginal	Poor	Remarks
Dry density (Mg/m^3)	> 2.9	2.6-2.9	2.3-2.6	< 2.3	(1)
Water absorption (%)	< 0.5	0.5-2.0	2.0-6.0	> 6.0	(1)
MgSO$_4$ soundness (%)	< 2	2-12	12-30	> 30	(2)
Freeze-thaw (%)	< 0.1	0.1-0.5	0.5-2.0	> 2.0	(3)
Methylene Blue Adsorption (MBA; g/100)	< 0.4	0.4-0.7	0.7-1.0	> 1.0	(4)
Uniaxial compressive strength (MPa)	> 200	100-200	50-100	< 50	ISRM
Fracture toughness K$_{ic}$ (Mpa.m^2)	> 2.2	1.4-2.2	0.8-1.4	< 0.8	(5)
Point load index (Mpa)	> 8.0	4.0-8.0	1.5-4.0	< 1.5	ISRM
AIV (% fines), wet	< 12.0	12-20	20-30	> 30	BS 812
Mill abrasion resistance Ks	< 0.002	0.002-0.004	0.004-0.015	> 0.015	(6)
Normalised velocity index I	< 1.2	1.2-1.5	1.5-2.0	> 2.0	(7)
Rock durability indicator; static RDIs	> 2.5	2.5--1	–1--3	<–3	(8)
Dynamic RDId	< 0.5	0.5-2.0	2.0-4.0	> 4.0	(8)

(1) Density and water absorption have been found to be useful indicators of material quality. WA is the single most important indicator of resistance against degradation and also a good indicator of weathering resistance. WA can be misleading for porous limestones with large free draining pores. (2) Indicates resistance to weathering and salt crystallisation growth. Important test for porous sedimentary rocks in hot dry climates. (3) Freeze-thaw and magnesium sulphate soundness both correlate with WA test. (4) This test is very useful to indicate the presence of swelling clay minerals in rock and soil and can be used in combination with petrographic examination (Stapel & Verhoef 1989, Verhoef 1992). (5) This test is a fundamental indicator of tensile strength of rock specimens. Correlates with abrasion resistance. (ISRM 1988). (6) Simulation test for abrasion of rock armouring stone (Latham 1991). (7) A method to predict possible breakage of large armourstone blocks, in research stage (Niese et al. 1990, Houwink & Verhoef 1991). (8) These indicators are proposed by Fookes et al. 1988.

such research is the development of a degradation model for armour stone by Latham (1991). These studies clearly show the complex interaction between the in-service behaviour of geomaterials and their properties. Usually the relation cannot be developed further than a qualitative identification of relevant properties. For example, it is quite well understood which properties determine the resistance of road aggregates against polishing in the Polished Stone Value test. However, giving a quantitative interpretation of this relation has so far not been possible. Neither has it been possible to relate the behaviour of the test material in a consistent manner to its behaviour in practice.

3.3 *The design of a laboratory investigation of rock materials based on standard practice*

Guidelines for rock quality such as given in Table 3 are only an indication. In fact, it is clear that a single test is not sufficient to indicate suitability. Evaluation of several test results, determining the physical, chemical and mechanical durability characteristics is necessary. One should also be sure that no unknown defects are present in the rock. Important in this respect are the presence of detrimental minerals and structures. There may be several reasons for testing rock or aggregate:
 – To assess the usefulness and quality of a new source of aggregate,
 – To compare quality of rock from different sources,
 – To assess sample variability,
 – To predict performance in service,
 – To predict durability,
 – To identify the causes of an unexpected failure or early deterioration.

3.4 *Basic philosophy*

The test program to be designed depends not only on the purpose, but also on the material to be tested. For most applications utilizing granular geomaterials, there are specific requirements regarding size and shape. Much of the information on potential size and shape of blocks that can be obtained from a quarry comes from special field studies, especially for large armour stone blocks. After adequate and representative sampling (Gy 1979) the work should start with a proper characterization of the material. Characterization implies determination of the geometry of the aggregate particles (size, shape, flakiness index, angularity), determination of grain size distribution, determination of physical properties (specific gravity, water absorption, porosity) and petrographic examination by microscopic techniques. The flow diagram of Figure 1 gives the outline of the investigation proposed.

The diagram emphasizes that before mechanical tests are performed, the mineralogical composition should be known. Apart from fine grained clay minerals and highly weathered minerals, most minerals can be determined using the petrographic polarizing microscope. Traditionally, clay minerals are determined using X-ray diffraction techniques and differential thermal analysis. In engineering geology one is not necessarily interested in the correct determination of a clay mineral. The question is whether such minerals are swelling when wetted and shrink when dried. Even if very small amounts of swelling clays are present in a rock, varying moisture conditions cause the swelling clays to act as small pumps from which cracks can grow, leading to degradation of the rock. Wetting and drying can, of course, occur when rock is used as coastal protection rip-rap. A less obvious example is the moisture variation that can occur within road base aggregate by the dynamic loading of traffic. Examples are known of argillaceous aggregate developing a thin clay layer by this process between sub-base and pavement, ultimately leading to failure of the road surface (Smith & Collis 1993). The simple method of staining the thin sections used for microscopic study with methylene blue and titrating ground rock or soil with a methylene blue solution (MBA test), is sufficient to identify and establish the swelling potential of clays (see Appendix). Since even 'fresh' rock can contain some clay, the MBA test should be standard procedure when performing a petrographic examination.

It has already been stated that most degradation processes that take place in engineering time are the result of the interaction of the new engineering environment with detrimental minerals or microscopic structures already present in the rock. Table 4 illustrates the forces that can be exerted by clay minerals or crystals growing in the pore spaces of rock.

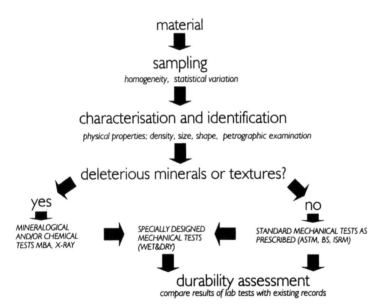

Figure 1. Procedure for the laboratory investigation of rock materials.

Table 4. Typical pressures exerted by physical processes on rock (Fookes et al. 1988).

Process	Exerted stress
Freezing (max. at −20°C)	200 Mpa
Crystallization of salts	2-20 Mpa
Hydration of salts	100 Mpa
Clay expansion	2 Mpa

3.5 *Design of mechanical tests*

The testing program should be designed after the characterization of the material is completed. If no special problems have been found, the standard tests as given in the requirements can be performed. However, if during the characterization potential detrimental features have been observed, an adjustment of the testing program is necessary. For example, the rock under evaluation may be transected by partially cemented cracks or may contain minor amounts of swelling clay. The testing program is adjusted to address the potential detrimental features. In the case of partially cemented cracks this should involve careful observation of the results of strength tests and the execution of magnesium sulphate soundness tests. These tests help to determine the susceptibility of geomaterials to degradation due to salt crystalisation. If clays are present that may have swelling properties, the mechanical tests should be carried out both wet and dry. For example, the difference between the results of the 10% fines test carried out under both wet and dry conditions gives an indication of the degradation due to the activity of clay minerals (Hosking & Tubey 1969, McNally et al. 1990).

3.6 *System and functional analysis*

The laboratory procedure just described does not comply with a rigid use of test result requirements such as those listed in Table 3. More and more it is recognized that for geomaterials used in construction another approach should be taken. Instead of using a suite of standard tests to assess the material, for each application an analysis of the system that completely describes the engineering environment should be made. This system analysis should describe as accurately as possible the stresses, loadings and environmental agents that operate on the rock material. The testing program then specifically addresses these features. This is akin to the systems approach used in tribological (wear and lubrication) engineering (Uetz 1986, Zum Gahr 1987).

Consider, for example, the design of a rock structure. During the construction phase certain loads are expected while for engineering life lower loads are foreseen (Fig. 2). If we knew the durability of the rock material with respect to the loading

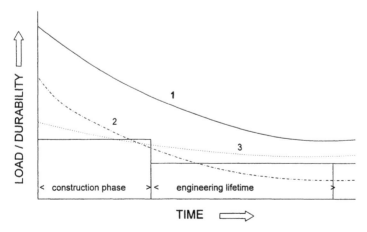

Figure 2. The calculated load on a rock stone structure during construction and engineering life compared with the durability (and related decrease in suitability) of three rock types.

conditions considered (static, dynamic, atmospheric, thermal, ...), we could select a more appropriate rock. In Figure 2 rock type 1 would give no problems and rock type 3 could possibly be used. However, some problems at the end of the construction phase are expected with rock type 3. Rock type 2 concerns a rock that is inferior, because it will not be durable both during construction and during the in-service period of the structure. If the system analysis has been carried out successfully, a solution to these problems might be found, or a reasoned adjustment of safety factors or construction procedure could be proposed. In this respect it is clear that a probabilistic approach, using the variability in measurements and observations, would be advantageous.

4 MATERIALS SCIENCE OF GEOMATERIALS: PETROGRAPHY

When new materials are developed and used for construction, the materials engineer plays a large role in its development, quality control and research. It is logical that, understanding the engineering behaviour of geomaterials, we should make use of rock material scientists. The petrographer is a scientist trained in the materials characterization of minerals and rocks. A petrographer with sufficient knowledge of the civil engineering requirements would be well suited for the job. Such people are scarce and have become knowledgeable in the field due to their personal career background. Yet it is generally agreed that a petrographic characterization is the essential core of a proper investigation into the suitability of geological materials. The importance of petrography in construction engineering is illustrated by Table 5, showing the dependence of mechanical properties of aggregates on petrographic parameters.

Table 5. Main petrographical features affecting engineering tests results (after Fookes et al. 1988).

Engineering test	Mineralogy[1]	Hardness[2]	Porosity	Texture[3]	Shape	Sec. Min.%	Micro cracks	Grain size	Packing
Water absorption			x			x	x		x
Density	x		x			x	x		x
Compressive strength		x	x	x	x	x	x	x	x
Tensile strength			x	x	x	x	x	x	x
Slake durability		x	x	x		x	x		
Wetting-drying			x	x		x	x		
Freeze-thaw			x	x		x	x		x
Sulphate soundness			x	x		x	x		x
Micro-deval (wet)	x	x		x	x	x	x	x	x
Los Angeles abrasion	x	x			x		x	x	x
Aggregate crush. val.	x	x	x		x		x		x
10% fines test	x	x	x		x	x	x		x
Aggregate impact val.	x	x	x		x		x		x
M.AIV	x	x	x		x	x	x		x
Polished stone value	x	x		x	x	x	x	x	x

[1] Excluding secondary minerals, [2] Indentation strength of composing minerals, [3] Including anisotropy.

The following paragraphs give examples of the use of petrography in construction engineering. Also some difficulties in the use of petrography are addressed.

4.1 *Swelling clays and the durability of rock aggregates*

The Methylene Blue Adsorption test has been mentioned previously. By comparing the MBA value of finely ground rock with the percentage of clay that is present in that rock, a good idea can be gained of the type of clay present. Rock intended for construction purposes can quickly be examined by adopting the following procedure (Verhoef 1992). First a part of the rock is ground to a grain size smaller than about 0.15 mm. About 2 g of this powder is titrated in the normal way to obtain the MBA value (Appendix). A thin section of the same rock is stained with an MBA solution (3g/l). The volume of blue colored minerals is estimated by point counting the thin section under the microscope. The thin section can also give important information on the microstructural position of clays. An estimate of the MBA of the adsorbing blue-stained minerals can be derived by applying the following formula:

$$MBA_{min} = \frac{MBA_{rock} \cdot \rho_{min}}{Vol.\% \cdot \rho_{rock}} \cdot 100 \, (g \, / \, 100g)$$

Commonly, the density of the MBA adsorbing minerals is nearly that of the rock. With regard to potential durability problems, criteria can be established concerning the MBA value of the rock or the MBA value of the minerals. In Table 6 the MBA values of some common rocks and minerals are listed. Using these values, the infor-

Table 6. MBA values for common rocks and minerals (Verhoef 1992).

Mineral	MBA [g/100g]	Mineral	MBA [g/100g]
Analcite	0.59	Sepiolite	7.20
Anhydrite	0.00	Sericiteschist	0.15
Antigorite	0.15	Serpentinite	0.18
Apatite	0.00	Serpentinite	0.15
Asbestos	0.14	Scapolite	0.15
Asbestos	0.23	Talc	0.38
Baryte	0.00	Wavellite	0.50
Biotite	0.15	Li-montmorillonite	11.00
Calcite	0.09	Na-montmorillonite	9.50
Cancrinite	0.00	K-montmorillonite	22.90
Chamosite	0.00	Rb-montmorillonite	11.40
Dolomite	0.00	Cs-montmorillonite	4.70
Epidote	0.00	Mg-montmorillonite	22.00
Garnierite	5.40	Ca-montmorillonite	21.20
Gypsum	0.00	Sr-montmorillonite	20.30
Hornblende	0.15	Ba-montmorillonite	17.50
Kaolinite	1.40	Palygorskite	14.60
Lazurite	0.00	Chlorite	0.60
Magnesite	0.00	Illite	2.50
Monticellite	0.10	Kaolinite	2.40
Muscovite	0.15	Serpentine	1.20
Nephelinite	0.00	Halloysite	1.30

mation obtained from a MB-stained thin section can be translated into quantitative information on the swelling potential of the rock under investigation. The use of the Methylene Blue Adsorption test is illustrated by the following examples.

The MBA test, the wet Deval test and a petrographic examination were performed on a suite of 25 rock types. The wet Deval test, also referred to as the 'Aggregate Attrition Value' can be used to evaluate rock for coastal defense structures, railway ballast beds and road constructions. Samples are rotated in a closed, inclined cylinder resulting in mutual attrition of the aggregate. The percentage of fines generated during the test is used to calculate the wet Deval Value (NF P18-577).

In the study the following criteria were used to indicate potential durability problems regarding swelling clays:
  – MBA rock > 1, or
  – MBA mineral > 10, if present in % vol.> 5, or
  – MBA mineral > 20.

If either of these criteria is met, the rock is considered suspect. Table 7 lists the results of the wet Deval test and the Methylene Blue Adsorption test. In Figure 3, the MBA value is plotted against the inverse wet Deval value. Filled squares indicate suspect rocks, open squares indicate rocks that were not suspect with regard to swelling clays. In the figure the inverse of the wet Deval values is used so that high val-

Table 7. MBA and wet Deval values of 25 rock aggregates.

Rock type	Vol.% stained	MBA (g%)	MBA min (g%)	Deval (wet)
Andesite*	1.00	0.35	35.1	9.7
Andesite*	1	0.59	58.5	9.1
Andesite*	5	1.29	25.7	6.0
Andesitic basalt*	1	0.35	35.1	9.6
Andesitic basalt*	1	0.82	81.9	9.0
Basalt	5	0.35	7.0	14.9
Dacite*	20	2.11	10.5	7.8
Granite	3	0.35	11.7	14.6
Granite*	4	0.70	20.1	12.1
Granite*	18	0.94	5.2	8.7
Granodiorite	10	0.59	5.9	15.0
Quartz diorite	22	0.35	1.6	20.1
Dolomite*	1	0.35	35.1	18.2
Limestone*	1	0.23	23.4	6.2
Limestone*	1	0.23	23.4	10.3
Limestone	5	0.35	7.0	9.3
Limestone	15	0.59	3.9	5.9
Quartz arenite	10	0.35	3.5	10.5
Quartz arenite*	1	0.47	46.8	6.2
Quartz arenite*	15	0.82	5.5	7.4
Graywacke	8	0.70	8.8	10.6
Metabasalt	5.00	0.47	9.4	13.7
Metagabbro*	45	1.17	2.6	5.5
Gneiss*	1	0.35	35.1	9.3
Gneiss*	1	0.59	58.5	21.5

*Considered suspect after petrographic examination of MB stained thin sections

ues indicate a poor performance (as in most aggregate tests). The figure also distinguishes marginal rock with MBA values between 0.7 and 1.0 and unsound rock with MBA values > 1. The results show that most of the suspect rocks performed poorly in the wet Deval test, including rocks that would pass the MBA < 0.7 criterion. The wet Deval test apparently causes the clays present in these aggregates to swell, with resulting failure, even when the Methylene Blue Adsorption is below the 0.7 threshold. Only by using the method of staining thin sections, minute amounts (< 1 vol%) of swelling clays can be detected. This method should therefore be included in the routine petrographic examination of geomaterials.

Figure 3 shows that some rocks did not behave as expected. It should be noted that the examination has the function of indicating potential hazards. Also, the micro structural position of the clay minerals was not considered. An example is the rock considered sound by both the MBA test and the petrographic staining, but with a poor performance in the wet Deval test. Here microscopic examination showed that the grains in this limestone are surrounded by a thin film of (non-swelling, low MBA) clay. The texture (fabric) of the rock explains the behaviour during the Deval test.

A second example of the use of the MBA method concerns the limestone rock that was used as rip-rap for the protection of the shores of Lake Tunis. Strong limestone blocks of a size of 350 mm were obtained from a nearby quarry to function as rip-rap. In Tunisia an arid, dry climate exists and apparently no degradational features of the rock were observed in the quarry. After Lake Tunis had been deepened and enlarged by dredging, in the late 1980's, the rip-rap was placed. After a few months it was observed that some of the blocks were breaking up, some even crumbled to rock powder. The rocks, coming from one quarry, were described to vary from white to gray in colour. Especially the gray rock seemed to be affected by the crumbling process. At places along the shores of the lake, more than 30 % of the rip-rap was completely degraded. Samples were sent to the laboratory to be examined (Fig. 4). The rock appeared by visual examination to be a fine grained layered, slaty lime-

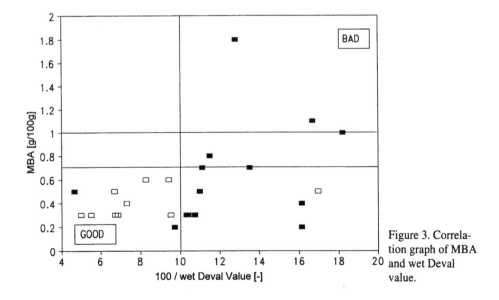

Figure 3. Correlation graph of MBA and wet Deval value.

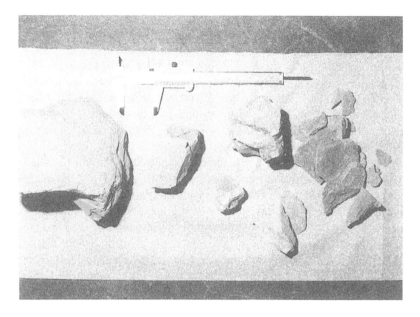

Figure 4. Rock samples from Lake Tunis showing degradation.

stone, suspect of containing clay (argillaceous limestone or calcilutite). Parts of the specimens were ground down, the rock powder was used for the MBA test and for X-ray diffraction, to obtain information on the mineralogical composition. Thin sections were stained with methylene blue and were examined by microscope. The results are given in Table 8. The swelling clay mineral Montmorillonite was found to be present in varying amount. The rocks that degraded had the swelling clay in thin layers or laminae (Fig. 5a). Rocks that had swelling dispersed in the matrix, or in patches, showed no degradation (Fig. 5b). Obviously the repeated wetting and drying of the rocks caused swelling and shrinkage of the clay seams, from which micro cracks could grow, leading to the degradation.

4.2 *Texture and mineralogy related to rock strength*

Limestone is sometimes used as cladding material for buildings or other engineering structures. However, in London it was found that after construction, parts of the cladding slabs started to break off, leading to dangerous circumstances. To find out the cause of the problem, the cladding slabs were examined.

It was observed that the limestone contained 'stylolites'. Stylolites are irregular suture-like boundaries that often develop in limestones. It is generally accepted that stylolites are formed by some kind of pressure controlled dissolution. Along this dissolution surface, insoluble residue is accumulated (Fig. 6a). As a result, stylolites form a clear mechanic weakness surface in the limestone. Figures 6b and 6c show schematically the development of stylolites. During the compaction of calcareous sands, usually because of the weight of overlying layers, local dissolution of calcite particles occurs due to high local contact pressures. Dissolution occurs in surfaces

Table 8. Results analysis samples Lake Tunis project.

Sample nr.	Rock type	MBA stained [%]	MBA rock [gr%]	MBA min. [gr%]
1*	Layered calcilutite	25	2.1	8.4
2	Layered calcilutite	31	0.6	1.9
3	Calcilutite	49	0.5	1.0

* XRD analysis: quartz, calcite, kaolinite, montmorrilonite, chlorite (?).

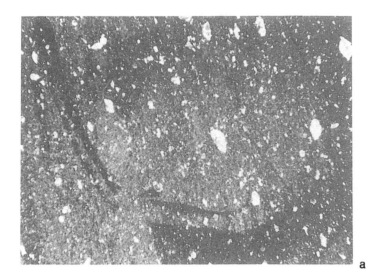

a

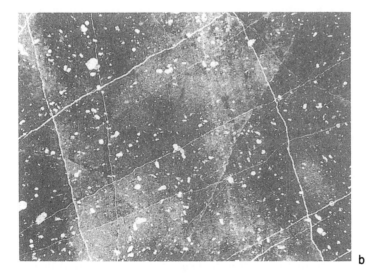

b

Figure 5. Thin section with MB staining The MB staining show up black in the photographs. a) Dispersed distribution of MB staining (swelling clays), b) Concentration of MB staining in certain areas leading to premature degradation of the rock.

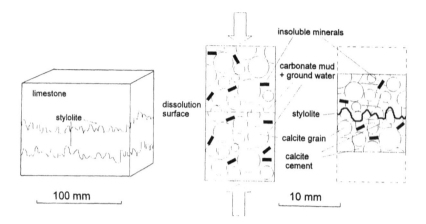

Figure 6. Development of stylolites in limestone.

perpendicular to the direction of the highest stress (usually horizontal). On the dissolution surfaces, insoluble minerals, organic remains or clay particles, accumulate.
The stylolites get their irregular shape due to the variation in dissolution rates. Figure
6c shows the result of the process: a stylolite, covered with insoluble particles and a
reduction of the thickness of the limestone layer. The dissolved calcite may be carried off through the still porous rock and be precipitated elsewhere, resulting in a
cementation of grains and reduction of available pore space. The result is a massive
limestone traversed by irregular lines or thin bands, showing through different colour
or texture.

Stylolites occur in most limestones. Their presence, however, does not automati-
cally imply that the rock is unsuitable as cladding material. The strength of the
weakness surface depends on the type and amount of insoluble residue deposited on
the dissolution surface. If the insoluble residue contains clay material, the plane will
have a very low tensile strength. The clay material, especially if it involves swelling
clays such as smectite, may cause swelling-shrinkage problems when exposed to
varying moisture conditions and freeze-thaw cycles.

The presence of stylolites can best be identified in the quarry. These features are
generally visible to the naked eye. If present, a petrographic study in combination
with a MBA test and possibly X-ray diffraction is necessary. If the insoluble residue
proves to contain clay or other undesirable material the rock should be rejected.

The failed cladding slabs contained stylolites with clay minerals. The clay-rich
stylolites showed clearly when the rock was wetted; the area around the stylolite remained moist for a longer period than the rest of the rock. Petrographic examination
pointed out that the area around the stylolite contained calcite, iron hydroxide, quartz
and 5-10% clay, stained blue with methylene blue dye. X-ray diffraction identified
the clay mineral as smectite, a swelling clay. The MBA value was 3 gr/100gr (values
over 1 gr/100gr indicate unsound rock). The findings of the study showed that the
reason for the failure of the cladding slabs could be attributed to the presence of stylolites with swelling clays.

4.3 *Difficulties in relating petrography to mechanical behaviour*

Although Table 5 shows that petrographic features influence test results, proving the relationships is not easy, especially when a group of petrographically different rocks is considered. Pieters (1992) attempted to find relationships between petrographic characteristics and engineering test results on a group of more than 60 aggregate types, but no obvious correlations were found. Sometimes useful correlations can be established if only one rock type is considered. For example, it has been attempted to establish a quantitative relationship between the petrographic characteristics of road aggregates and their behavior in the Polished Stone Value (PSV) test.

The PSV test is used to quantify the durability of the surface roughness of aggregates in road surfaces. The test involves the polishing of aggregates and the subsequent measurement of the friction between a rubber slider and the aggregate surface. Several rock properties influence the developed friction after polishing, of which the packing proximity is one. This parameter is defined as the number of grain-to-grain contacts of the hard minerals divided by the total number of contacts (grain-to-matrix contacts). In Figure 7 the PSV is plotted against the packing proximity for a wide range of rock types. The graph clarifies one of the problems in using single petrographic characteristics for the explanation of the mechanical behaviour of rocks. In this case only a single rock type, quartzite, shows good correlation of PSV with the packing proximity.

Present and future research is therefore concentrating on a quantification of petrographic characteristics that can be used more rigorously. Since linking of specific test results to the petrographic characteristics proved difficult, an approach using classification systems based on the petrography is attempted.

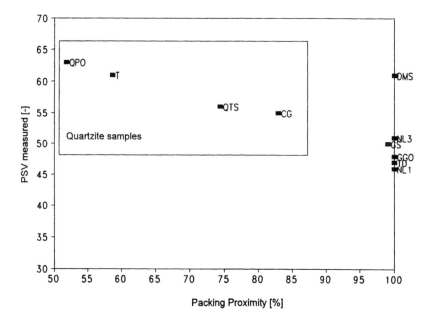

Figure 7. Correlation graph of Polished Stone Value and packing proximity.

4.4 *A classification system using petrography*

The Polished Stone Value of aggregates is an extensively researched topic as the test is used as a major criterion for the suitability of aggregates as road surface material. Several petrographic characteristics have been found to influence the Polished Stone Value (PSV), of which the most important one is the mineralogy (Tourenq 1971, van de Wall 1992). Because of the complex interactions of the petrographic characteristics, attempts to correlate these characteristics to the PSV have generally not come beyond a qualitative stage. In such cases, a classification system may be useful. For the Polished Stone Value a system can be set up using the mineralogy and the structure of the rock.

The mineralogy can be expressed in a term depicting the average hardness of the rock particle and a term depicting the average differences in hardness of the minerals present in the rock particle, the rock differential hardness (Tourenq 1971). The structure of the rock can be classified as porphyritic, i.e. large grains embedded in a finer matrix, or non-porphyritic, i.e. equal grainsize. The PSV can be classified as low, medium or high, indicating respectively a PSV lower than 48, 48-53 and higher than 53. Table 9 shows a possible definition of the classes with a suggested rating number. The PSV is classified with regard to the specifications for road construction in the Netherlands. Road class 1 and 2 require a PSV > 48 and road class 3, 4 and 5 a PSV > 53. An assessment of the suitability of a rock is obtained by summing all ratings.

The classification has been applied to 24 rock samples with varying Polished Stone Values. If the suitability ratings for the samples are compared with their Polished Stone Value, it can be seen that the rating is quite adequate in its identification of the suitability of the samples (Table 10).

Based on the results in Table 10 it can be stated that a rock having a suitability rating lower than 6 is likely to have a low Polished Stone Value. Rocks with a suitability rating between 6 and 9 can be expected to have a medium Polished Stone Value, while the Polished Stone Value for rocks with a suitability rating higher than 9 is probably high. Table 10 shows that 2 samples deviate (SR-PSV = 9-46 and SR-PSV = 11-48). These deviations may be explained by the fact that the classification involves only a few petrographic characteristics. The table also shows that a PSV of 53 seems to fit better in the medium group than in the high group.

Table 9. Definition of classes for the polished stone value.

Diff hardness	Class	Rating	Mean hardness	Class	Rating
< 100	Low	1	< 300	Low	1
100-250	Medium	4	300-700	Medium	2
> 250	High	8	> 700	High	3
Structure	Class	Rating	PSV	Class	
Porphyric	Yes	2	0-48	Low	
Non porphyric	No	1	48-53	Medium	
			> 53	High	

Table 10. Cross tabulation of suitability rating and polished stone value, SR: Suitability rating, PSV: Polished stone value.

PSV	SR	3	6	8	9	11	12
	42	1					
Low	46				1		
	47	1					
	48					1	
Medium	50			3			
	51			3			
	52		1	2			
	53			3			
	55						1
	56					1	1
High	57					1	
	61					1	
	63						2
	65					1	

5 CONCLUSIONS

Durability and degradation are properties that are dependent on the nature of the construction material, the engineering environment in which it is applied, the expected life time of the engineering structure and the function that the construction material is to perform.

Sometimes it is advisable to perform a functional analysis (sometimes called system analysis) to estimate the loads that will be imposed on the construction and to define an optimal testing program.

An intelligent study of rock materials, including petrography and mineral analysis techniques can support the testing program for quality assessment. Classification systems including petrographic characterization data can be used for this purpose. This does call for a continuing study of methods to quantify the petrographic characteristics of rock materials.

Quantitative data on the properties of geological construction materials obtained in the laboratory will rarely be directly applicable to the in-service behaviour of the construction material and the engineering structure.

Proper study of rock materials, before these are used as construction material, through appropriate laboratory testing and by petrographic analysis, performed by a qualified rock material scientist, can prevent or minimize problems with the construction later on.

ACKNOWLEDGMENTS

The support of the General Directorate for Public Works and Water management, Division for Road and Hydraulic Engineering (Rijkswaterstaat, DWW, The Netherlands) is gratefully acknowledged. The comments of two referees on the first draft of this paper are appreciated.

APPENDIX: THE METHYLENE BLUE ADSORPTION SPOT TEST

Scope of the test

This test is used to quantify the effect of clay minerals (smectite group) present in an aggregate and hence to indicate the soundness of the aggregate. If a significant amount of methylene blue is adsorbed by the soil or ground rock material, this may indicate the presence of swelling clay minerals, although there exist substances that also may adsorb methylene blue. Further study, for example using specialised X-ray diffraction techniques suitable to identify smectite clay minerals, is necessary to assess the true nature of the adsorbing substance. Low MBA (methylene blue adsorption) values, however, nearly always indicate absence of significant amounts of swelling clay minerals. If a petrographic study is carried out concurrently, the adsorbing minerals may be stained, by immersing the thin sections with the methylene blue solution. In this way the volumetric amount of adsorbing minerals can be estimated and an educated guess of the nature of the clay mineral may be made. The methylene blue test is an alternative way to determine the cation exchange capacity of a soil or finely ground rock.

Apparatus and reagents

- 25 ml burette mounted on stand
- 250 ml Erlenmeyer flasks
- 100 ml volumetric flask and stopper
- 250 ml beakers
- Glass stirring rod
- Magnetic stirrer
- Small clock glasses
- Sample containers
- Spatula
- Whatman No. 40 filter papers (12.5 cm diameter)
- Distilled or de-ionised (demineralized) water
- Methylene Blue
- Chemical waste tank
- Analytical balance; aggregate crusher; grinding apparatus; riffle boxes; scoop; metal trays; an oven.

Preparation of the Methylene Blue solution

The crystalline Methylene Blue (3.9-bis-dimethylamino-phynazothionium-chloride; $C_{16}H_{18}N_3ClS)nH_2O$ is hygroscopic. The solution can be prepared by dissolving a certain weight of methylene blue crystals in distilled water and determine the normality of the solution. To determine the hygroscopic water content of the MB, a sample of the crystalline dye is dried at 105 °C and the loss of the weight, x, is determined (a typical weight loss is about 12.34 %). The normality of a MB solution can then be calculated:

$$N = (c \times 1000)/319.9 \times (100.00 - x)/100 \text{ [meq/l]}$$

where: c = concentration methylene blue solution [g/ml], x = weight loss of crystalline dye by drying, in percentages [%], A 3 g/l solution is commonly prepared for the method used in the way described here[1]. The methylene blue solution must not be kept more than one month maximum.

Preparation of the aggregate suspension

If the MBA value of a soil has to be prepared, the soil has to be sieved. The test has to be performed on the fine fraction of the soil (at least smaller than 63 μm: silt and clay fraction). The fines content f of the soil has to be determined. If the MBA content of a rock has to be determined, a representative portion of the rock has to be ground down to smaller than 0.063 mm.

Prepare from the rock or soil aggregate a very fine grained powder by drying, crushing and sieving, as described above. The powder must be representative of the aggregate or rock composition or the fines fraction of the soil. The grain size must be less than 63 μm.

Take a part of the sample, weigh it, and dry it in the oven at 105°C to determine the water content of the sample. (Remember that the MBA test has to be performed at least twice; keep enough undried sample for this purpose).

Make a suspension of the undried aggregate powder (the amount should correspond to about 2 g dried aggregate powder) in the Erlenmeyer flask with 30 ml distilled or demineralized water.

The crushed rock or soil may come from a small sample or hand specimen. If a representative aggregate sample is available, the following procedure is recommended (Wimpey Laboratories, UK, see Higgs 1986).

To prepare the test sample, riffle the sample of aggregate down to a portion of about 1 kg and crush it in a jaw crusher to passing the 6.3 mm sieve. Then riffle the portion down to about 100 g and, using the jaw crusher, crush it as finely as possible (say passing 0.425 mm). Follow this by grinding it in a mechanical pestle and mortar until nearly all of it passes the 0.063 mm sieve. The material retained on the 0.063 mm sieve is ground to passing this sieve using a hand agate pestle and mortar. A portion of the sample has to be used for the determination of the water content (see above).

Procedure of the test

From the Methylene Blue solution 0.5 ml is added to the aggregate suspension by means of a 25 ml burette. Shaking during the addition is necessary, for which a magnetic stirrer may be used. To perform the titration, add successive volumes of 0.5 ml

---

[1] Note that an alternative method is being developed by LCPC (Paris, France), where a 10 g/l solution is used to test larger quantities of soil/rock (about 30 g)

of the methylene blue solution to the Erlenmeyer flask. After each addition, agitate the flask for 1 minute and remove a drop of the dispersion with the glass rod and dab it carefully on a sheet of filter paper. Initially, a circle of dust is formed which is coloured dark blue and has a distinct edge, and is surrounded by a ring of clear water.

When the edge of the dust circle appears fuzzy and/or is surrounded by a narrow light blue halo, agitate the flask for 1 more minute and do another spot test. If the halo has disappeared, add more blue. If there is still a halo, agitate the flask for a further 2 minutes and do another spot test. Whatever the outcome of this test, add more blue, agitate for 2 minutes, do a spot test, then agitate for a further 2 minutes and do another spot test. This sequence, with a total of 4 minutes of agitation is repeated until there is a definite light blue halo. It is recommended to note down the sample number and the amount of methylene blue added below each spot on the filter paper.

To determine the end-point, hold the filter paper up to daylight while it is still damp, and compare the dust circles made after 4 minutes of agitation. It should then be possible to see where the halo first appears and thus where the end-point is. The corresponding volume of methylene blue solution added is noted down.

This procedure is called the 'spot' method. After completing the test, pour the remaining methylene blue solution and the titrated suspension in a chemical waste container, specially determined for this purpose.

Calculation of MBA and CEC

The Methylene Blue Adsorption value is – in the literature -normally expressed in grams Methylene Blue adsorpted by 100 g of sample material, given as g% or g/100g.

$$\text{MBA} = \{(X \text{ g}/Y \text{ ml}) \cdot p \text{ ml MB}\}/(A \text{ g} / 100 \text{ g}); [\text{g\%}]$$

It is preferred to express the MBA in milliequivalents adsorbed per 100 g of sample material:

$$Mf = (100 \cdot n \cdot p \text{ ml MB})/A \text{ g}; [\text{meq}/100\text{g}]$$

where: X = weight of dried methylene blue crystals, Y = volume of diluted methylene blue solution, p = volume of methylene blue solution added, A = weight of dry soil or rock powder i.e. a correction for the water content of the sample should be made (see Section 4), n = normality of the MB solution (0.0094 meq/l for a 3 g/l methylene blue solution).

If a soil sample is tested, the MBA and Mf are determined as follows: When the test is carried out on a (fine) size fraction (*o/d*), then the Mf of the total soil (*o/D*) can be found by:

$$Mf_{o/D} = Mf_{o/d} \times C_d/100$$

where: C_d = the weight percentage of size fraction o/d in the soil with size distribution *o/D*.

The $Mf_{o/D}$ or the similarly determined $\text{MBA}_{o/D}$ can act as an index which quantifies the effect of clay present in the soil. M_f can be regarded as a good approximation of cation exchange capacity (CEC).

Reporting of results

At least two determinations of the MBA adsorption value should be done on each sample; the average is reported.

The M_f value should be reported; the cation exchange capacity should be given with an accuracy of 0.1 meq/100g. To compare the data with literature also the MBA value should be calculated with an accuracy of 0.1 g/100 g.

It is advisable to keep record of:
– The amount of soil or ground rock used (with 0.01 g accuracy),
– The water content of the sample,
– The amount of methylene blue solution added (with accuracy of 0.5 ml).

Together with the result of the calculations, the filter paper, on which the sample number and the amounts of MB (in ml) are noted down, should be kept in the laboratory files.

Interpretation of results

Limiting values used in the UK (Wimpey Lab., see Higgs 1986) are 1.0 for basaltic rock and 0.7 for coarse grained sandstone. Cole & Sandy (1980) also give a boundary for unsound basaltic aggregate, which (recalculated) is 1.5 g/100 g MB. That value is also given as a boundary value for rock by Tran Ngog Lan (1985). The values of Wimpey Lab. are as follows:

Indication and durability	MBA (g/100 g)	Mf (meq/100)
Acceptable	< 0.7	< 1.9
Marginal	0.7-1.0	1.9-2.7
Unsound	> 1.0	> 2.7

Some characteristic MBA values for soils and rocks are given in Table 9.

REFERENCES

Brown E.T. (ed.) 1981. Rock characterization, testing and monitoring. *ISRM suggested methods.* Oxford: Permagon press.

CIRIA/CUR 1991. *Manual on the use of rock in coastal engineering, CIRIA Special Publication 83*, CUR Report 154. Rotterdam: Balkema

Cole, W.F. & Sandy, M.J. 1980. A proposed secondary mineral rating for basalt road aggregate durability. *Ausralian Road Research* 10(3):27-37.

Fookes, P.G. Gourly, C.S. Ohikere, C. 1988. Rock weathering in engineering time. *Quarterly Journal of Engineering Geology* 21(1): 33-57.

Fookes, P.G. 1991. Geomaterials. *Quarterly Journal of Engineering Geology* 24(1): 3-17.

Gy, P.M. 1979. *Sampling of particulate materials; theory and practice.* Amsterdam: Elsevier.

Higgs, N.B. 1986. Preliminary studies of methylene blue adsorption as a method of evaluating degradable smectite-bearing concrete aggregate sands. *Cement and concrete research* 16: 525-534.

Hosking, J.R. & Tubey, L.W. 1969. *Research on low grade and unsound aggregates. Report LR 293.* Crowthorne (UK): Road Research Laboratory.

Houwink, H. & Verhoef, P.N.W. 1991. Detectie verborgen scheuren in waterbouwsteen. *Memoirs of the Centre of Engineering Geology no. 88.* GEOMAT.01.

ISRM 1988. Suggested methods for determining the fracture toughness of rock. *Int. J. Rock Mech. Min. Sci.& Geomech.* Abstr. 24:71-96.

ISRM: see Brown, E.T. (1981)

James, A.N. & Lupton, A.R.R. 1978. Gypsum and anhydrite in foundations of hydraulic structures. *Geotechnique* 28: 249-272.

Latham, J.P. 1991. Degradation model for rock armour in coastal engineering. *Quarterly Journal of Engineering Geology* 24(1): 101-108.

McNally, G. Williams, W. & Stewart, I. 1990. Soil & Rock Construction Materials. Key Centre of Mines Short Course, 27-29 November. Kensigton: University of New South Wales.

NF P18-577 1979. *Granulats: essai deval* Norme Francaise enregistrée.

Niese, M.S.J. Van Eijk, F.C.A.A. Laan, G.J. & Verhoef, P.N.W. 1990. Quality assessment of large armourstone using an acoustic velocity analysis method. *Bull.* IAEG. 42: 55-65.

Smith, M.R. & Collis, L. (eds) 1993. *Aggregates: Sand gravel and crushed rock aggregates for construction purposes (2nd edition).* London: The Geological Society.

Stapel, E.E. & Verhoef, P.N.W. 1989. The use of the Methylene Blue Adsorption test in assessing the quality of basaltic tuff rock aggregate. *Engineering Geology* 26: 233-246.

Tran Ngog Lan 1985. Deux nouveaux essais d'identification des sols argileux. Paris: Laboratoire des Ponts et Chaussees.

Tourenq, C. & Fourmaintraux, D. 1971. Propriétés des granulats et glissance routière, *Bulletin Liaison Laboratoire des Ponts et Chaussees 5.*

Uetz, H. 1986. *Abrasion und Erosion.* Mhnchen: Hansen Verlag.

Van de Wall, A.R.G. 1992. The polishing of aggregates used in road construction. *Memoirs of the Centre for Engineering Geology in the Netherlands 96.*

Verhoef, P.N.W. 1992. The Methylene Blue Adsorption test applied to Geomaterials. *GEOMAT.02, Memoirs of the Section for Engineering Geology in the Netherlands 101.*

Verhoef, P.N.W. Van de Wall, A.R.G. & Van Mier, J.G.M. 1994. *Geological construction materials, course notes MP3720.* Delft: Delft University of Technology, Faculty of Mining and Petroleum Engineering, Section Engineering Geology.

Zum Gahr, K.H. 1987. *Microstructure and wear of materials.* Elsevier.

Assessing the environmental impacts of sand harvesting from Kenyan rivers

JOHN S. ROWAN & J. JOHN KITETU
Institute of Environmental and Biological Sciences, Lancaster University, Lancaster, UK

1 INTRODUCTION

Like many other developing countries, Kenya faces serious environmental problems including exhaustion of natural resources. Environmental degradation makes it difficult to provide for basic needs (water, food, fuelwood, etc.) that negatively affects quality of life issues and may eventually constitute an obstacle to development. Throughout the world, rivers have long been exploited as sources of sand and gravel, however the impacts of mining on their stability and ecology has only recently been recognized (e.g. Lagasse et al. 1980, Collins & Dunne 1990). The harvesting of sand from river beds in Kenya now provides approximately 90% of annual consumption, but little is known of the local or cumulative effects of the extraction process. The United Nations Environment Program based in Nairobi, Kenya, have recorded sand and gravel extraction projects as one of the most pressing issues currently on the international agenda for environmental action (UNEP 1990).

Commercial river sand harvesting was first recognized in the early 1950s (Baker 1954) and has since grown to be an important element of the national economy. As early as 1982 the Kenyan Government acknowledged that uncontrolled river mining was causing widespread damage to river ecosystems (Diang'a 1992). However, indiscriminate gathering of sand and gravel continues to be a major issue of public concern. National newspapers frequently refer to conflicts between miners and local peoples (e.g. Musyoka 1983). On occasion disputes have led to violence and rural roads and tracks have been sabotaged to prevent the sand lorries gaining access to the river systems. Vehicles have also been burned by villagers seeking to protect water wells sunk in the channel-sand aquifers (e.g. Odongo 1987, Senda & Mulinge 1988).

A critical aspect in the mismanagement of the Kenyan river sand industry has been the failure to recognize the compound effects of many small but intensive harvesting operations. Effective regulation does not take place because of the absence of appropriate agencies. The problem is partially related to ownership laws which grant mineral rights to riparian landowners and gives them freedom to sell sand without consultation. The involvement of local authorities is restricted to issuing operating licenses to individual sand-haulage lorries. This scheme is ineffective as a control measure because its primary rationale is taxation rather than regulation. The sand harvesters are therefore free to decide where to collect sand, how much to extract, and how to get it to market. Mining river-beds near urban areas greatly increases profit margins for as little as 30 km of transport can more than double production costs (Bull & Scott 1974). Thus, commercial harvesting of river sand is most developed in those districts which are close to Nairobi and other main industrial urban centres.

A priori consideration suggests that in-stream mining operations may significantly impact channel stability and sediment transport processes. Nuisance impacts such as noise, dust, increased traffic and risk of road accidents may also occur. This not only makes the activity a 'bad neighbour', but can lead to economic repercussions such as the long term decline in resource base productivity. In semi-arid environments such as Kenya, seasonally-dry rivers are also key sources of water supply for the local populace. Other important qualities and values of river ecosystem which are not considered 'economic' include use for scientific research and the natural beauty of unregulated rivers. These qualities may exceed the market value of extracted aggregate and justify the protection of natural areas against encroachment from sand harvesting interests.

Such potentially serious problems require systematic study. Hitherto, no comprehensive investigation has been carried out on the quantitative effects of sand harvesting from Kenyan rivers. This paper seeks to identify the magnitude and significance of both specific and compound environmental effects at a range of scales spanning from local to national level. These findings offer a framework to inform policymakers and environmental planners about the need for the long term conservation and sustainable use of the natural resource base.

2 ENVIRONMENTAL EFFECTS OF RIVER MINING: A REVIEW

Sand and gravels are normally extracted from channel systems as part of navigation or flood-control programs; the commercial sale of the aggregate often being a secondary issue. Though a common place activity there are few reported instances where such projects are guided by environmental assessment procedures (Walker 1994) i.e. the construction of locks and weirs on the upper Mississippi River since 1930 has induced extensive sedimentation in the weir pools (Kesel & Yodis 1992). Similarly, Lagasse et al. (1980) speculated that gravel harvesting has caused an increase in channel avulsions on the lower Mississippi.

Sand and gravel mining often creates large holes in river beds and floodplains and has been shown to cause substantial channel degradation (e.g. Kostourkov 1972; Wayne et al. 1985). Removal of bed material in excess of the natural replenishment by sediment transported from upstream areas causes channel lowering which can propagate upstream and downstream from the extraction point. This may lead to the destabilisation of engineering structures such as bridge piers and pipe-line crossings. In an extreme example, Bull & Scott (1974) recorded the collapse of three highway bridges and destruction of seven houses following the failure of an off-channel excavation pond during a large flood on Tujunga Wash, California USA.

The removal or reduction in size of in-channel structures such as sand and gravel bars can cause adjacent banks to erode more rapidly and alter the geometry of meander bends (Wayne et al. 1985; Collins & Dunne 1989). Lowering of the main channel can also result in lowering of the local base level of tributary channels, thereby initiating headward erosion of the tributary system (Galay 1983). River diversions can occur when large amounts of material are removed from flood chutes or large off-channel ponds are excavated across the necks of meander bends. To prevent river degradation, it is essential that the rate of extraction of sand and gravel from the bed or other transitory storage sites such as bars and benches does not exceed the rate of replenishment. This factor is particularly important where bed controls such as armoured bars or subsurface cobbles and boulders are removed. Lagasse et al. (1980) have outlined the importance of armoured gravel bars in preventing excessive bed scour and channel avulsion.

In addition to important changes to channel morphology, sand and gravel removal can also generate indirect impacts to the hydrology and ecology of river systems. A potentially useful attribute of the mining activity is to enlarge the channel section thereby reducing the risk of overbank flooding and floodplain inundation because of increased conveyance capacity (c.f. Kesel & Yodis 1992). However, over-deepening may also lead to reduced rates of groundwater recharge to important floodplain aquifers. In the case of the Cache Creek, California, gravel removal caused the channel elevation to lower by 9 m. This loss of head effectively reversed the piezometric gradient thus the channel became a drain rather than a source of groundwater (Collins & Dunne 1990). Such adjustments can have significant impacts on potentially sensitive riparian ecological systems.

The response of planning authorities to the problems of river-based aggregate harvesting has varied in different areas of the world. Once recognized, the most common course of action has been to introduce bans on mining activity. These bans can either be time-based as in the case of a 5 year preventative ban in the River Gaula, Norway (Dahl & Tesaker 1992), or may involve the selective banning of mining from specific reaches i.e. near sensitive features such as bridges (Kira 1972). The purpose of this review is to illustrate that the phenomenon of river sand and gravel harvesting is widespread, but also to show that it has received relatively little attention in the international literature. Moreover, the majority of studies have been based in industrialized Western countries and involve mainly perennial river systems. The difference in the present study is that relatively little is known about the hydrology and hydraulics of ephemeral Kenyan rivers and that the target mineral is primarily sand.

3 STUDY APPROACH

Environmental considerations are becoming increasingly important components for sustainable development planning and improving natural resource management. Many countries, following the lead of the USA's Natural Environmental Policy Act of 1969 (NEPA), have introduced legislation requiring environmental impact assessment (EIA) to be included in the planning and operation of large development projects. While many recognize the need to adopt these procedures at the policy and sectoral level, in practice most assessments have been targeted to individual projects (c.f. Lee 1995). For this study, a systematic evaluation of the effects of numerous small-scale, but widespread, itinerant mining operations was needed. In such cases the cumulative impact can be greater than the sum of the component parts.

The study scheme reported here involved the systematic collection of data relating to the sand harvesting industry and its impacts. Information on the natural replenishment rates of sand and gravel from upstream areas was obtained by reference to the literature and the use of empirical sediment transport equations. Data on harvesting techniques and extraction rates were collected from local government registers, interviews with mining operators and by field survey. To evaluate the physical impacts on channel systems and riparian habitats a range of measurements were taken during field seasons in 1993 and 1994. These involved surveying of channel cross-sections, measuring sediment properties, groundwater depths and vegetation patterns. The timescales investigated were extended using aerial photographs and pre-existing monumented sections found at bridge and pipeline crossings. The 'significance' of different environmental impacts were evaluated from analysis of questionnaires distributed to harvesting companies, local residents and government officials.

4 BACKGROUND TO THE STUDY AREA

A generalized map of the geology of Kenya is shown in Figure 1a. This illustrates the main lithological groups (c.f. Dunne 1979). Precambrian rocks are widespread in western Kenya in the form of folded volcanics with numerous granitic intrusions. To the east of the Rift Valley, Precambrian rocks consist mainly of gneisses, schists and magmatites (metamorphic rocks which have been subjected to various degrees of granitization). After the Precambrian the region behaved as a cratonic block characterized by broad warping and eperiogenic movements during the Cenozoic (Baker 1970). To the east, marine sediments of Mesozoic and Cenozoic ages cover the Precambrian basement. The extensive faulting and volcanism which began in the Miocene and continued through the Quaternary produced the spectacular Rift Valley region which is bounded by a series fault escarpments reaching 3000 m above sea level. Cenozoic volcanism produced great thicknesses of basic and intermediate lavas and pyroclastic rocks formed extensive plateaux and hill masses around the margins of the Rift Valley. For example, the Yatta Plateau is composed of Tertiary volcanics (Kapiti phonolites).

The distribution of commercial sand and gravel harvesting operations is determined by geological controls on resource availability and an appropriate market for the product. In Kenya these factors combine to concentrate river mining within the districts situated to the south-east of Nairobi. Historically this area had the best developed transport infrastructure, and the geology is dominated by gneiss, granitoid and schistose basement rocks which weather into sandy soils and produce sand-bed rivers containing deposits highly suited to cement manufacture (see grade curves Fig. 1d). Most of the remaining area around Nairobi, especially to the north and west, is underlain by volcanic deposits which weather to silt and clay substrates unsuitable to the construction industry.

For environmental assessment purposes, the most suitable study would be inclusive of the western headwaters of the 38,000 km^2 Athi River basin. By contrast

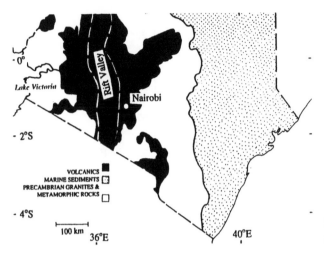

Figure 1a) Generalised geology of southern Kenya.

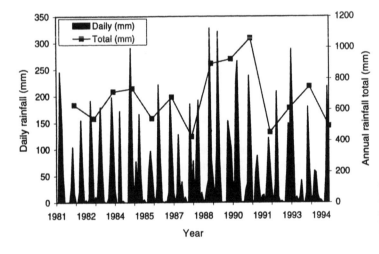

Figure 1b) Rainfall patterns, Ukai Farm Machakos District (1981-1995),

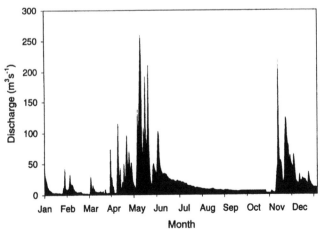

Figure 1c) 1980 discharge variations, Athi River (Station 3F02),

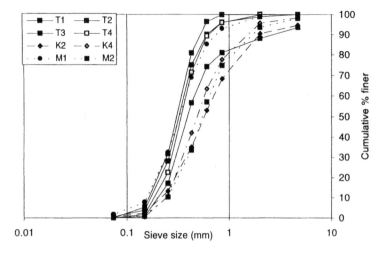

Figure 1. Continued. d) Bed sediment grain-size distribution in study rivers (T = Thwake, K = Kiati, M = Muooni).

nearly all the relevant data available was based in local government administrative units (Districts). Fortunately nearly all of Machakos and Makueni Districts (combined areas 11,000 km^2) fall within the Athi basin, giving these districts some functional relevance. These two districts (unified until 1992) supply approximately 90% of the national demand for construction sand (MRD 1992). Rivers closest to Nairobi have been exploited for the longest period, but as demand has grown and readily available local supplies have been exhausted, the industry has progressively extended in a southerly direction.

The topography and drainage of the study area strongly reflects the geological framework. Elevations range from 600 to 2150 m above sea level. Approximately 80% of the region lies within the arid and semi-arid zones defined by Mortimore (1991) with a highly seasonal annual rainfall between 500 mm to 1000 mm. More humid conditions occur in the central hill mass, characterized by steep slopes and intensive cultivation. The seasonality of the system is revealed in Figure 1b which illustrates rainfall to be concentrated into two main periods known as the 'short rains' (March-May) and the 'long rains' (October-November). The Tana River drains the northern part of Machakos District, while the Athi River and its tributaries Thwake, Muooni and Kaiti drain south-east and are the focus for the present investigation (Fig. 1c). These rivers are deeply incised with extensive sandy deposits along the channel bed and between rock ledges (Fig. 1d). The lower reaches exhibit meanders and transport sufficient gravel to maintain occasional point bars. Soil erosion has been recognized as a major problem for over a century, but as a result of the seasonality of stream flow, silt, sand and gravel are transported only intermittently by flood flows.

Some perennial springs offer the potential for gravity-fed water schemes for domestic and livestock water uses, but generally surface water is scarce. Groundwater supply is also limited by the low water-holding capacity of the basement complex rocks underlying the region. Instead, subsurface water is mainly obtained from the sandy river beds which are critical sources of water, particularly in rural areas (c.f. Gezahegne 1986). The sandy river beds offer the potential of a reliable piped water supply (by construction of sand or subsurface dams); however, such projects are increasingly jeopardized by uncontrolled sand mining.

5 OVERVIEW OF THE KENYAN AGGREGATE INDUSTRY

Prior to the present study there was no authoritative database on the Kenyan aggregate industry. A key aspect of this project was therefore to collate and synthesize data from a range of sources as shown in Table 1. The need for comparative data was highlighted by the contradictory nature of much of the evidence obtained. Again all output figures must therefore be treated as provisional.

A sand lorry registration scheme in the Municipal and County Offices of Machakos District provided several years of records reporting 'official' numbers and sizes of lorries operating in the period 1987 to 1992. However, the records were very incomplete and grossly underestimated the true activity levels observed by independent traffic counts. A comprehensive survey was made of sand operators at their head offices and at key locations such as sand markets. This provided information on important aspects such as the main locations where river sand was worked, the extraction methods used, haulage vehicle size, transport routes and number of trips per day. Additional information was obtained from a local 'sand co-operative' and by direct monitoring of extraction rates within selected study reaches.

Sand harvesting is widespread throughout the western tributaries of the Athi basin and in the southern tributaries of the Tana River (Fig. 2). The most recent reliable official figures date from 1991 when an annual yield of 6.5 Mt was estimated to have been mined in Machakos District (MDAO 1991). The findings of this study suggest that this total had increased to approximately 9.5 Mt by 1994. This rise in production by over 30% indicates that demand for construction sand has continued to grow in recent years, in spite of economic recession.

5.1 *Sand harvesting techniques*

Sand extraction methods are simple and can be defined into two groups. The first is non-mechanized and involves teams of labourers who scoop and gather sand into heaps in the channel bed where it is sold to an agent or driver (Fig. 3a). The drivers take the lorries into the river bed after paying a fee to the land owner. The lorry is

Table 1. Data sources for information on the distribution and intensity of sand harvesting.

Data source	Production yield (t)	Extraction sites	Lorry trips	Timing of extraction	Haulage routes	Extraction method	No. of employees
Operator interviews	x	x	x	x	x	x	x
District Council records	x	–	–	x	–	–	–
Field traffic survey	x	–	x	–	x	–	–
Yiika Sand Co-operative	x	–	x	x	x	x	x
Field channel survey	x	x	–	–	–	x	–

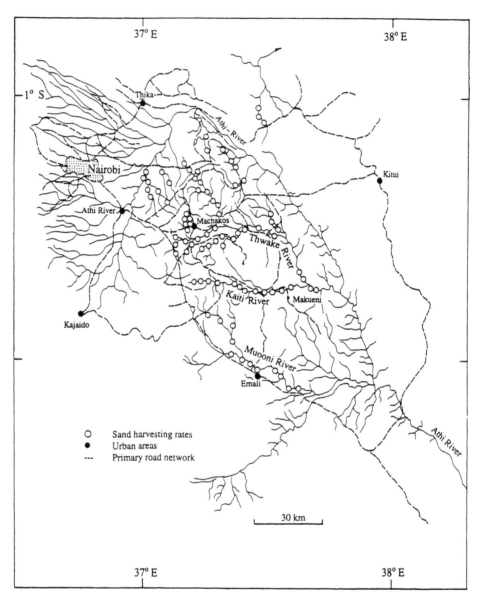

Figure 2. Distribution of major sand harvesting sites within the upper Athi basin.

then loaded by 3-5 people using hand shovels and transported to market. Manual labour is used throughout the study area, especially in the Thwake and Kaiti catchments. In the remoter reaches of the Kaiti River and where lorries cannot gain access to the river bed, sand is dragged to river banks before loading. These methods are slow and time consuming and typically cater for small < 10 tonne lorries. Because these methods are labour intensive, they provide widespread employment for locals as sand gatherers, loaders, off-loaders and drivers.

Where the deposits are of sufficient thickness and access is suitable, these traditional techniques are increasingly being replaced by mechanized shovels. These are driven into the channel where they scoop sand and load it directly onto lorries parked on the river bed (Fig. 3b). The lorries then transport the sand to collection depots or directly to urban construction sites. Further economies of scale are produced because mechanized operations also generally involve much larger vehicles up to 50 tonnes. In 1993 and 1994 these methods were restricted to selected sites on the Muooni and Iuuma Rivers. Although these methods are quicker and more economical they also result in greater damage to the river bed and channel banks. Employment opportunities for the local people are also reduced.

a

b

Figure 3. a) Sand harvesting on the Thwake river using traditional hand shovel methods. Note number of individuals involved and the lorries driving into the channel bed, b) Mechanised harvesting methods are becoming increasingly common. The effects of these machines is to increase the intensity of the impact and regularly involves complete stripping of the sand layer from the channel bed.

5.2 *Temporal and spatial patterns in harvesting activity*

A seasonal pattern is evident in harvesting activity. Figure 4a shows two production peaks during the months of November-January and April-May. These peaks immediately follow the two wet seasons which are periods of replenishment when soil erosion washes sediment into the channel network and fresh sand is deposited on river beds. The mining peaks lag the rainy seasons because during these times the earthen roads used by sand lorries to access the river beds often become washed out and impassable. As the rains recede the harvesters quickly return to exploit the fresh deposits. Intense competition means those rivers closest to the urban centres are rapidly depleted. As easily recovered sand becomes scarce many casual miners return to other forms of employment leaving a smaller core of full-time operators to continue harvesting and journey progressively farther from Nairobi.

There is a strong spatial dependency in the mining industry with proximity to urban areas and the quality of the road network being key determinants of harvesting intensity. In general, sand production rates declined with increased distance from Nairobi and other urban centres (Fig. 4b), and was greatest where the road ran parallel to the river or was bridged. Around Machakos town, many relatively small headwater streams including the Kithaayoni, Maluva, Ikiwe, Mutituni, Mwania, Love, Mavoloni and Ikondoni are devoid of sand most of the year because removal rates exceed the natural replenishment rates and fresh deposits are quickly exhausted. The response of the mining operators has been to extend their harvesting activities farther south coupled with a greater use of mechanized techniques. For example, Figure 4c illustrates a steady increase in output from the river Muooni since 1989 when the first mining took place.

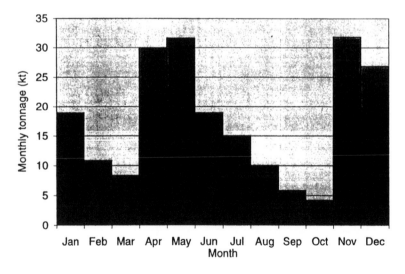

Figure 4. a) Monthly harvesting rates from Municipality Council rivers 1991.

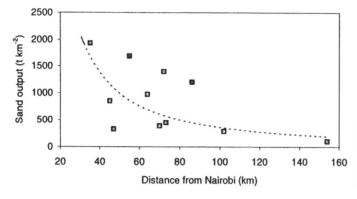

Figure 4. Continued. b) Distance decay effect of sand production by Division in Machakos District 1991

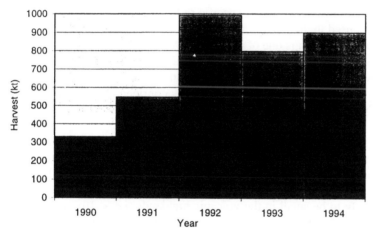

Figure 4. c) Growth in sand production from Muooni river.

6 IMPACTS OF SAND HARVESTING ON RIVER SYSTEMS

Sand harvesting causes a wide range of impacts. These can be classified according to a range of alternative schemes such as on-site and off-site, reversible or irreversible or isolated and cumulative. Impact intensity varies according to the type of technology used, the extent and timing of the harvesting and the geological and geomorphological characteristics of the mined area.

6.1 *Harvesting rates in relation to natural replenishment*

Channel mining impacts are heavily dependent on the rate of extraction of bed material relative to its replenishment from upstream sources. Where the harvested volume is small in relation to the influx, the effects are likely to be minor and short-lived. However, where extraction rates are high or exceed the replenishment rates, then significant and potentially irreversible changes to hydraulic performance and channel stability may result. Determination of natural sediment yields is therefore an important element in assessing the potential for degradation and formulation of management strategies.

Sediment yields are generally divided into two components. A fine-grained suspended fraction, mainly composed of silts and clays (< 63 μm) supported in the water column by turbulent eddies, and a coarse-bedload fraction which is transported by a combination of sliding, rolling and bouncing along the channel bed. The distinction is an artificial one, particularly in sand-bed rivers where the sandy substrate can be mobilized by relatively small flow and the bed contains a large proportion of suspendable material. Collins & Dunne (1990) described various methods for estimating bedload including 1. Direct measurement, 2. The use of empirical equations, and 3. As a proportion of published suspended sediment yields. This last method was adopted here because of the paucity of available hydrological and hydraulic data for the study rivers. Accordingly a regional bedload sediment yield was derived by reference to previous studies (Table 2). Reach specific sediment yield calculations were undertaken, but are not used in the present analysis.

The main sand harvesting area is one of the most severely degraded environments in Kenya. Soil erosion is widespread with agricultural land and unmetalled roads recognized as the two principle sediment sources (Dunne & Dietrich 1982). Sediment transport is highly episodic and runoff is generally confined to a small number of flash floods which experience high transmission losses due to infiltration into dry channel beds.

Dunne (1979) generated a series of regression equations for estimating mean annual suspended sediment yields based on catchment runoff, land use and topographic factors. Applying this method gives a regional estimate in the range of 200-4000 tonnes km^{-2} yr^{-1}, but it does not account for bedload. Dunne suggests that the bedload component is likely to be small in the volcanic highlands (i.e. < 10%), but will be much higher in Precambrian and metamorphic lithologies. Confirming this idea, Thomas et al. (1981) measured bedload to comprise 90% of the sediment yield of the Iuuni experimental catchment near Machakos town. It was observed that most flows were capable of transporting medium to coarse sand size (0.2-2.0 mm), while larger particles up to cobble-size (128 mm) were transported only during the highest flows.

Wain (1983) provided a reassessment of previous studies and reported the suspended sediment yield for the entire Thwake catchment to be 1265 tonnes km^{-2} yr^{-1}, with much of this consisting of sand-sized particles. This figure is in general

Table 2. Selected sediment yield data in the Athi basin, Kenya.

Catchment	Area (km²)	Sediment yield (t km⁻² yr⁻¹)	% Bedload	Author
Iiuni	11.3	535	90	Thomas et al. (1981)
Kalunda	25	1075	50	Dunne & Ongweny (1976)
Maruba	–	1500	50	Edwards (1979)
Thwake/Kaiti	4610	1265	–	Wain (1983)
Upper Athi	5590	156	–	Wain (1983)
Mukugodo*	2.5	1080	–	Ondieki (1995)
Athi basin	38000	200-4000	10	Dunne (1979)

* Catchment outside limits of Athi basin but situated on Precambrian basement complex.

agreement with the recent findings of Ondieki (1995). Taken together, an average annual suspended sediment yield for the Kenyan basement complex was estimated at 1100 tonnes km^{-2} yr^{-1}. Edwards et al. (1979) noted that bedload accounted for 50% of the sedimentation at Maruba reservoir. Accordingly the mean annual bedload was established to equal the mean annual suspended sediment, giving a tentative bed replenishment flux of 1100 tonnes km^{-2} yr^{-1}. It is acknowledged that this regional value carries a high degree of uncertainty.

6.2 *Cross sectional changes*

Sand harvesting causes major morphological changes which influences the hydraulic performance of the channel network and sediment transport processes. A comparison between surveyed mined and unmined 'control' reaches is shown in Figure 5a, and illustrates that the channel cross-sectional areas of mined reaches were much greater, often twice as large, as unmined reaches. Unmined reaches typically show only minor morphological changes, or limited amounts of net deposition. By contrast, mined reaches evidenced considerable deepening with increases of 1-2 m yr^{-1}. Where the harvesting is carried out by traditional hand methods, the most common approach is to excavate a series of trenches down the centre-line of the otherwise flat channel bed. Point-bars are also favourite sites for excavation. Where mechanized shovels are used the procedure generally involves the systematic removal of the sand across the entire channel width. This is particularly important in removing vegetated lateral bars which otherwise protect the bases of the high channel banks and is a major factor in accelerated bank collapse.

An insight into the longer term response of the channel network to the mining activity was gained from an analysis of historical aerial photographs. Bankfull widths were compared for a selection of representative reaches (Fig. 5b). In general, all the reaches examined evidenced a net increase in width over the observation period. Mined reaches showed the greatest rates of change, and this was particularly true of sites on the Thwake River. The Thwake has the longest history of mining and several reaches have more than doubled in width. Moreover, the evidence suggests that the rate of widening has accelerated, reflecting the growing demand for construction sand in recent decades. Overall the least amount of change was found in the unmined reaches of the Muooni River (< 10 year history of commercial mining). On inten-

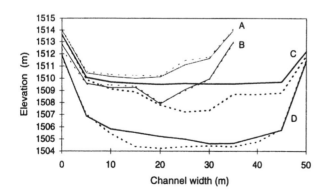

Figure 5. a) Changes in channel cross-sectional area of mined reaches (A, B) and unmined reaches (C, D) in the Thwake river (— 1993, --- 1994),

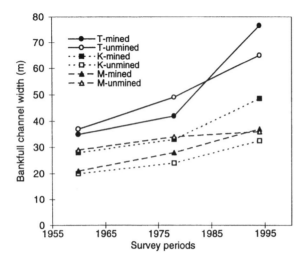

Figure 5. Continued. b) Channel bankfull width observed from field survey and historical air photographs (1960-1994). (T = Thwake, K = Kiati, M = Muooni).

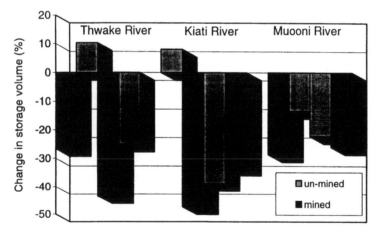

Figure 5. c) Observed changes in channel sand storage volumes for the three study rivers (1993-1994).

sively mined rivers such as the Thwake and Kaiti, many unmined reaches also experienced significant increases in bankfull width over the observation period. These patterns are tentatively linked to instabilities propagated upstream and downstream from harvesting sites (e.g. Collins & Dunne 1990).

6.3 *Channel sediment budgets*

Major cross-sectional changes also have important implications for sediment storage within channel systems. Sand depth was systematically measured along a series of study reaches varying in length between 250-1000 m. For each reach, the volume of sand was determined in 1993 and again in 1994. The results of the survey are presented in Figure 5c which expresses change in storage volume over one year. Two important results can be concluded from the analysis. Firstly, almost all sites evidenced a decline in the storage of sand including the nominally unmined areas. In some areas of the Thwake and Kiati, almost half of the original volume of sand was removed, and in several places this meant stripping the sand to bedrock. With the exception of some sites on the Thwake and Kaiti Rivers, most unmined sites also revealed net sand storage losses. The second conclusion is that in all the mined sites the rate of extraction was exceeding the natural sediment replenishment. In fact in mined reaches the channel bed dropped at approximately 0.43 m yr^{-1}, whereas unmined reaches showed a positive mass balance during the same period with a net deposition of 0.21 m. More detailed sediment budget dynamics for individual reaches have yet to be calculated, however; it is clear from these examples that the rates of sand mining in many parts of the river system are at least 3 times greater than the annual rate of replenishment.

6.4 *Implications of channel change for engineering structures*

The increased channel capacity has the advantageous effect of reducing the risk of flooding within the study area. However, the deepening and widening of channels has important indirect effects on engineering structures such as pipe-lines and bridges which cross the channels. Exhumation of the bridge piers and their foundations increases maintenance costs and the risk of accidents. The region has experienced human tragedies associated with bridge collapses during floods which are at least partially related to mining activity. The official response to these problems has been to introduce sand harvesting bans within a distance of 1-2 km upstream and downstream of bridge crossings (MRD 1992). Unfortunately, the policy is not based on a functional understanding of the geomorphological behaviour of these channels and hence is of uncertain value. Moreover, such bans are poorly enforced meaning that sand removal around bridge piers is still active.

Table 3 provides an indication of the growing threat that exists for further bridge failure within the study area. Two scour rates are shown, the first relating to a mean post-construction scour rate and the second to losses observed between 1993-1994. The long term rate of scour was reconstructed from sand marks identified on the original structures which indicated the maximum depth of sand loss approached 4 m on the river Thwake (Fig. 6a). In all cases, the most recent figures suggest that the scour rates have accelerated as a result of increased mining activity. Moreover, even

Table 3. Scour effects of sand mining on selected bridges sites in the study area.

River	History	Bridge type	Built	Scour depth (m)	Scour rate (mm yr^{-1})	1993/94 scour (mm yr^{-1})
Thwake	Unmined	High	1975	3.7	185	560
	Mined	Low	1960	2.5	71	380
	Mined	Low	1963	2.3	74	370
	Unmined	High	1982	1.3	108	260
Kaiti	Mined	High	1978	2.9	181	450
	Unmined	Low	1965	1.5	52	130
Muooni	Mined	High	1980	2.7	193	610
	Mined	Low	1965	1.9	66	210
	Unmined	Low	1971	1.1	48	120

a

b

Figure 6. a) Thwake Bridge illustrating the undermining of the central pier foundation. If scour continues at the present rate the projected time to failure is less than 10 years, b) An illustration of the importance of channel sand aquifers for well provision on the river Muooni, Machakos District. Within a two month observation period the water level in this well dropped by more than 1 m relative to unaffected sites upstream.

on sites lacking direct evidence of mining, the minimum scour depth exceeded 1 m and implies the propagation of scour both upstream and downstream of mining sites in order for the long profile of the river to achieve grade.

6.5 *Hydrological considerations*

Previous work in the USA has shown that mining-induced channel deepening can lower regional water tables due to reduced groundwater recharge (Bull & Scott 1974; Collins & Dunne 1990). In dryland countries such as Kenya, subsurface water from aquifer storage in sandy river beds provides a key source of water to rural communities. Shaw (1989) reported 33% of Botswana's groundwater is now derived from sand bed rivers. Wells dug into the river beds provide water for domestic use and livestock. The number of people who routinely depend on subsurface water is staggering; for example, a recent report estimated that wells on a 5 km stretch of the river Thwake provided the essential water requirements to approximately 10,000 people (MRD 1992). Clearly any threat to this water source is a major concern.

The impact of sand harvesting is to reduce the depth of the channel aquifer and often results in exposure of the groundwater to the atmosphere. Even a minor reduction in water storage can be critical in the marginal semi-arid environment of the study area. Evaporation from sand beds is a function of the water table depth below the bed surface, the grain size distribution and porosity of the aquifer. Hellwig (1973) showed that a water table 0.3 m below the surface of a fine-grained sand bed reduced evaporation rates by 50% relative to open surface water. Field observations confirmed that sand harvesting was depleting the channel bed depth and exposing the subsurface water to accelerated evaporation. Monitoring carried out during 1993 indicated that within two months of sand extraction that groundwater levels in mined reaches of the River Muooni dropped by 1.5 m relative to unmined reaches. Mechanized harvesting is widespread in the Muooni and locally sand is totally stripped to the underlying bedrock. Within the more populated Thwake and Kiati rivers the mining reaches also demonstrated groundwater depletion 0.8-1.0 m greater than in unmined sections.

The differences between mined and unmined reaches are shown in Table 4. The extreme lowering of water table levels throughout the study area during 1993-1994 are testimony to erratic rainfall totals in 1994 which were significantly below the long term mean annual totals. However, the greater reductions in water table depth within the mined areas clearly illustrates these channel reaches to be more susceptible to drought and imply greater losses of this key resource. An example of a river bed well showing evidence of severe drawdown due to upstream mining activity is illustrated in Figure 6b. In an extensive questionnaire survey, approximately 90% of respondents reported diminished water supply to be the most significant environmental impact of mining and the primary cause of conflict between mining and local interests.

Table 4. Mean water table reductions within the study rivers 1993-1994.

River	Lowering of ground water depth (m)	
	Unmined	Mined
Thwake	− 0.83	− 1.85
Kiati	− 0.90	− 1.60
Muooni	− 0.85	− 1.85

Reduced water supply impacts may be thought of in terms of three types of increased costs to the local people. Firstly, extra travel distances for people and animals to reach alternative sources of water, secondly, lost income stemming from lower production of vegetables, fruits, good pasture and occasional activities such as brick manufacture and thirdly, the costs associated with the need to establish new sources of water supply. When the channel aquifer dries, especially as a result of sand harvesting, 85% of the local community have to look for other sources. Most of them (mainly women) take as long as six hours on one trip, covering a distance of 5-10 km to reach an alternative water supply. Significant indirect impacts on the general development of the area result from the lost time spent searching for water, particularly during the long dry season between June and September.

The impact of the mining activity upon water quality was also assessed. Water samples were tested for pH, conductivity, salinity and turbidity. Turbidity was the only factor which consistently distinguished mined and unmined sites, explained by the effects of digging and lorry trafficking, that produced sediment concentrations above WHO 'desirable standards' for drinking water. Of potentially greater concern was the observation that where sand harvesting lorries and mechanized shovels were working in the channels, there was evidence of leaking gasoline and engine oil. In some reaches layers of oil were clearly visible on the water and sand surfaces. Moreover, where the water table was exposed the temporary pools act as drinking holes which concentrate cattle with increased risk of faecal-coliform contamination. The potential for these pollutants to damage drinking water supplies is obvious. Locals blamed high rates of stock mortality (i.e. anecdotal reports of 5000 head of cattle lost between 1993-1994) on water quality problems.

6.6 *Impacts of sand mining on riparian ecosystems*

River and riparian ecosystems are amongst those most frequently damaged by human activities (Allan 1995). Removal of channel substrate and attendant bank erosion processes causes widespread destruction of habitat. In semi-arid areas like Central Kenya important foodchain organisms such as locusts and grasshoppers breed in the seasonally dry channel beds. Vegetation is susceptible to a range of direct impacts such as abrasion from sand lorries and indirect effects resulting from heightened physiological stresses due to soil compaction and falling groundwater levels in the riparian zone. Dramatic reductions of species diversity and percentage cover in mined reaches are shown in Table 5. A quantitative analysis, using the cluster analysis and ordination techniques within the VESPAN III program (Malloch 1994), confirmed significant differences between semi-natural and mined sites but these results will be fully described elsewhere.

Table 5. Analysis of riparian vegetation community structure.

Vegetation parameter	Unmined reaches	Mined reaches
Percentage ground cover (%)	90	10
Floristic diversity (number of species)	23	17

Loss of cover and changing vegetation patterns have important feedback effects by accelerating erosion in riparian areas. All farmland and rural roads exposed to trafficking from the sand lorries are also adversely affected by increased erosion. Vegetation can influence the erosion resistance of a soil by 1-2 orders of magnitude (Kirkby & Morgan 1980), and may enhance the stability of channel banks with respect to mass failures (Thornes 1990). Dunne & Deitrich (1982) have suggested erosion and particularly gullying of roads may be the key factor promoting landscape degradation in drylands such as Kenya. Clearly the extensive sand harvesting traffic has the potential to aggravate this problem.

6.7 *Impacts on local communities*

In order to assess the impact of sand harvesting on local communities an extensive interview and questionnaire program was carried out principally directed towards chiefs, assistant chiefs and village leaders. As spokespersons for their respective communities these individuals were expected to articulate the wider views of the peoples in the region. The questionnaires were answered by over 175 respondents, but because of the hierarchical nature of the society this may reasonably represent coverage of 30,000 people.

The purpose of the questionnaire was to gain an insight into the views of the local communities; to learn what their main concerns were and what solutions to the problems might be forwarded. As part of a larger questionnaire a number of potential impacts were listed. The respondents were asked to score the impact on a 1-5 scale, ranging from insignificant 1 to very significant 5. This approach is similar in form to a scoping survey used to identify the broad range of potential impacts and a preliminary means to focus resources and subsequent investigations only on those impacts deemed 'significant'.

Two methods were used to compare responses. In the first a numerical 'impact index' was derived as the mean score from the answers given, where 1 is the lowest possible score and 5 is the highest. This impact index was then used to rank the impacts shown in Table 6. In the second, scores of 4 and 5, expressed as a percentage, were considered to represent 'significant impacts' to the local people. The results

Table 6. Results of scoping questionnaire by local resident groups.

Environmental parameter	% response in each catagory					Impact index	% significant
	1	2	3	4	5		
Water supply	1	3	8	20	68	4.5	88
Water quality	3	3	11	32	52	4.3	84
Dust	1	3	29	47	20	3.8	67
Land degradation	5	8	31	36	20	3.6	56
Vegetation damage	5	23	27	20	25	3.4	45
Conflict with miners	12	13	24	40	11	3.2	51
Channel instability	9	16	28	43	4	3.2	47
Stock problems	8	43	24	19	7	2.7	25
Traffic	15	31	40	11	4	2.6	15
Noise	47	29	5	5	13	2.1	19

clearly highlight water supply and quality to be the primary concerns. Dust and damage to farmland and crops from sand lorries were the next most important topics. By contrast increased traffic problems and increased noise were considered to be insignificant.

7 MANAGEMENT IMPLICATIONS AND PERSPECTIVE

The Kenyan Government has clear policies on environmental protection for sustainable development. However, itinerant sand harvesting falls outside these provisions, hence the industry is unregulated and out of control. Predicting the level of growth in the national economy is difficult but growing demand for construction sand seems certain. In order to meet both present and future demands for river sands, whilst ensuring environmental protection, there is a need for new and more effective mechanisms of control on the mining operations.

One of the main reasons why environmental degradation continues to grow and remains unchecked is the sectoral nature of Central and Local Government. Water resources are the primary remit of the Ministry of Water Development, but other Ministries including Environment and Natural Resources, Agriculture, Tourism and Wildlife, and Power and Communications are also concerned with water and environmental issues. Regional planning is undertaken through the District Development Committees but again environmental responsibility is blurred between subcommittees (Fox 1988). The complex range of on-site and off-site environmental impacts span a range of sectors from transport, minerals, industry, water and agriculture (Fig. 7). With so many effects and so many organizations with vested interests no single government agency has assumed the necessary responsibility to tackle the problem. Even TARDA (the Tana and Athi River Development Authority) has to balance competing demands between HEP, irrigation and domestic water supply, and by necessity often has to tolerate undesirable compromise (Rowntree 1990).

Over harvesting is encouraged by the lucrative market for sand in Nairobi. A key observation of the present study is that sand is being extracted at rates significantly greater than the rates of natural replenishment by fluvial transport processes. Serious degradation of the riparian environment was revealed by channel scour approaching 4 m, and channel widths more than doubling in the past 30 years. In response to a few specific cases of intense local opposition nominal harvesting bans and site specific river restoration schemes have been proposed. However, the successful implementation of such schemes has been limited (MRD 1992).

Local authorities must play a more direct role in the planning and management of natural resources within their jurisdiction. The current licensing procedures are viewed only as a means of raising revenue from the sand lorry permits. A thorough and systematic appraisal of sand and gravel resources in the rivers needs to be established. A code of practice should be issued to harvesters and sand merchants providing guidance on conservation needs. Only a voluntary scheme appears to have any future. Specific information on the distribution of activities and harvesting returns is also needed. These figures can then be used as a means to guide harvesting rates. However, little is known about the sediment dynamics of these rivers and research is needed into transport rates and the time scales of sediment storage and redistribution.

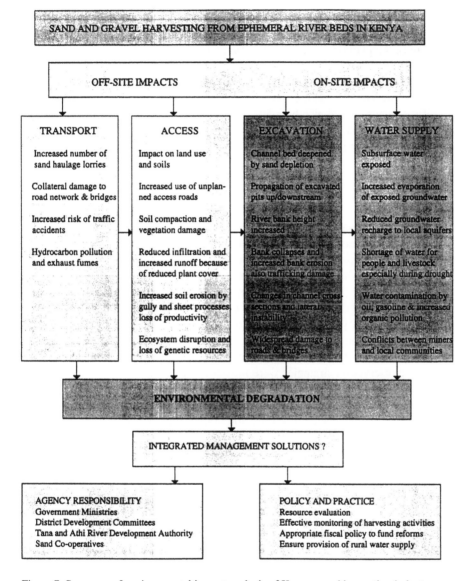

Figure 7. Summary of environmental impact analysis of Kenyan sand harvesting industry.

The location and extent of workable deposits is difficult to infer from geological maps. Establishing effective survey and monitoring procedures to evaluate sand and aquifer characteristics within seasonal rivers is therefore a priority. At present more than 95% of the local population receive their water for domestic and livestock use from rivers where sand is harvested. This is the central cause of conflict between harvesters and the local community. Commercial sand extraction should be controlled in such a way that it does not jeopardize water supplies from channel aquifers. One method is to construct checkdams across channel sections to provide subsurface water storage dams. These structures serve the dual purpose of stabilizing the channel course, controlling degradation and providing a more predictable supply of sand filtered water for rural communities and their livestock.

In rural areas, the planning for, and good maintenance of, access roads used by the heavy sand vehicles is essential as it would minimize trespassing and associated land degradation. Traffic surveys in the area should be conducted to assess levels and composition of present use, taking into account, daily, weekly and seasonal variations. Careful surveys should also be made of the present road network and the routing of sand lorries should be controlled. Ultimately the success of such schemes will be dependent on the administrative, political and legal framework in which it operates.

7.1 *Perspective*

The harvesting of river sand and gravel has reached unsustainable levels in Kenya. Machakos District has been the traditional source of sand to the Nairobi markets, but as the most readily accessed sand deposits (i.e. those near urban centres and surfaced roads) are depleted, the industry and its associated environmental degradation are systematically spreading. River sand is principally derived from soil erosion on agricultural land which is perhaps an even more pressing environmental problem which threatens the sustainability of the Kenyan resource base. Soil conservation strategies, linked to catchment management planning is required. However, there is a paradoxical dependency between the sand industry and the soil erosion problem. If effective soil and water conservation strategies were introduced then replenishment of sand to the channel network would fall, thus it is clear that solving one environmental problem would exacerbate another. This situation highlights the complexity of environmental processes and the need for a fully integrated approach to natural resource management.

ACKNOWLEDGEMENT

John Kitetu gratefully acknowledges the receipt of a overseas scholarship from the Ministry of Education, Kenya. We would also like to thank those individuals who helped with the fieldwork in Kenya, particularly Mr Muchiri, and to Dr Karanja for making available laboratory facilities in the Department of Agricultural Engineering, Egerton University, Kenya.

REFERENCES

Allan, J.D. 1995. *Stream Ecology*. Chapman and Hall. London.
Baker, B.H. 1970. Tectonics of Kenya's Rift Valley. Unpublished PhD thesis, University of East Africa, Nairobi, Kenya.
Baker, R.G. 1954. Geology of the Southern Machakos District. *Geological Survey of Kenya Report No. 27*, GSK, Nairobi.
Bull, W.B. & Scott, K.M. 1974. Impact of mining gravel from urban stream beds in the Southwestern States. *Geology* 2: 171-174.

Chen, Y.H. & Simons, D.B. 1979. Geomorphic study of Mississippi River, *Journal of the Waterway Port Coastal Ocean Division Proceedings American Society of Civil Engineers* 105: 313-328.

Collins, B.D. & Dunne, T. 1989. Gravel transport, gravel harvesting and channel-bed degradation in rivers draining the Southern Olympic Mountains, Washington, USA. *Environmental Geology and Water Science* 13: 213-224.

Collins, B. & Dunne, T. 1990. *Fluvial Geomorphology and River Gravel Mining. A Guide for Planners*. Seattle, Washington.

Dahl, T.E. & Tesaker, E. 1992. Monitoring of the effect of gravel mining on the morphology of Gaula river; Norway. In Bogen, J., Walling, D.E. & Day, R. (eds), *Erosion and Sediment Transport Monitoring Programmes in River Basins*, IAHS Publ. No. 210, poster volume, pp. 54-57.

Diang'a, A. 1992. Dangers of wanton sand harvesting. *The Daily Nation,* April 24, 10.

Dunne, T. & Ongweny, G.S.O. 1976. A new estimate of the rate of sediment rates on the Upper Tana River in Kenya. *Kenya Geographical Journal* 2: 109-126.

Dunne, T. 1979. Sediment yield and land use in tropical catchments. *Journal of Hydrology* 42: 281-300.

Dunne, T. & Dietrich, W.E. 1982. Sediment sources in tropical drainage basins. In Lal, R. (ed.), Soil Erosion and *Conservation in the Tropics*. ASASSSA, Madison. pp. 41-55.

Edwards, K.A. 1979. Regional contrasts in rates of soil erosion and their significance with respect to agricultural development in Kenya. In Lal, R. & Greenland, J. (eds.), *Soil Physical Properties and Crop Production in the Tropics*. Wiley, Chichester, pp. 441-454.

Fox, R.C. 1988. Environmental problems and the political economy of Kenya. *Applied Geography* 8: 315-335.

Galay, V.J. 1983. Causes of river bed degradation. *Water Resources Research* 19: 1057-1090.

Gezahegne, W. 1986. Sub-surface flow dams for rural water supply in arid and semi-arid regions of developing countries. Unpublished MSc thesis, Tampere University of Technology, Finland.

Hellwig, D.H.R. 1973. Evaporation of water from sand, IV: the influence of the depth of the water-table and the particle size distribution of the sand. *Journal of Hydrology* 18: 317-327.

Kessel, H. R. & Yodis, E.G. 1992. Some effects of human modifications on gravel and sand-bed channels in Southwestern Mississipi USA. *Environmetal Geology and Water Science* 20: 93-104.

Kira, H. 1972. Factors influencing the river behaviour – river-bed variation due to dam construction and gravel gathering. *Transactions of the 8th Congress of International Commision for Irrigation and Drainage*. Varna Bulgaria 5: 405-432.

Kirkby, M.J. & Morgan, R.P.C. (eds). 1980. *Soil Erosion*. Wiley, Chichester.

Kostourkov, G. 1972. Exploitation of regulated and non-regulated rivers with gravel pits in the river bed. *Transactions of 8th Congress of International Commission for Irrigation and Drainage*, Varna, Bulgaria 5: 1-10.

Lagasse, P.F. Winkley, B.R. & Simmons, D.B. 1980. Impact of sediment of gravel on river system stability. *Journal of the Water Port Coastal and Ocean Division* 106: 389-404.

Lee, N.I. 1995. *Environmental Impact Analysis*. Routledge, London.

Machakos District Agricultural Office. 1991. *Sand Harvesting Report on Machakos District*. Soil and Water Conservation Section, Machakos, Kenya.

Ministry of Reclamation and Development of Arid, Semi-Arid Areas and Wastelands. 1992. *Environmental Action plan* (Draft) *for Machakos District*. Nairobi, Kenya.

Malloch, A.J.C. 1994. VESPAN III: a computer package to handle and analyse multivariate species data handle and display species distribution data. University of Lancaster. Lancaster.

Mortimore, M. 1991. *Environmental Change and Dryland Management in Machakos District, Kenya: Environmental profile*. Ministry of Reclamation and Development of Arid, Semi-Arid Areas and Wastelands, Nairobi, Kenya.

Musyoka, C. 1983. MP calls for end to sand row. *Daily Nation,* August 12, 5.

Ondieki, C.M. 1995. Field assessment of flood event suspended sediment transport from ephemeral streams in the tropical semi-arid catchment. *Environmental Monitoring and Assessment* 35: 43-54.

Ondongo, T. 1987. New clashes over sand harvesting. *Daily Nation*, August 17, 4.

Rowntree, K. 1990. Political and administrative constraints on integrated river basin development: an evaluation of the Tana and Athi Rivers Development Authority, Kenya. *Applied Geography* 10: 21-41.

Senda, K. & Mulinge, P. 1988. *Daily Nation*, February 8, 13.

Shaw, P. 1989. Fluvial systems of the Kalahari a review. *Catena Supplement* 14: 119-126.

Thomas, D.B. Edwards, K. A. Barber, R.G. & Hogg, I.C.G. 1981. Runoff, erosion and conservation in a representative catchment in Machakos District, Kenya. In Lal, R & Russell, E.W. (eds), *Tropical Agricultural Hydrology and Watershed Management*. Wiley, Chichester. pp. 395-417.

Thornes, J.B. (ed.). 1990. *Vegetation and Erosion*. Wiley, Chichester.

Wain, A. S. 1983. Athi river sediment yeilds and significance for water resources developments in soil and water conservation in Kenya. *Proceedings of 2nd National Workshop, Universty of Nairobi Occassional Paper 42*. pp. 274-293.

Walker, H.J. 1994. Environmental impact of river dredging in Arctic Alaska (1981-1989). *Arctic* 47: 176-183.

Wayne, D.E. Geary, P.M. & Outhet, D.N. 1985. Potential impacts of sand and gravel extraction on the Hunter River, New South Wales. *Australian Geographical Studies* 23: 71-86.

United Nations Environment Programme. 1990. Enivironmental guidelines for sand and gravel extraction projects. *Environmental Guidelines No.20*. UNEP, Nairobi.

Mineral extraction and the conservation of geotopes – Experiences in the Netherlands

GERARD PIETER GONGGRIJP

Institute for Forestry and Nature Research, Wageningen, Netherlands

1 INTRODUCTION

Since mankind started its career on earth, the products of geological processes have been won and used. Clay and loam are used for pottery, making ovens, and brickstones and plastering reed walls of houses or facing dykes. Sand is used in the brick and glass industry, in roads, iron-foundries and concrete. Gravel is useful as drainage and necessary in concrete and roads.

Erratic boulders were important for building graves or as monuments in prehistoric times. They formed the ballast in ships or were used in armouring dykes and in road building. Chalk and limestone in the eastern (Fig. 1) and southern part of the Netherlands has been cut as building stone, burnt for lime or ground for agricultural purposes. Sandstone and quartzite in South Limburg was processed in housebuilding. Browncoal was an effective fuel. Peat, covering large parts of the Pleistocene and Holocene areas in the Netherlands, was found an attractive fuel. In coastal areas peat was sometimes burnt to extract salt. Now it is dug for peat-dust to cover our gardens. In the formerly undrained eastern part of the Netherlands, bog ore could be exploited in the brook valleys every 40-50 years for industrial purposes.

All these extractions, serving many interests, were important in the process of progress. But there is a reverse to the coin: these extractions cause wounds in the landscape. Landforms, geological and pedological sequences and phenomena are being affected or even disappearing. The only esker system in the Netherlands from the Saalian has been dug away for an extensive part.

Along with the substratum, parts of the vegetation and related fauna and cultural-historical features like artifacts and historical patterns are lost. Hydrological patterns change, influencing ecological situations locally but also in more distant areas. For example the extensive German brown coal mining industry not only has its impact even in parts of the Netherlands despite being relatively far from the border .

Initially most products were extracted from local resources for local need. But even in ancient times rare and valuable materials like flint and gold, used as a basis for tools and ornaments, were transported over large distances. Now a one day distance is not a problem at all. For example Norwegian granite is exported all over the world as building stones and Polish peat finds its way to the Dutch market. This ex-

Figure 1. Limestone winning in the province of Gelderland. Mineral extraction results in irreversible change of landscapes. Therefore decisions should be made after a sound weight of all interests, including geomorphology.

pansion of trade means that locally the mining of products used for export may have a tremendous influence on the landscape.

2 MINING AND LANDSCAPE

2.1 *Mining policy*

The exploitation of minerals is necessary to maintain or improve the quality of society. But untouched nature and landscape is also an essential part of this quality. To keep the balance a set of legal regulations has been developed, though the existing regulations are often obsolescent and no longer sufficient. Recent changes in Dutch policy, aimed at a more environmentally friendly approach, have been worked out on a national scale in:

– The draft note 'Gegrond ontgronden' (Earth Removal Note) by the Ministry of Transport and Public Works (1988),

– The 'Natuurbeleidsplan' (Nature Policy Plan) by the Ministry of Agriculture, Nature Management and Fisheries (1990),

– The 'Nota Landschap' (Landscape Note) by the Ministry of Agriculture, Nature Management and Fisheries (1992),

– The 'Structuurschema Groene Ruimte' (Structure Scheme Green Space) by the Ministry of Agriculture, Nature Management and Fisheries and the Ministry of Housing, Planning and Environment (1993),

– The 'Structuur Schema Oppervlaktedelfstoffen' (Structure Scheme on Surface Mineral Supply) by the Ministry of Transport and Public Works (1993).

The draft Earth Removal Note confirmed in the recent Structure Scheme on Surface Mineral Supply points out the national government's responsibility for ensuring that sufficient minerals are available, when needed. The licensing authorities (provinces) are responsible for regulating the location and method of extraction and for indicating potential mining localities in their regional plans. The general starting point for the national policy on extraction is that mining should be socially acceptable. That means that extraction of minerals should be carefully weighed against other social interests. In general, this can be translated to the use of more alternative products and recycling and with respect to nature conservation that mining in valuable nature and landscape areas should be avoided.

The Nature Policy Plan, the Landscape Note and the Structure Scheme Green Space give insight into the categories of nature and landscape areas. Ecologically valuable areas, that form part of the so-called National Ecological Network, earth-scientific, culturally and historically valuable sites should be kept free of mining. When mining is allowed, the conditions will include terms for reclamation of the exploited area. A comparable policy arose in the United Kingdom, where a strategy for the incorporation of sustainable development principles into minerals planning is being developed (Scott 1994).

2.2 *Geotopes in the landscape*

In order to maintain or even improve landscape quality a list of these ecological, earth-scientific, culturally and historically important areas is essential for nature and landscape conservation. In 1969 the working group Gea was established to make an inventory of all important geological, geomorphological and pedological sites (geotopes). The criteria used were: rarity, soundness, representativity, scientific and educational importance. Important in this respect is the non-replaceability of most of the sites, because they are the results of geological processes in the past (Gonggrijp & Boekschoten 1981). Sub-tropical soil profiles from Tertiary age or ice-pushed ridges from glacial times will certainly not be formed in the near future in the Netherlands.

More than eight hundred sites have been selected, of international, national as well as provincial importance, and published in provincial reports. On behalf of the Nature Policy Plan 119 internationally and nationally important geomorphological sites were listed (Gonggrijp 1989, Ministry of Agriculture, Nature Management and Fisheries 1990). The policy plan allows also for the safeguarding for research and education of for example, rare drumlins or coversand hills etc. which are otherwise ecologically uninteresting. Small sites likes outcrops and erratic boulders have been excluded, not because of their lack of importance but purely because of their size.

3 MINING AND OUTCROPS

3.1 *Geological interest in exposures*

Surface mining creates outcrops in formerly hidden geological formations, which are of interest to geologists for scientific and educational reasons (Figs 2-3). In Dutch

358   *Gerard Pieter Gonggrijp*

Figure 2. Excursion on the type locality of the Tiglien for discussing on
protection of this international scientificly important site (province of Limburg).

Figure 3. Opening of an educationally
important Geological monument in a
former sand pit, showing cemented
Early Pleistocene beach sediments near
the Belgium border (province of Zee-
land).

legislation the owner or the mining company is responsible for the security of the pit or quarry. However, during the period of exploitation, scientists and amateurs can usually request permission to visit the pits and quarries at their own risk.

The opportunity to do research is very important because a lot of geological data becomes available, spread in time and space, giving the three-dimensional picture of the local geology, and contributing to our understanding of its genesis. Also exposures in pits and quarries allow all kind of detailed research that otherwise is almost impossible to carry out by drilling or seismic methods.

In a lowland country like the Netherlands in which the number of natural outcrops is very rare the artificial ones made by mining are of great importance. The understanding of the origin of several landforms had to be revised after detailed studies in man-made exposures. For example, after a detailed study of just one temporary outcrop, features geomorphologically interpreted as eskers from the Saalian glacial period, turned out to be meandering longitudinal dune systems of Weichselian age.

In the Netherlands natural outcrops are present only in river valleys in the eastern and southern part. In the Pleistocene half of the country nearly all existing outcrops are artificial: road cuts, temporary excavations for houses and service-pipes and mineral exploitation. In the Holocene part even the artificial exposures are exceptional. Sides of ditches and special activities like tunnel building allow a glance into the earth. This stresses the importance of pits and quarries for geological research especially in the Dutch situation.

In the past there existed a rather dense network of small pits and quarries in Pleistocene and older parts of the Netherlands. This was a rather favourable situation for earth scientists to become informed about the general geology at the surface. But due to a change to a more environmentally friendly policy in the 1970s these excavations were not allowed anymore and therefore closed. At first they were replaced by a smaller number of large and often very deep pits, which were if possible concentrated in less valuable areas. In general this tendency of concentration had some positive influence on nature and landscape protection as a whole, as fewer areas were effected by mining activities. For study and education in geology, however, it had pros and cons. Instead of exposures in many different geological features, important for an overview of geological history, there were a few very enormous exposures of a limited number of features, providing possibilities for detailed studies. However, these large exposures were often difficult to access and dangerous because of steep walls and excavation below groundwater level. Nevertheless the continuous increase of demand of minerals lead to a further enlargement of the excavations and loss of natural landscapes. This caused not only direct damage but also indirect by changing the hydrological situation in neighbouring areas.

At the end of the 1980s, extraction became more and more subject to social pressure which, among other things, led to the 'Earth Removal Note'. This meant that existing exposures in pits and quarries should be carefully selected and conserved for future generations of earth scientists and the public for research and education. Of course this tendency of loss of exposures is not restricted to the Netherlands (Bridgland 1994, Schlüchter 1994). In several European countries a policy of recycling, offshore mining and alternative products is increasingly preferred.

3.2   *Restoration of pits and quarries: Mind geotopes*

For a conservationist of earth-science sites the extraction of minerals poses a significant dichotomy between conserving geological landscapes on one side and creating and preserving new outcrops on the other. To avoid being accused of opportunistic behaviour the geoconservationist should stand up for the conservation of the threatened landscape in the first place. If the decision to allow mining has been taken, the conservationist has to ensure that geologically important sites in the pits and quarries remain accessible for research and education after extraction ceases. This will necessitate co-operation with officials and mining companies to ensure the safeguarding of geological heritage in future pits or quarries. This means that *licenses should include restoration conditions in favour of conservation of geological sequences and phenomena.*

Rather common at the moment is restoration in favor of nature development. This means reclaiming former pits and quarries in such a way that they serve an optimum natural development, which of course does not have to exclude the geological aspects. An active policy in conserving parts of pits and quarries should put an end to continuous loss of geological sites by infilling (Fig. 4), replantation or neglect.

Earth-science conservation, if relevant, should be an obligatory part of the restoration plan on which a license is granted. The new regulations in the Netherlands make it much easier to attach conditions to a license in order to protect these interests. They include 'the restoration of the area when quarrying is completed, co-responsibility for plans to modify the uses to which the area surrounding the remaining site can be put and making the license-holder bear part of the capitalized management costs of the area following quarrying'. It remains to be seen how effective they are implemented.

Figure 4. A great deal of geologically important exposures disappear by refuse dumping. In this case a trench has been kept open (province of Limburg). Neglect and finishing off also cause loss of outcrops.

Sometimes during the extraction process, unique features become exposed which should be preserved from a scientific point of view but which may not be possible, because of the perforce continuation of mining and the cost of possible compensation. However, under favourable exploitation conditions there may be a solution, by extending the allowed mining area in exchange for the establishment and preservation of the geotope. Sometimes the pit or quarry can be deepened locally and filled in later with waste materials or the walls can be restored in such a way that more material can be extracted. Redefining the license conditions demands willing co-operation from both industry and the authorities (Fig. 8).

4 OUTCROP STABILITY

Saving an outcrop in a pit or quarry from covering, plantation or land fill is one thing, but arranging and maintaining it safely for research and education is quite an other (Fig. 5). Each excavation produces an unstable situation. Pit and quarry faces are subject to degradation by *weathering* (especially frost action), *erosion* (especially run off), *mass movements* (falling as well as sliding and flowing) and *human activities* (collecting samples, digging, cutting and walking).

How effective these degradational processes are depends largely on:

– External conditions such as *climatic circumstances* (frost, precipitation), *face exposure* (sun-exposed or not), *slope characteristics* (angle and height), *trembling natural as well as artificial* (earthquakes, blasting and heavy traffic),

– Internal properties like *rock type* (Tables 1-2), *structure* (weak zones, dipping slope) and *hydrological properties*.

Figure 5. Cretaceous chalk layers near Valkenburg (province of Limburg) cleaned by a local nature organization with financial support of the provincial authorities.

Table 1. Average slope values for bedrock excavations (Baker & Gray 1960, p. 11-35). The table shows slope angles in various types of hard rock, based on experiences in quarries. In general, faces in most hard rocks can be rather steep without losing too much stability. Some rock types like platy or clayey sedimentary rocks may cause problems. But in comparison with unconsolidated rocks there is still a great difference.

Rock type	Horizontal slope	Vertical slope	Degrees
Igneous			
– granite, trap, basalt and lava	1/4:1	1/2:1	76°-63°
Sedimentary			
– massive sandstone and limestone	1/4:1	1/2:1	76°-63°
– interbedded sandstones, shales and limestones	1/2:1	3/4:1	63°-53°
– massive claystone and siltstone	3/4:1	1:1	53°-45°
Metamorphic			
– gneiss, schist and marble	1/4:1	1/2:1	76°-63°
– slate	1/2:1	3/4:1	63°-53°

Table 2. Soil properties and their relative influence on stability. The conservation of unconsolidated rocks is much more difficult. Properties like texture (size, sorting and rounding) and structure (packing and cohesion) influence the slope stability for a great deal.

Soil properties		Influence on stability	
		Positive	Negative
Texture	Size	Fine	Coarse
	Sorting	Poorly sorted	Well sorted
	Roundness	Poorly rounded	Well rounded
Structure	Packing	Compact	Loose
	Cohesion	Cemented	Not cemented

In summary, most igneous, metamorphic and consolidated sedimentary rock faces are relatively stable unless the following is present:
– Weak zones, joints, etc. or sensitivity to frost action, causing falling and sliding; potential slip faces like clayey beds causing sliding,
– Penetration of relatively large quantities of water causing sliding and flowing,
– Beds dipping towards the face causing sliding.
Most unconsolidated sediments are unstable and suffer from falling, sliding and flowing caused by gravity, frost action (Schenk 1965), and erosion (rain and groundwater).

5 OUTCROP ARRANGEMENT AND MANAGEMENT

For scientific or educational reasons it may be necessary to maintain even relatively unstable faces in hard or soft rock. In that case measures have to be taken to stabilize the face. Of course it is difficult to give general applicable rules for arrangements of rock faces. The Nature Conservancy Council (1990) in Great Britain produced a geo-conservation strategy report with an appendix containing a lot of very illustrative examples of outcrop arrangements.

In general the same principles can be used in hard and soft rocks:
- High faces which are dangerous or unstable can be lowered by infilling, supplying material at the foot of the face or by creating terraces, depending on the local situation. Terraces can be used if complete sections should be conserved,
- Faces which produce falling material, for example by frost action, can be provided with a gutter at the foot of the face or with geofabric covering the face itself,
- Faces which suffer easily from erosion or water infiltration in the slope behind the face should be protected by planting water-retaining vegetation or construction of drains leading the water away,
- Very sensitive (unconsolidated) rocks or rocks with fossils can be sheltered by a shed or even a closed construction. In special cases exposed fossils can be prepared by penetrating lacquers, etc.
- Stabilization of unconsolidated sandy (not too fine) rocks with high educational qualities (frost wedges, cryoturbation, etc.) can be realized by penetrating water-glass or plastics into the pores (Kutzner 1969, Fiedler & Czerney 1963). However this restricts further research at that spot,
- Small exposures in soft (unconsolidated) rocks with much room behind can be cleaned regular. But in the case of little room or vulnerable exposures they should be buried and only exposed when necessary. One of the Dutch Weichselian strato types is simply protected in that way (see below),
- Quarries and pits which are filled up with waste and in which a part of the excavation is kept free of waste often need special arrangements to prevent gas and leachate problems (Wright 1990, MacKirdy 1994).

The arrangement of a quarry pit for earth-science interest is an important aspect of earth-science conservation. But when a site has educational values facilities such as displays, pamphlets, etc. should be produced, depending on the educational importance and the vulnerability of the location. Where several sites are present in a relatively small area, trails can be laid out. The financial consequences of these special arrangements could be part of an arrangement between authorities, extracting companies and future owners.

6 SOME DUTCH EXPERIENCES

6.1 *Usselo: Conservation by covering*

In the early 1940s an archeologist discovered an archaeologically and geologically interesting exposure in the vicinity of the small village of Usselo near Enschede in the province of Overijssel. In this exposure, Palaeolithic artifacts were related to periglacial eolian sands, the so-called coversands, that could be dated. The layer in which the artifacts occurred was a bleached horizon with a high charcoal particle content. In the exposed profile the bleached layer, interpreted as the A2 horizon of a podzol, passed into a peat layer of Allerød age, the last interstadial of the last glacial period, the Weichselian. After this discovery near Usselo, the bleached layer with the charcoal particles was found on several places in an area from Belgium to Poland. The charcoal, mainly from *Pinus* was dated somewhat younger than the Allerød.

Figure 6. Scientificly important site exposing the 'transition', of the Usselo Layer into an Allerod peat layer. The site situated near Usselo (province of Overijssel) was dug up for the third time and covered again to protect the sensitive layers.

After the relatively warm conditions, the Allerød, during which soil and peat formed, a new cold and dry period started. The *Betula-Pinus* woods that covered a large belt south of the ice margin died producing an enormous quantity of dead dry wood that could easily set on fire by lightning or hunting tribes. As in 1905 in Siberia when an area as large as Europe was set fire, it is not surprising that traces are found from Belgium to Poland. The charcoal particles were transported by bioturbation into the A2 horizon of the podzol. The traces of beetles and their digging have been found. This site became the type locality for the 'Layer of Usselo'.

After the research was completed the site was covered with sand to protect it. In 1949 the site was re-opened, searched and covered again by a geologist specialized in palaeobotany. Twenty six years later (Fig. 6) the same procedure took place. New specializations and new methods in the field of geology ask always for new investigations. A few years later the site was officially protected by law.

This way of protecting a sensitive site is very effective. The layers are perfectly conserved by the sand infill and out of sight. Of course this not a site for educational use and it should only be opened for scientific research and for important international excursions. However, the Usselo Layer with charcoal can be seen permanently in the next example (the Zândkoele).

6.2   *Heetveld: The Zândkoele, more than just an exposure*

In 1981 when this site was noticed by the author (Gonggrijp 1994), it was a former communal sandpit used as a storage yard. One of the small walls showed eolian coversands of Weichselian age on top of Saalian till. It is a typical profile for the northern, Pleistocene part of the Netherlands, nothing spectacular, nothing very special, but seldom exposed.

At the time the municipality of Brederwiede (province of Overijssel) answered an appeal for protection of the site for educational purposes (1982), the site was already promised for use as a cycle-cross racing track for the local youth. However, the physical properties of the till proved unsuited to the needs of this sport and conservation was given a chance. Serendipitously the preparation of the racing track had opened fresh faces to supplement the original exposures. These showed much more geological detail than before, like frost wedges, a desert pavement with ventifacts, cryoturbation, an 'Usselo layer' etc.

In 1983, in a report ordered by the local authorities, several alternative designs for the lay-out for a monument, varying in complexity and costs, were presented. The most advanced, and expensive one was selected. The design included protection of the exposure and the realization of a large map of Scandinavia on which erratic, ice-carried boulders found in the north of the Netherlands were placed at a location on the map of Scandinavia representing their original sources (Fig. 7), so that visitors could have an idea what had happened some 150,000 years ago. For convenience, a viewing mound was provided, and there are two interpretative displays related to the map and to the exposure to be seen in the pit. In 1984, the geological monument was opened. In 1994 the site was included in a geological cycle trail, explaining the geo-

Figure 7. The geological monument 'de Zândkoele' near Heetveld (province of Overijssel), was founded in 1984. The wall in de background shows boulder clay covered with cover sand. For educational purposes a 'boulder map' of Scandinavia was constructed on the floor of the former sand pit to provide visitors with information on ice ages and boulders.

logical genesis of the surroundings, the solution for an isolated site. The money for the project was provided by the owner, the municipality and a general fund and the work was mainly carried out by unemployed volunteers.

As almost everywhere vandalism was a problem. But after a session at the locality with pupils and teachers of the nearby schools, the problems disappeared for a while. But every now and then vandalism starts again. In fact involving the locals, for example by adoption by a local society can be a guarantee to prevent vandalism and to keep a site in good condition. In this case the main problem was that although there was money for developing the monument there was no budget for the yearly maintenance.

6.3 *Hattem: Saved from 'landscape restoration'*

In 1974 when the exploitation of a sand pit in one of the ice-pushed ridges of Saalian age was well under way, the working group Gea started negotiations with the owner of the pit, the municipality of Hattem in the province of Gelderland, in order to establish a geological monument.

The pit showed impressive sections of fluvial Pleistocene sediments pushed by the Scandinavian ice sheet during the Saalian. Formations, normally covered by 50 m of sediments, were here exposed at the surface. The pit more than twenty meters deep, should have followed the same procedure as so many other pits and quarries, being filled in with rubbish after extraction ceased. This normal 'landscape restoration' procedure takes many years. During the negotiations it became clear that the municipality was willing to establish a monument, but it had to be compensated for the loss of the landfill site. The proposal was to keep a trench (100 m in length, 3 m high and 5 m wide) open and free of landfill. The provincial authorities responsible for the permitting of the mining, approved this plan. Normally the extension of excavation needs long and difficult procedures, but in this case, as it was just for the creation of a monument, it needed no official permission. Only a relatively small extension of the pit was needed and the slight loss in dumping was financially compensated by the value of extra sand.

This agreement, reached in 1975, was implemented in 1988. In the same year the site (Fig. 8) was presented at the first meeting of the European Association for the Conservation of the Geological Heritage (ProGEO) to an international expert group. Until now this site has not been subsequently cleaned. The wall is partly covered with talus, but, because layers are tilted, all formations are still visible in the upper meter. It has been proposed to clean only small parts for special occasions and to cover them again after use.

6.4 *South Limburg: National Landscape geotope project*

At the end of the seventies the concept of National Landscapes was developed in the Netherlands. These landscapes were to get a status more or less comparable with the English National Park system. This idea has now been abandoned. However, at that time there was room and money for all kinds of projects supporting the ideals of the National Landscapes.

Figure 8. Geological site in a former sand pit used for scientific and education purposes. The pit near Hattem (province of Gelderland) shows ice-pushed estuarine and fluvial sediments, which are nowadays seldom exposed. The deep pit is completely filled up by refuse, except for the trench with the geological profile. In case of research and excursions parts of the 3 meter high wall are being cleaned.

From 1982-1985 one such project included the restoration of a series of pits and quarries for science and education in geology. In a rather small area several Carboniferous, Cretaceous, Tertiary and Pleistocene outcrops are present, which is unique in the Netherlands.

A working group composed of representatives of the provincial authorities, provincial nature conservation agencies (private as well as official), the Geological Survey and the Research Institute for Nature Management (now Institute for Forestry and Nature Research) worked out a plan for the restoration of neglected exposures (Gonggrijp & Felder 1988). The budget for the restoration of the sites was priced and sent to the provincial authorities for approval and financing. After approval the project was carried out in three phases to spread the costs over several years.

After three years of restoration, 12 sites (Figs 5 and 9) were cleaned from vegetation and debris and made safe where necessary by fences on top of the walls. Two sites with boulders were cleaned of vegetation. Several boulders were dug up and added to the others. Planning and preparation was conducted very well except there were no funds available for maintaining the sites. The hard rock profiles are still in perfect condition, but the soft rock exposures urgently need some maintenance. However, there is hope for the future. Since the eighties, the Netherlands have provincially organized foundations, employing unemployed people, and have become active in maintaining landscape elements. First these foundations were mainly concentrated on elements like hedges, trees and ponds for amphibians. Recently they have extended their activities and included geological outcrops and geomorphologi-

Figure 9. Carboniferous sediments forming a thrust fold are exposed in a former quarry near Cottessen (province of Limburg). The quarry is being cleaned as part of a National Landscape Geotope Project and is rather easy to manage.

cal landforms. In the province of Limburg a scheme has been developed to restore and maintain the important geological sites. The next step will be negotiations with the various owners to persuade them to offer financial support for the restoration activities.

7 CONCLUSION

The extraction of aggregates is necessary for the continuation of all kinds of social activities. However, during mining, landscapes are being affected or even destroyed. Mining should be confronted with all natural values including geological ones during environmental assessment studies. Careful considerations should form the basis for final decisions. When, in principle, extraction has been approved the establishment of geological sites together with other environmental aspects have to be embedded in a license and a restoration plan. The final development of geological sites depends largely upon the local situation and of course upon the rock properties.

REFERENCES

Baker, F.R. & H. Gray 1960. Design of foundations, embarkments and cut slopes 11-1/11-80. In K. Woods (ed.), *Highway Engineering Handbook.* New York, Toronto, London.
Bridgland, D. 1994. The conservation of Quarternary geology in relation to the sand and gravel extraction industry. In D. O'Halloran, C. Green, M. Harley, M. Stanley & J. Knill (eds), *Geological and Landscape Conservation*: 87-91. Geological Society, London.

Fiedler & Czerney 1963. Möglichkeiten der Bodenstabilisation mittels Kunstharzen. Zeitschr. f. Landw. *Versuchs- und Untersuchungswesen* 9(4/5): 427-443.

Gonggrijp, G.P. 1989. Nederland in Vorm. Achtergrondreeks Natuurbeleidsplan nr. 5. SBU 's-Gravenhage, 141 pp.

Gonggrijp, G.P. 1994. Two geological monuments in the Netherlands. In D. O'Halloran, C. Green, M. Harley, M. Stanley & J. Knill (eds), *Geological and Landscape Conservation*: 323-328. Geological Society, London.

Gonggrijp, G.P. & G.J. Boekschoten 1981. Earth-science Conservation: No science without conservation. *Geol. en Mijnb*. 60: 433-445.

Gonggrijp, G.P. & W.M. Felder 1988. Mergelland, een geologisch 'buitenmuseum'. Natuurh. Maandblad, 77/ 7-8, 129-137 (with English summary/Mergelland, a geological 'outdoor museum').

Kutzner, C. 1969. Baugrundverfestigung durch Injektionen. *Strassen- und Tiefgbau* 7: 644-650

Mac Kirdy, A.P. 1994. Technical solutions to conservation problems for natural and artificial rock sections. *Mém. Soc. géol. France* 165: 221-226.

Ministry of Agriculture, Nature Management and Fisheries 1990. Nature Policy Plan (Abridged English version) The Hague.

Ministry of Transport and Public Works 1988. Gegrond ontgronden Den Haag.

Ministry of Agriculture, Nature Management and Fisheries 1992. Nota Landschap.

Ministry of Agriculture, Nature Management and Fisheries and the Ministry of Housing, Planning and Environment 1993. Structuurschema Groene Ruimte.

Ministry of Transport and Public Works 1993. Structuur Schema Oppervlaktedelfstoffen.

Nature Conservation Council 1990. Earth-science conservation in Great Britain: A Strategy. Appendices: A handbook of Earth-science conservation techniques.

Schenk, E. 1965. Der Zusammenbruch von Baugruben und Böschungen infolge der Frosteinwirkung. *Felsmechanik und Ingenieurgeologie* VIII (3-4): 103-112.

Schlüchter, C. 1994. A Model of consensus in aggregates mining and landscape restoration: science-industry-conservation. In D. O'Halloran, C. Green, M. Harley, M. Stanley & J. Knill (eds), *Geological and Landscape Conservation*: 39-42. Geological Society, London.

Scott, M.M. 1994. Sustainable development: the implications for minerals planning. In D. O'Halloran, C. Green, M. Harley, M. Stanley & J. Knill (eds), *Geological and Landscape Conservation*: 9-11. Geological Society, London.

Wright, R. 1990. Of rocks and rubbish. Earth-Science Conservation NCC. 27, 9-11.

A review of mineral aggregate production and operating conditions in Alberta, Canada

W.A. DIXON EDWARDS
Alberta Energy and Utilities Board, Edmonton, Alberta, Canada

1 INTRODUCTION

Sand and gravel has been produced in Alberta (Fig. 1) since the turn of the century. Concern that supplies from existing sources were running short was reported as early as 1946: 'Since the demand for gravel is increasing year by year, and since many of the gravel deposits are being worked out, it is imperative that new sources of this material be found'. (Research Council of Alberta 1946). More supplies were found: 99% of Alberta mineral aggregate needs in 1991 still were met through sand and gravel.

Total Canadian production of mineral aggregate in Canada in 1991 was 304 million tonnes worth $1,281 million (Vagt 1994). Alberta's mineral aggregate production in 1991 was 45.8 million tonnes worth $156.8 million (Edwards 1995). Sand and gravel is the preferred source of mineral aggregate in Canada comprising seventy-one per cent of Canadian production in 1991. Sand and gravel has a lower cost of production per tonne than the main alternative, crushed stone; respective average costs in Canada in 1991 was $3.43 and $6.15 (Vagt 1994). Average cost of a tonne of sand and gravel in Alberta in 1991 was $3.37 per tonne (Edwards 1995). As sand and gravel deposits become exhausted and constraints sterilize replacement reserves the industry is forced to turn to bedrock sources.

Mineral aggregate, unlike other minerals used in Alberta, is not imported, save for small amounts of specialty products. Alberta has large, widespread supplies of sand and gravel, major potential bedrock sources, and has considerable control and flexibility in the development of the resource. Supplies are dwindling rapidly, however, and concerns about both prevention of development and the potential harms of development are expressed from all parts of Alberta. These concerns are concentrated in the more densely settled areas of the province, particularly the Calgary region at this time. Concerns expressed revolve around a few key issues: the conflict between the need of the operators to develop close to the market and the desire of residents to defend their quality of life; protection of natural areas (particularly water courses) versus the desire of operators to maximize recovery from deposits; and the concern of all parties about the nature and enforcement of laws and regulations.

It is within the regions of Alberta, that is the Counties, Municipal Districts, Improvement Districts, cities, and towns, that the direct management of the resource

Figure 1. Location of the province of Alberta, Canada.

Figure 2. Location of the largest cities in Alberta.

takes place, interaction between producers and competing interests occurs, and industry competes. To understand these forces and to effectively manage the resource it is necessary to have detailed information about the resource at the regional level. Regional sand and gravel production or demand studies have only been done for the Edmonton region (City of Edmonton 1978), for the combined Edmonton-Lloydminster region (Edwards et al. 1985), and for the St. Paul-Bonnyville region (Edwards & Fox 1980). Public knowledge of sand and gravel supplies is also incomplete. Detailed maps identifying potential resources are available for only about 18% of Alberta (Edwards & Chao 1989) and there has been no new mapping since 1990.

This paper describes the first attempt in Alberta to accumulate data for all the regions of Alberta. A recent survey by the Alberta Geological Survey (AGS) collected data from private and public sector mineral aggregate producers for 1991 and attributed this production to regions. These regions were defined as: cities with over 50,000 population together with the surrounding Counties or Municipal Districts (Calgary, Edmonton, Lethbridge, and Grande Prairie regions, Fig. 2), and rural Counties, Municipal Districts, and Improvement Districts. In total producers in sixty regions were surveyed. Producers from all regions did not respond to all questions so the total regional responses in the following figures and tables are variable and do not total sixty.

The recent survey data are used in this paper as the basis for describing: the pattern of production of mineral aggregate in Alberta; transport distances; geological, environmental and legislative conditions which affect production; and environmental concerns.

2 PRODUCTION

Alberta mineral aggregate production data gathered by Natural Resources Canada (NRCan) for 1991 are reported as 38,722,000 tonnes, of which 38,303,225 tonnes were sand and gravel (Vagt 1994). An evaluation of the NRCan data in 1981 (Edwards 1989a) suggested that NRCan data are inaccurate for Alberta. In 1991, data were gathered in Alberta by NRCan from 172 survey respondents (Table 1). The AGS identified another 128 producers and an additional 7,181,611 tonnes of sand and gravel production (Table 1). It is assumed that the production number (45,484,836 tonnes) generated from a combination of the AGS and NRCan surveys (Table 2) still represents a minimum amount of sand and gravel as some producers did not respond to the AGS survey.

The AGS survey recovered more data from smaller producers than the NRCan survey (Table 1). Seventy-seven percent of the AGS survey data was recovered from operations producing less than 50,000 tonnes in 1991 whereas the NRCan survey recovered 62% of its data from operators producing in excess of 50,000 tonnes. The smaller producers represent the rural regions of Alberta and are quite as important to these regions as the larger producers are to the large urban centres. The production data recovered by the AGS were recorded by region so that for the first time a provincial wide estimate of production by region was available for evaluation.

Table 1. Mineral aggregate production from AGS and NRCan surveys for 1991.

	Amount of production (10^3 t)						
	> 500	> 250	> 100	> 50	> 10	< 10	Total
NRCan survey							
# reporting production	10	16	47	33	48	18	172
% of respondents	6	9	27	19	28	11	100
AGS survey							
# reporting production	3	3	13	11	37	61	128
% of respondents	2	2	10	9	29	48	100
Total distribution							
# reporting production	13	19	60	44	85	79	300
% of respondents	4	6	20	15	29	26	100

Table 2. The estimated amounts, values, and unit values of sand and gravel products in Alberta for 1991.

	Sand	Gravel	Crushed gravel	Other	Total
Production amount					
(t)	4,467,634	6,070,748	26,838,076	8,108,378	45,484,836
(%)	10	13	59	18	100
Production value					
($)	17,723,871	15,563,877	107,770,556	12,168,385	153,226,689
(%)	12	9	71	8	100
Value per ton					
($/t)	3.97	2.56	4.02	1.50	3.37

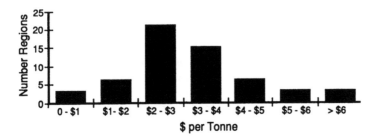

Figure 3. Regional costs per tonne of sand and gravel at the pit in Alberta for 1991.

The average cost of sand and gravel at the pit produced in 1991 was $3.37 per tonne and varied widely in the regions from $0.48 to $9.51 per tonne (Fig. 3). When the regions are divided into groups based on population, the average cost per tonne of sand and gravel exhibits an overall pattern which reflects the population (Fig. 4). In general, regions with smaller populations have lower unit costs and the highest costs are for the large Calgary and Edmonton markets.

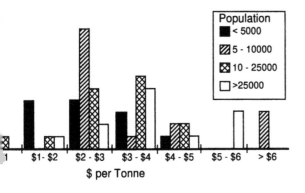

Figure 4. Costs per tonne of sand and gravel at the pit for regions of different populations (1991).

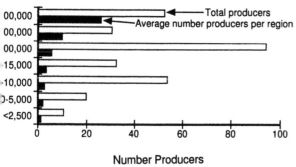

he total and average number of producers operating in each region during 1991.

eatest total production of sand and gravel is concentrated around Calgary onton, the two regions with the greatest population (Fig. 2). Calgary and have the largest average number of producers per region (Fig. 5) and the mpanies.

erta, 4% of the producers (13) mine more than 500,000 tonnes of sand and ually (Table 1) and produce 53% of the total provincial production (Table large producers are all located around the Calgary and Edmonton regions) population each, Fig. 5). In total, the Edmonton and Calgary regions have ers or one producer for every 30,200 people. Annual per capita production equal to consumption) for the Edmonton and Calgary regions is virtually t 11.9 tonnes (Fig. 6).

ix percent of the provincial production comes from only 10% of the opera- that mine over 250,000 tonnes annually). These producers are all located gary and Edmonton regions plus the Red Deer, Lethbridge, and Grande ions, each with populations > 50,000 (Fig. 2). The mineral aggregate re- s of these five regions are similar, the infrastructure in these populated re- ighly developed, most roads are paved, and there is a larger industrial and construction component than in the rural regions. There is relatively little in the per capita consumption for these regions because each has sufficient on to be little affected by a single construction event. Annual per capita

Table 3. Sand and gravel production for 1991 categorized by producer size.

	Amount of annual production by producer size (10^3 t)						
	> 500	> 250	> 100	> 50	> 10	< 10	Total
Production	23,867	6,141	9,780	3,131	2,102	464	45,485
% total production	53	13	21	7	5	1	100

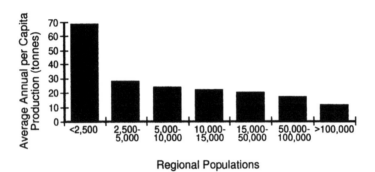

Regional Populations

Figure 6. Average annual per capita productions (tonnes) of sand and gravel for Alberta regions in 1991.

production in the Red Deer, Lethbridge, and Grande Prairie regions is 17.0 tonnes (Fig. 6).

Regions with intermediate populations between 5000 and 50,000 (Fig. 5) typically have been settled for a long period of time, may have an agricultural base, and contain one or more small to medium sized towns. The transportation infrastructure in these regions is well established although up-grading of roads is usually underway. The greatest number of producers are active in these regions (Fig. 5) because there are more regions of this size (41 of 60 regions) but the average number of producers per region decreases with decreasing regional population (Fig. 5). Data from regions with less than 50,000 population had a large range but overall the average per capita production for 1991 increased with decreasing regional population (Fig. 6).

Regions with populations less than 5000 (14) are scattered across Alberta, occurring in the central and southern plains as well as in northern Alberta. Any amount of infrastructure development or upgrading in these regions will produce a large per capita production value and considerable annual variation is expected for a given region. In these sparsely populated regions there was an average per capita production for 1991 of 67.5 tonnes and one producer for every 1091 people.

The survey data for 1991 shows production differences for regions of different size and several patterns. The average number of producers per region declines with decreasing regional population, from ~26 (Calgary or Edmonton) to ~2 for regions of < 2500 (Fig. 5). The most populated regions have the largest producers. The greatest annual per capita production of sand and gravel occurs in the least populated regions (Fig. 6) and decreases with increasing regional population, from 67.5 tonnes in regions with < 2500 people to 11.9 tonnes for the Edmonton and Calgary regions.

Land use and resource strategies should reflect these differences in amount and style of production.

Ideally mineral aggregate reserves should be determined from geological mapping. However this information is not available in Alberta. Consequently in the 1991 survey some idea of supply was gathered from public and private sector producers. Producers would not reveal their volume of reserves but did provide an estimate of the number of years a supply of sand and gravel is available to them. Responses were identified by region. Using this method it was possible to establish when the first producer in a given region depleted his sand or gravel supplies and when all producers in that region depleted their supplies. Information on gravel supplies was received for 49 regions and for sand supplies for 48 regions. This method of reporting over-estimates years of supply if producers fail to anticipate increases in regional demand.

Initial depletion of gravel is predicted for at least one producer in 34 of the 49 regions reporting in ten years or less and in half of the municipalities in 12 years (Fig. 7). Gravel supplies of all producers are predicted to give out within 10 years in 15 of 49 municipalities and in half of all regions in 20 years or less. Initial depletion of sand is predicted within 10 years in 24 (half) of all regions reporting (Fig. 8). Supplies held by all producers in the regions are predicted to expire within 10 years in 10 regions and in 21 years for half of all regions reporting.

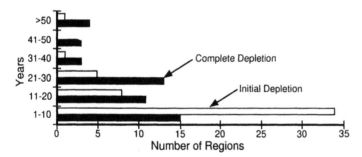

Figure 7. Years to initial depletion of gravel supplies by one producer (clear bar) and the depletion of all supplies (solid bar) for each of 49 regions.

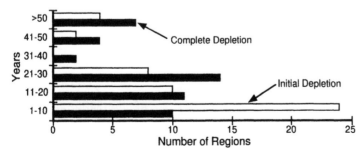

Figure 8. Years to initial depletion of sand supplies by one producer (clear bar) and the depletion of all supplies (solid bar) for each of 48 regions.

3 TRANSPORTATION

The costs for sand and gravel cited above are for the mining and processing on site of the mineral aggregate. They do not include costs due to the transportation of the aggregate from the site for further processing, use as a raw material in other products, or for direct use. The transportation cost in many cases is greater than the cost of the mineral aggregate at the pit.

The only aggregate currently moved by rail is for use on the rail lines although markets near existing rail lines continue to monitor the related costs of truck and rail transport. A few places in Alberta are located on navigable waterways (Fort McMurray) and barge transport has been considered. The primary method of transport in Alberta is by truck. Public and private sector mineral aggregate producers reported truck haul distances for 51 regional municipalities. Average maximum haul distance is 42 km with a range from 8 km to 140 km (Fig. 9). Haul distance is a factor which indicates depletion or scarcity of the resource. Three of the seven longest hauls in Alberta are for regions with the largest populations and greatest demands for aggregate. Edwards (1989b) reports a maximum haul in Alberta of 100 km for 1983. Beyond this single report there is little historical data on haul distance for comparison, however, it does point to a trend of increasing depletion. In many situations, especially in urban settings, haul route is a more important development factor than haul distance. In a land use conflict northwest of Calgary one of the primary issues was the number of trucks and the routes they would take (Edwards 1995).

Region Populations
>500,000 (Edmonton, Calgary)
50-500,000 (Red Deer, Lethbridge, Grande Prairie)
10-50,000 (solid bars)
   <10,000 (hatch bars)

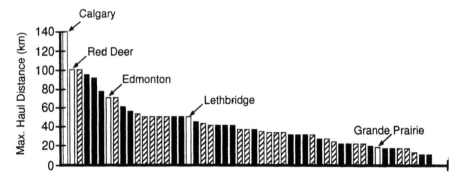

Figure 9. Maximum haul distance (km) for sand or gravel for 51 regions.

4 OPERATING CONDITIONS

Factors which impact on the ability of private and public sector producers to operate include physical conditions, regulatory requirements, and environmental and land use issues. The physical conditions include the geology and distribution of the deposits, available technology, climate, and the value of the local product compared to substitutes or imports. Any operation in Alberta is required to follow various regulations. Provincial legislation is outlined in this report. Environmental and land use issues are becoming increasingly important as they can curtail or stop potential development, reducing aggregate resources. These will be addressed at some length as it is a relatively new consideration in Alberta and an area that is not adequately documented or described from the resource perspective.

4.1 *Physical conditions*

The geology of some aggregate deposits in Alberta require special extraction. Preglacial deposits often require significant overburden removal and extraction may require dredging or pumping water from the pit as they also may be aquifers. Kame deposits commonly are poorly sorted and require selective recovery. Without due care operations in kames can create a 'moonscape' appearance to the pit. If colluvial fans with intermittent streams are excavated, changes to dry channels can alter the water flow during spring run-off. Such diversions can endanger other land uses on the fan. Glaciofluvial deposits commonly have a high percentage of fine sand. Processing is required to separate the sand and extra costs ensue for discarding the excess. Settling ponds are incorporated in operations on many types of deposits but they are especially important in alluvial deposits near river courses. It is essential that sediment does not escape into natural water bodies (Edwards 1995).

The technology required for excavation and processing of sand and gravel for mineral aggregate is well established. Mining almost always takes place in the summer and fall to avoid problems associated with the cold winter months. This requires that the annual supply of aggregate is produced in half the year. Any difficulties during the summer and fall which curtail the mining process can have serious effects on the operation for the entire year. Processing also proceeds during the summer and fall but may be carried on into the winter in larger operations using stockpiled materials.

5 LAND USE LEGISLATION AND REGULATIONS

Legislation which regulates aggregate mining and the lands disturbed by aggregate mining was initiated because of concerns about the effects of resource exploration or testing. Much of the early impetus (1963-1973) for provincial regulation of mineral activities in Alberta came as a result of oil and gas activities. Later, regulations became more specific to aggregate mining. A current outline of regulations governing aggregate mining is presented in Badke (1994) and in a pamphlet entitled 'Aggregates and Our Environment' (Alberta Sand and Gravel Association 1994).

5.1   *Environmental Protection and Enhancement Act 1992-present*

Early legislation in the form of the Surface Reclamation and Land Surface and Rec-lamation Acts was revised and expanded in the Environmental Protection and En-hancement Act introduced in 1992. This is the main body of legislation which cur-rently regulates the reclamation and environmental activities of the aggregate industry on private and Public Lands. It is administered by Alberta Environmental Protection. There is an established application procedure with the possible require-ment of an Environmental Impact Assessment. An appeal procedure is in place for citizens or applicants (Environmental Appeal Board) and public participation is en-couraged.

5.2   *Conservation and reclamation regulations 1993-present*

These regulations established the environmental assessment process which includes the submission of detailed development, conservation and reclamation plans (reclamation to equivalent capability), posting of a security deposit equal to the cost of reclamation, penalties for environmental offenses, and regulation of pollution or emissions from pits or processing plants.

5.3   *Public Lands Act*

The Public Land Act is the primary legislation dealing with the management of sand and gravel resources on Public Lands. The resource is managed and regulated on primarily privately held areas by Alberta Agriculture, Food and Rural Development (Public Lands) and on primarily Crown areas by Alberta Environmental Protection (Land and Forest Services). Authorization to remove sand or gravel is by a variety of licenses or leases within the Surface Materials Regulations. A royalty of $0.60 per yd^3 or $0.47 per tonne is collected for sand and gravel removed.

5.4   *Planning Act – Currently under revision*

The Planning Act is major legislation affecting aggregate producers through munici-pally generated and enforced land use and development regulations. Provides for creation of Regional Planning Commissions which create Regional Plans that include guidelines for municipalities on land use and development issues and requires each municipality to create General Municipal Plans that provide broad guidelines for land use and development, Area Structure Plans for parts of the municipality, Land Use Bylaws for specific land use districts (there is no appeal mechanism for refused re-zoning applications but a 6 month re-application), and require that a Development Permit is obtained for all developments (applicants can appeal to Development Ap-peal Board).

6 ENVIRONMENTAL CONDITIONS

Increasingly important aspects in the ability of a company or municipality to open, develop, and operate a mineral aggregate pit or quarry are land use and environmental approvals. About one quarter of the mineral aggregate producers in Alberta say they have encountered situations where their efforts to mine an aggregate deposit were curtailed or prevented by land use or environmental restrictions. Producers surveyed were from the private sector (companies and individuals) and the public sector (Counties, Municipal Districts, Improvement Districts, Special Areas, cities, and towns). The breakdown of this response is shown in Table 4. The restrictions which producers cited can be categorized into residential opposition, environmental restrictions, regulatory restraints, and conflicting land use issues (Table 5). Private and public sector producer responses were not separated as they cited similar issues and in similar proportion.

The restrictions due to residential opposition to mineral aggregate mining centre around concerns such as truck traffic, noise, and dust. Public protest can result in the

Table 4. Summary of a survey to identify mineral aggregate producers who encountered situations where their efforts to mine an aggregate deposit were curtailed or prevented by land use or environmental restrictions.

	Number responding	Number encountering restrictions
Private sector	80	20
Public sector	82	22
Total	162	42

Table 5. Restrictions to mineral aggregate mining identified by producers. Regions are separated according to population.

	Number of situations reported per region (population 10^3)								
	< 2.5	2.5-5	5-10	10-15	15-25	25-50	50-100	> 100	Total
Residential opposition				1	1			5	7
Environmental restrictions unidentified [1]			3	1					4
– due to wildlife and vegetation [1]					1	2			3
– natural areas	3	1		1	1				6
– environmentalists			1						1
– reclamation costs [2]			1						1
Water course proximity [1]	1		3		4	3	2		13
Current regulations [3]	2	2	3		2				9
Conflicting land use [4]			2					1	3
Total	6	3	13	3	9	5	2	6	47

1. Restriction resulted from regulations requiring buffers or denying development, 2. Reclamation costs are a requirement of all operations but may affect operations differently, 3. Regulations may be environmental in nature, may concern various levels of authority, 4. Conflict with pipelines, road allowances, and urban residences which cover deposits.

delay or rejection of an application to develop an aggregate operation (for example in the Municipal District of Rocky View as described in Edwards 1995). Such protest also can result in the implementation of bylaws or regulations which require the attention of all subsequent applicants. There is a link or continuum between the category of residential or lobby group opposition and restrictions listed in Table 5 under restrictions resulting from current regulations or conflicting land uses. The majority of restrictions due to residential opposition were reported from regions with high population density (Calgary region) or a long history of settlement (Red Deer region) (Table 5).

Environmental restrictions include four basic aspects. One is the opposition to mining by an individual or group. This 'environmentalist' opposition is cited in only one response for an environmentally sensitive area in the foothills.

A second type of environmental restriction is exclusion of areas, or regulations intended to protect fauna, flora, or natural areas. Specific exclusions to mineral aggregate are noted for protection of wildlife, vegetation, natural or sensitive areas, and protection of land through the formation of parks or recreation areas. These restrictions are through particular land designation or zoning and can be considered as land use management issues. These restrictions are noted by operators in central and northern Alberta, outside of the major urban centres.

The third aspect of environmental restriction to mineral aggregate mining is the cost of reclamation. One response is specific to this topic but others in the current regulations category probably include reclamation. The intention of reclamation is to return the land to a useful or natural state after mining. Most operators pay reclamation costs and did not respond that this cost was a restriction to development. It should be recognized, however, that reclamation costs can be a major disincentive to development if the requirements are not reasonable or fairly applied, and that reclamation costs can affect operations differently; reclamation costs can become a block to development.

The fourth and most common environmental restriction to mining is proximity to a water course. Restrictions cited are primarily for development near, but not in, a river or lake. The restrictions usually are in the form of a buffer between the operation and the water course. This issue is raised by operators in all parts of Alberta and appears to be a result of geographic and geologic circumstance and not due to population or settlement.

Current regulations are mentioned in a significant number of responses. The regulatory issues mentioned include the enforcement of national, provincial, or municipal regulations to prevent possible damage to wildlife or vegetation, to restrict possible damage to natural systems (rivers and lakes), to reclaim the site, and to ensure the safety and quality of life of citizens. These responses came primarily from the rural regions.

Some restriction to development comes through the loss of land to competing land uses. These alternate land uses include pipelines, road allowances, and construction of dwellings on land which could have produced mineral aggregate. These restrictions occur in both urban and rural areas.

The mining and transportation of aggregate can have a definite impact on the environment and the quality of life of residents in a region. These impacts ultimately result in restrictions to mineral aggregate development. Municipalities were asked to identify situations where mineral aggregate mining or transport had an impact on the environment or quality of life. Nineteen of 81 municipalities responding cited situations where mining is perceived to impact on the residents or the environment (Table 6). The situations cited reflect the rural or urban nature of the region.

The situations documented are divided into four basic areas of concern: economic, quality of life, concerns for the natural environment, and concerns with existing controls on mineral aggregate resource development. The greatest number of situations reported involve concerns that the quality of life of residents is being disrupted by mining operations and the transport of aggregate materials by truck. All the types of concerns cited in this survey were identified by opponents to mineral aggregate development in the Calgary region.

Some concerns identify the economic effect on the land by an aggregate operation. For example, an unreclaimed pit cannot be used for agriculture, and soil loss affects reclamation and ultimately the post-mining value of the land. These concerns reflect rural concerns that land developed for aggregate has not been returned to a viable state for agriculture and concerns that aggregate operations are lowering the value of adjacent lands. Loss of residential property value close to aggregate operations, presumably as a result of the decline in the quality of life factors, was cited by urban respondents.

Concern for the effects of mineral aggregate on the natural environment are reported mainly, but not exclusively, from the rural areas. These concerns include the impact on wildlife, flora, natural areas, and on lands already set aside for the conservation of natural areas. These concerns overlap with those maintaining that current

Table 6. Environmental and quality of life concerns resulting from mineral aggregate mining.

	Number of situations reported per region (population 10^3)								
	< 2.5	2.5-5	5-10	10-15	15-25	25-50	50-100	>100	Total
Economic concerns:									
– top soil loss			1						1
– unreclaimed pits			1				1		2
– devaluation of property								1	1
Quality of life:									
– resident/ ratepayer concerns[1]			1	1	1	1	1	3	8
– impact on recreation area			1						1
Natural environment concerns:									
– impact on natural areas		1							1
– elevation changes	1								1
– vegetation change	1								1
– effect on wildlife, vegetation	1								1
– damage, change to river			1				1		2
Concerns with lack of controls:									
– poor regulation enforcement			1				1		2
– poor environmental controls							1		1
– lack of data	1								1
Total	4	1	6	1	1	1	5	4	23

1. Concerns include noise from crushing operations, hours of operation, truck traffic, dust.

regulations are not adequately controlling mineral aggregate development. Concerns about the lack of control over aggregate development contrast markedly with the view expressed by many operators in the survey that controls hamper their opportunity to develop pits.

7 SUMMARY AND CONCERNS

National and provincial statistics are useful for provincial comparisons and the identification of long term trends in production but have relatively little application at the local level. Described here is the first attempt to identify market regions in Alberta. It appears that this could be a valuable way of tracking the use of the resource, providing important clues to the state of the resource, resolving decisions which would affect more than one jurisdiction, and determining long term trends.

This initial survey provides clues to regional differences. High consumption takes place in the urban areas. It is in these areas where mineral aggregate producers come into the greatest contact with residents. Conflict can result over differing visions of land use. These areas also are centres which have required aggregate for long periods of time and now may have limited sources of supply without long haul distances. These regions require very intense land use assessments of the need and value of future mineral aggregate resources.

Regions with lower populations have different, but no less important, requirements for management of the mineral aggregate resource. They are in the process of developing the basic transportation and industrial infrastructures. Careful selection and use of mineral aggregate at this stage of development is extremely important and has more flexibility than in the urban areas. It is also in these regions that the provincial government has a greater ability to manage the aggregate resource and plan for the future.

Procedures are in place to restrict development, based on environmental concerns, and to reclaim sites after mining. There is little information on the economic value and benefit of resource development going into the evaluation process and the province needs an effective resource conservation strategy. But signs are that a long term strategy must be designed soon if we are to sustain our mineral aggregate supplies. First, it is essential to inventory our resources as most other provinces have done. These resource data must be followed with a resource conservation strategy which identifies those resources to be preserved for the future. The long term impacts of resource exclusions or approvals must be considered and a working relationship needs to be established and maintained between the various stakeholders in the mineral aggregate resource sector.

REFERENCES

Alberta Sand and Gravel Association 1994. *Aggregates and our environment.* Pamphlet produced and distributed by the Alberta Sand and Gravel Associatio, 4 pp.

Badke, D.A. 1994. *Report on legislation and statistics relating to the gravel mining industry.* Prepared for the Alberta Sand and Gravel Association by D.A. Badke Enterprises Ltd., 33 pp.

City of Edmonton 1978. *Edmonton regional aggregate study.* Edmonton Engineering Department, City of Edmonton, 65 pp.

Edwards, W.A.D. 1995. *Mineral aggregate commodity analysis.* Alberta Geological Survey Open File Report 1995-08, 54 pp.

Edwards, W.A.D. 1989a. *Current and future demand for aggregate in the Edmonton region.* CIM Bulletin 82: 118-123.

Edwards, W.A.D. 1989b. *Aggregate and nonmetallic Quaternary resources.* In R.J. Fulton (ed.), Chapter 11, *Quaternary Geology of Canda and Greenland,* Geological Survey of Canada, Geology of Canada 1: 684-686.

Edwards, W.A.D. & Chao, D.K. 1989. Bibliographic index and overview of aggregate resource publications. Alberta Research Council Open File Report 1989-13, 90 pp.

Edwards, W.A.D. & Fox, J.C. 1980. *Sand and gravel resources of the Cold Lake area, Alberta.* Alberta Research Council Open File Report, 1980-8, 45 pp.

Edwards, W.A.D., Scafe, D.W. & Hudson, R.B. 1985. *Aggregate resources of the Edmonton/Lloydminster region.* Alberta Research Council, Bulletin 47, 64 pp.

Edwards, W.A.D., Scafe, D.W., Eccles, R., Miller, S., Berezniuk, T. & Boisvert, D. 1994. *Mapping and resource exploration of the Tertiary and preglacial formations of Alberta.* Alberta Research Council Open File Report 1994-06, 123 pp.

Research Council of Alberta 1947. Annual report of the Research Council of Alberta 1946, Report 50.

Vagt, O. 1994. Mineral aggregates. In *Canadian Minerals Yearbook.* Mining Sector, Natural Resources Canada: 32.1-32.14.

An environmental perspective on quarrying for the construction industry in Lebanon

MOHAMAD R. KHAWLIE
National Centre for Remote Sensing, Beirut, Lebanon

1 INTRODUCTION

A country coming out of a crippling crisis and undergoing reconstruction and development is in dire need for construction materials. Lebanon now typifies this case. After 17 years of internal strife and destruction, governments since 1990 have set down a number of priorities. They reflect the spirit of the 1990's where the Earth Summit at Rio de Janeiro (United Nations Conference on Environment and Development) emphasized linking development to environmental concerns.

If one reviews the status of the construction industry in Lebanon from the point of view of resources, i.e. quarries and their operations, one can be easily shocked (Khawlie 1995). Whether it is the distribution, or the handling or other technical and managerial aspects, the local supply for construction materials in Lebanon has been an environmental curse. Wherever they occur, and under whatever program they operate, quarries have contributed to further environmental deterioration. If Lebanon intends to catch up with the rest of the word, even with the neighboring countries, proper environmental practices and law enforcement must be implemented.

What is interesting about this problem is its economic – environmental character. In as much as it typifies a case of modern versus past views of such character linkages, it also presents a case where environmental technology can play a big role. Innovations here will not be necessarily linked to the technical aspect although that may help, rather they are related to an improved management approach. It focuses on resource environmental management in safeguarding both the environment and the production line.

The current state of affairs is chaotic and must be regulated. Whether securing construction materials from natural sands, crushed rocks, or even from the marine environment, there is an increase in negative environmental impact. An environmental assessment, creating the required database, and following a well planned framework covering all steps in the quarrying operation must be followed. Total quality management techniques are available and can easily contribute to that end.

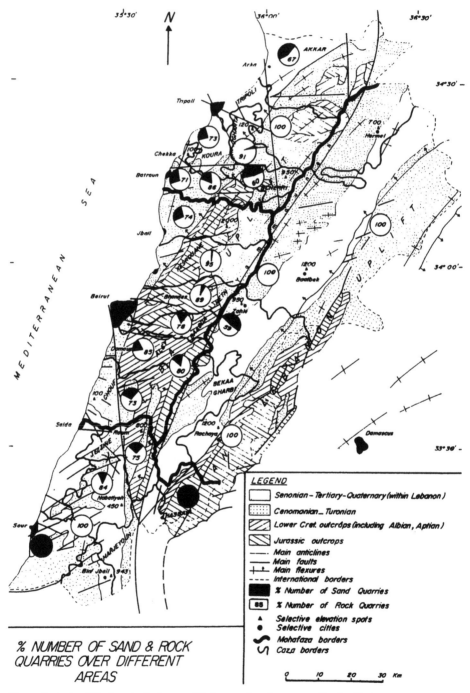

% NUMBER OF SAND & ROCK
QUARRIES OVER DIFFERENT
AREAS

Figure 1. A simplified geological map of Lebanon showing percentile distribution of sand and rock quarries over administrative districts.

2 DEVELOPMENT PROJECTS AND THE CONSTRUCTION INDUSTRY

It is natural for a country coming out of a devastating war to undergo redevelopment. The upgraded government redevelopment plan 'Horizon 2000, or the 10-years plan' includes 136 projects covering 15 sectors. All will require huge amounts of construction materials. Notable among these are projects for upgrading the major seaports, Beirut international airport, the southern and northern suburbs of the capital, country highway network and the Arab Auto-Route. In addition, the reconstruction of Beirut down town by itself will bring about a heavy burden on demand for construction materials.

The above gives the perspective from the public sector, but equally important are the construction requirements in the private sector. As stated by Bechara (1994) the real estate sector increased tremendously in 1994, with construction investments estimated at $3.3 billion. The building permits increased by 49.66% in 1994 compared to 1993, and the cement delivery also increased 16.20 %, which is equivalent to about 3.5 million tons of cement, with built up surfaces estimated at 17 million m^2.

Other aspects which pushed the construction industry forward are prices of materials. The selling price of aggregates increased from $9 to $13 per m^3 due to the government's decision to close some quarries that were environmentally problematic. Cement prices increased about 12%, from $60 to $68 per ton because of increasing government tariffs and taxes.

A United Nations Development Program study (UNDP 1994) estimated the per capita need for aggregates at about 5 tons/year (in Europe it is 6-10 t/yr). The estimate is about 16 million tons annually, or 30,000-60,000 m^3/day. The average of 45,000 m^3 is divided into 5000 m^3 sands and 40,000 m^3 crushed limestone subdivided as follows: 5000 m^3 riprap, 27,000 m^3 aggregate and 8000 m^3 sands. The projected estimated needs for the coming few years will increase by about 30% to an average of about 20 million tons annually.

With a relatively dense population distribution in Lebanon, 320/km^2 compared to 104/km^2 in France (of course in cities it may exceed 15,000/km^2), the general land-use pattern is indeed crucial. The erratic distribution of quarries needed to supply nearby centers of demand, and their inevitable closeness to residential and commercial areas, and the associated use of heavy explosives (ammonium nitrates) has led to closing some of them. But the ongoing reconstruction meant larger needs for construction materials The supply-demand picture faced a gap which led some contractors to start dredging marine sands, both from the beach and the near-shore (Khawlie 1995). Of particular concern here is not only the quality of these marine sands, rich as they are in salts, but also the detrimental environmental impact that their dredging leaves on the marine ecosystem. It must be noted that Lebanon's mountainous rocky areas are very well exposed and huge potentials exist for supplying construction materials from crushing rocks.

The problem is with supplying natural sands. This source could be supplied with some proper investigation (Hamad et al. 1995). Interestingly, from a resource conservation point of view, the huge amounts of construction debris resulting from the destroyed Beirut structures during the war are estimated at 500,000 m^3 of used concrete. No doubt the use of the recycled construction materials should be a welcome 'environmental ' source, and it is available where previous destruction took place. Its use also contributes in several ways by reducing debris, and the new demand on materials.

3 EXISTING AND POTENTIAL PROBLEMS

In trying to project the environmental technology applicable for quarrying operations in Lebanon, one has to understand the problems these operations are facing or inducing. If the amounts of needed construction materials are between 30,000-45,000m^3 per day, then Lebanon is at risk because of the amount of uninspected explosives used to produce these materials. One kg of dynamite produces about 1m^3 of material, over a working period of more than 250 days per year, this means about 3000 tons of explosives are used annually. The declared or 'licensed' amount is less than one third this value, therefore a huge amount of explosives go unchecked.

Obviously, many problems are at hand, for example, the handling and utilizing of explosives for the construction industry, relate to improper practice, endangering the safety of workers and the public. Scrutiny on health issues, such as those coming from pollution of air, land and water is also a concern. Of particular importance is the ill-planned logistics of a quarry, its location, design of working face, mobility and maintenance of machinery, the timing, distribution and amounts of explosives per hole, etc.; all are leading to reduce the long-term function of the quarry. Because current operations care only for quantities of production regardless of environmental outcome, the rate of short-term production is up.

Other problems result from improper administering of the quarrying sector by the authorities in charge. First, officials who are supposed to give working licenses are either not aware of what is actually on the ground, or they are 'blind' to the facts. For example, the distribution of about 780 quarries implies an average of one quarry every 10 km^2. Mount Lebanon area alone takes 42% of total, making the potentials of environmental problems increase many fold in that area (Fig. 1). As was shown by Khawlie (1995), other areas in the country have the following distribution: Beirut area 12%, South Lebanon 14% and 21% in each of the North and Beka'a. An estimated figure on the total area coverage of quarries from early 1980's to the mid 1990's shows an increase from less than 0.1% to about 0.2% of total territory (UNDP 1994). This by itself is alarming and, from an environmental perspective, implies land degradation spreading all over the country.

Both the geological character and seismicity of the terrain add further difficulties to the quarrying operations. Surficial instability and earthquakes are common features in Lebanon. The consequences of improper quarrying operations, notably the use of explosives could, and did, lead to landsliding or other earth movements. The mislocation of quarries and the current dangers due to the inevitable encroachment of urban complexes is not taken seriously. It is the lack of land-use planning that has led to this situation. At any rate, there are people living in the vicinity of quarries that are under daily danger because of the frequency and intensity of tremors, with possibility of earth movements along the many and dense faults in the country.

In most areas the basic scientific information on the quarrying operations and their impacts on the environment are lacking. This is definitely a serious problem. Forgetting about it is leading to intensify and diversify the difficulties of quarrying. Not securing the proper database, including detailed scientific investigations on the site, the materials, the surrounding natural and human potentials is resulting in delaying the problems not eliminating them. Environmental deterioration in the natural ecosystem and endangering people's safety is common place wherever a quarry exists in Lebanon.

Many quarries are abandoned and their sites not rehabilitated or reclaimed, therefore they constitute spots of land and habitat degradation. Implementation of environmental legislation and enforcement of codes could have contributed to saving the land and the people from many of the mentioned problems. Unfortunately, implementation was either improper or, in some instances lacking altogether. A prominent point emerging from this picture is the lack of cooperation between local communities and the people in charge for quarrying, i.e. quarry owners and officials. Quarrying is known to impinge a heavy environmental toll, therefore cooperation among these groups is a must.

4 ENVIRONMENTAL TECHNOLOGICAL PERSPECTIVES

It is not only a matter of technological innovations that quarrying operations need, but also environment-friendly applications. For a country like Lebanon, coming out from 17 years of war and economically devastated, the need actually is for proper resource planning and management. The authorities in charge of quarrying must know what areas require urgent attention versus others that are less environmentally deteriorated. Likewise, what areas, or problems, can be dealt with through preventive approaches versus others needing remediation. In some cases where quarrying is still at an early stage, then relying on incentives, tax relief, facilitation of transactions for the quarry owner can help. Short and long-term environmental impacts should be identified and steps to undo their effects be taken accordingly.

The present author is calling for solutions that combine some advanced and some simple technologies to help reduce environmental problems, while securing the availability of construction materials. This is crucial due to the increasing demand with ongoing growth, and expected expansion in redevelopment projects. The technologies apply to different stages of the operation as shown in Table 1.

The above concerns require that several steps be taken starting with the availability of reliable databases. Most data on quarrying operations in Lebanon are lacking and outdated. They need upgrading and full coverage of all sites, i.e. both active and inactive quarry sites. Since these sites are spread all over the country, a quick and accurate means to secure such data is through studying and analysis of satellite imageries. Remote sensing can give immense information on the distribution of quarries, extent of operations, levels and types of environmental impact, plus proximity to human communities and their infrastructures.

The next technique to apply is aerial photo interpretation (Khawlie & A'war 1988). This gives further details on the character of the land, its nature and main uses. Therefore, the terrain can be evaluated in view of several criteria that are crucial to quarrying operations. Accordingly, a systematic aerial analysis divides the land into classes and hierarchies showing, for example, where environmental deterioration has taken place, its extent, the urgency of the problem, etc. This kind of study

Table 1. Environmental technological perspectives on quarrying.

Stage	Environment-friendly technology	Purpose	Assessment
Ia	Satellite imagery	Secure quick, accurate and total country database	Deterioration Localisation Monitor patterns of change
Ib	Terrain evaluation and classification (aerial photogeology)	Determine urgency of environmental problem Designate new safe quarrying areas	Extent of deterioration Level of deterioration Landform characterisation
Ic	Seismology & faults activity	Determine seismic risk and potentials of land failure	General safety Monitor pattern of use of explosives
IIa	Geotechnical evaluation of materials	Secure safety of people and site stability	Monitor material and slope character
IIb	Technical management	Assure proper operation	Monitor quarrying steps
IIIa	ISO-9000 & TQM	Environmental management and optimisation	Monitor resource & habitat systems
IIIb	Non-market valuation methods	Environmental accounting	Monitor cost of environmental deterioration

will also serve to designate new quarry areas that are geotechnically suitable and environmentally safe. It is unfortunate that the new study completed by consultants for the United Nations Development Program did not cover such details. Rather, it allocated quarry areas by relying simply on very old and inaccurate small-scale geological maps considering only the general distribution of lithologies and faults (UNDP 1994). Another type of crucial study, whose technological application is necessary, is seismology which relates to earthquakes, active faults and the effect of the use of explosives on the terrain. The nature of earthquake pattern and relevant data will assure reducing hazards on both long and short terms. Geological delineation of active faults in Lebanon is important for safety measures as the terrain is seismically active (Tabet 1993).

These techniques cover the first stage for studying the quarries, i.e. Items Ia, Ib, and Ic in Table 1. They must be followed by indicating the requirements for proper working at the quarry site, i.e. the actual daily operations covering detailed steps that make the complete operational process, including: the quality of the materials, safety of people and stability of the site. To achieve this it is crucial to follow an environment friendly approach, that is to have proper quarry management. This management must assign steps that will identify all needed resources, human, material and technical including a market study (Woakes 1978, Thurrell 1981), and safety requirements of quarrying and blasting (Lester 1981, Grimshaw & Poole 1983). It is important to prepare a proper design and planning of the quarry (Groom 1982, I.Q. 1969), even minor but important details focusing on reducing environmental impacts such as dust suppression (Reilly 1964) – this is not yet practiced in Lebanon – flow of bulk materials (Dick & Carson 1989), plus implementing laws and regulations (D & M 1976, MMAJ 1992, UNDP 1994) including land reclamation.

The third group of environmental applications relate to indirect advances that have emerged which, if properly implemented, could greatly benefit the construction industry and the environment. These are government and quarry management approaches, regulatory policy options and actions that encourage the quarrying sector as well as other productive sectors to abide by environment – friendly practices. The approaches are inherent in ISO-9000 and TQM methodologies which are internationally followed standards and total quality management procedures that can be implemented at enterprises of all sorts to secure upgrading of production and sustainable growth. Their framework is optimization in an environment and economically feasible manner. This includes: environmental labelling, environmental management systems, environmental performance evaluation, environmental auditing, life cycle analysis, and environmental aspects in product standards (ISO- IEC 1993, US President's Committee 1993).

Approaches for valuing environmental change by non-standard methods are useful indeed. Such is the use of non-market values to modify income accounts so that they reflect improvements and declines in environmental resources. Through the use of standard methods, a quarry could destroy the resource base, the environment, and even the workers plus community at and near the site, and yet shows an increase in its wealth (Hoehn & Walker 1993). Economic and environmental concerns are not yet given equal importance. This leads to furthering environmental deterioration on the short-term, while on the long-term may lead to forcing cessation of quarrying.

The World Bank Environmental Department has been producing a number of important relevant studies. An example is that by Anderson (1990) which reviews policies of taxing pollution and the alternatives of regulating it. What is the role of public and private investment in environmental improvement. In a way, losing a scenic and ecologically fertile land in Lebanon by improper quarrying practices, is very much like losing the rich forests in a country like Costa Rica. It was Costa Rican government intervention, by setting policies and forcing application of sound management, that helped stop the deterioration (Lutz & Daly 1990). Land use and land capability mapping was a major concern that helped the Costa Rican government face the above deterioration together with use of incentives and regulations.

Recently, Lebanon witnessed the inception of its national environmental strategy. It included an assessment of the state of the environment and identification of policy options – to which the current author was a major contributor (ERM-Jouzy & Partners 1995) under sponsorship of the World Bank. This should lead to increasing hopes as proper implementation of the policy options and actions delineated in the strategy would increase environment-friendly trends. The policies related to construction materials and quarrying reflect to a considerable extent the purposes outlined in Table 1 of the present study.

5 CONCLUSIONS

Securing construction materials at any price, even for a country undergoing redevelopment and reconstruction, is not acceptable. This paper is a case at hand where the dilemma of rebuilding war-torn Lebanon required huge amounts of construction materials. The problem is that this is occurring at the expense of the environment.

The quarrying operations, both old and current, have been going on with minimal concern to environmental requirements. The siting of the quarries, their everyday processes and the lack of proper official control led to further deterioration. The present study emphasizes the need for securing databases and upgrading both technical and management aspects. Simple and advanced procedures, as well as legislative enforcement, can be applied to arrive at securing the required construction materials and saving the environment.

ACKNOWLEDGMENTS

The National Council for Scientific Research has always been keen on supporting applied research in Lebanon. The Ministry of the Environment, though newly established, is pushing to solve the problem of quarrying in Lebanon. Both institutes helped in many ways.

REFERENCES

Anderson, D. 1990. Environmental policy and the public revenue in developing countries. *World Bank Envi. Dept. Env. Working Paper No. 36.*

Behcara, H. 1994. *Bulletin de la construction et du batiment*, No. 101 Beirut.

D &M 1979. *A digest of environmental regulations pertinent to open pit mining in Canada.* Dames & Moore, Canmet Rept. 76-16.

Dick, D. & Carson, J. 1989. New developments help characterize and improve the flow of bulk granular solids. Mining Engineering 41: 3 : 163-165.

Grimshaw, G. & Poole, G. 1983. Blasting techniques for surface extraction in the U.K. In *Surface Mining & Quarrying.* Inst. Mining & Metallurgy (SMQ-IMM), 39-48.

Groom, A. R. 1982. Planning conditions and the quarry manager. *Quarry Management & Products*, July.

Hamad, B., Yassine, M. & Khawlie, M. 1996. A survey study on geology and location of major sand resources in Lebanon, Eastern Mediterranean. *Bull. Intl. Assoc. Eng. Geol.* 53:39-48.

Hoehn, J., & Walker, D. 1993. When prices miss the mark: Methods for valuing environmental change. Policy Brief, EPAT/MUCIA No.3.

I. Q. 1969. Quarry planning (law business, production*). The Inst. of Quarrying, Symp*. Proceed, Reprint, 95-179.

ISO-IEC 1993. Report of the 4th meeting of the ISO/IEC strategic advisory group on environment (SAGE). Toronto, Canada.

Khawlie, M. 1995. The Lebanese coastal environment : natural regime and man's interference. Proceed, *UNESCO symp*, pp.71-86, Arabic.

Khawlie, M. 1995. The potentials of environmental hazards resulting from quarrying operations in Lebanon. *Proceed.* NCSR Symp. (in press) Arabic.

Khawlie, M. & A'war, R. 1988. *Terrain analysis for development studies: typical mountainous area along the eastern Mediterranean.* Bull. Intl. Assoc. Eng. Geol. 38: 95-104.

Lester, D. (Comp). 1981. *Quarrying and Rock Breaking.* Intermed. Tech. Pub. London.

Lutz, E. & Daly, H. 1990. *Incentives, regulations, and sustainable land use in Costa Rica.* World Bank, Env. Dept. Working Paper No. 34

MMAJ 1992. Preventive activities of mining-related pollution in Japan. Metal Mining Agency of Japan, Tokyo.

Reilly, R. H. 1964. *The use of water sprays for dust suppression in quarry plant.* The Inst. of Quarrying, Reprint.

Tabet, C. 1993. A possible seismic gap along the northern segment of the Yammouneh fault in Lebanon. Seminar Earthquake Hazards, Cairo.

Thurrell, R.G. 1981. *Quarry resources and reserves: The identification of bulk mineral resources.* Quarry Management & Products, March, Reprint.

UNDP 1994. Projet PNUD LEB 87/002 – Impact des carrières sur l'environnement du Liban. UNDP, Beirut.

US Pres. Comm. 1993. Total quality management, framework for pollution prevention. *US Presidential Committee on Environmental Quality.* Washington, D.C.

Woakes, M. (comp.) 1978. *Mineral resource management in developing countries.* AGID Symp. Sydney.

Aggregate resources of the Greater Vancouver and Lower Mainland market, B.C., Canada: Problems and future outlook

Z.D. HORA

B.C. Ministry of Employment and Investment, Victoria, Canada

1 INTRODUCTION

Sand and gravel is, by volume, the largest mineral commodity used and produced in British Columbia. In 1994, about 42 million tonnes were produced in the province, of which half was used in Greater Vancouver and the Lower Mainland. In terms of value, sand and gravel is the most important industrial mineral commodity in the British Columbia economy. In 1990 in British Columbia, this industry employed between 4000 and 5000 people directly in the mining and processing of aggregate and employed many more in transportation to construction sites. Total direct value of the industry was estimated at $370 million (Thurber 1990). The study area covers the westernmost 100 km of the Fraser River valley ('Fraser Lowland') with the adjacent slopes and tributaries to the north and south. The southern limit is the International Boundary following the 49th parallel (Fig. 1). It is the most densely populated area of the province with the City of Vancouver and its suburban areas ('Greater Vancouver'), where the river empties into the Strait of Georgia, and a number of municipalities scattered along the river to the east. The valley is also the most agriculturally productive part of the province and most of the undeveloped land has been placed into the Agricultural Land Reserve (ALR) to protect it from residential and industrial encroachment. The analysis of the aggregate supply and demand situation was initially done between 1978 to 1980 to provide basic data for development of aggregate resource management policy for the province. The information was updated by collecting new data for the Aggregate Forum, held in Richmond, B.C., in March 1995.

2 QUALITY OF SAND AND GRAVEL

Nature does not always provide deposits containing particles ideally sized and sorted for industrial requirements including pavement, road base, concrete and drainage fill. This is especially so for the most abundant surficial deposits, products of deglaciation, which are characteristically poorly sorted. As a result, most of the aggregate that enters the Greater Vancouver and Lower Mainland area markets is, to some degree, pre-processed. Smaller producers usually employ only simple screening, leaving the boulders as waste, and produce only a few types of construction aggregate or

397

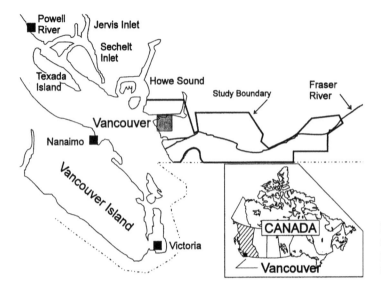

Figure 1. Location of study area, Lower Mainland, British Columbia, Canada.

fill. Larger operators, however, have crushing, screening and washing facilities and are capable of supplying many types of aggregate products for a variety of uses.

Fortunately, the Lower Mainland gravel deposits do not contain appreciable amounts of deleterious components such as chert, glassy volcanic rocks and weathered rocks. In other areas, these attributes limit the use of natural aggregate by lowering the final quality of concrete. The only significant deleterious components of the sand and gravel deposits in the study area are silt and clay, both of which are easily removed by screening and washing.

In general, Quaternary granular sediments of the Lower Mainland consist of mostly isometric, rounded pebbles, cobbles and boulders of mainly granitic and some highly metamorphosed rocks. A notable exception is the Holocene alluvial fan of Chilliwack River, a relatively small source area with a rather distinct geology resulting in a higher proportion of flat and elongated pebbles.

3 ORIGIN AND DISTRIBUTION OF SAND AND GRAVEL DEPOSITS

Sand and gravel resources of the southern coastal region of British Columbia may be linked to various episodes of Wisconsinan glaciation. The distribution of sand and gravel deposits in the Fraser Lowland and along the coast is controlled by a number of factors. During the Quaternary Period, the province experienced several glacial-interglacial cycles. Major glaciations were accompanied by isostatic and eustatic changes in sea level of up to 200 m. As a result, low-lying areas were, at times, covered by the sea.

Since the Fraser Lowland is bounded to the north and south by high mountain ranges, western glacier margins would have occupied the sea at certain times. During deglaciation, meltwater from the ice produced widespread and extensive deposits of sand and gravel along the coast, throughout the Fraser Lowland and adjacent areas.

The interaction of waves and changing sea level positions, resulted in the widespread accumulation of gravely beach deposits up to a few meters thick at elevations between 0-200 m. (Hora & Basham 1981).

4 LOCAL GEOLOGY

Gravel-bearing formations are present throughout most of the Lower Mainland in a variety of stratigraphical positions and lithological units (Fig. 2). The largest accumulations of gravel include deltaic deposits in North Vancouver (Capilano sediments), a complex of units in the Coquitlam Valley, as well as Sumas drift and Fort Langley Formation sediments near Langley and Abbotsford (Fig. 3). Alluvial fan sediments of the Chilliwack River (Salish sediments) cover a large area, but economical parts are restricted by largely uneconomic thicknesses of deposits. North of Vancouver, along the coast of the Strait of Georgia and its inlets, are numerous Capilano deltaic fans. In flat and gently sloping areas, local patches of raised beach gravel and sand can be found (Armstrong 1977).

5 DEPOSIT CHARACTERISTICS

Deposits of sand and gravel vary in size, shape and granular composition. The producing deposits in the study area range from small fans, a few hundred meters across and only several meters thick, to areas of more than 50 km^2 underlain by up to 50 m of gravel. Deposits south of the Fraser River occur in generally flat terrain and contain well sorted gravel clasts, whereas deposits north of the Fraser River and along the coast occur on sloping terrain, are unsorted and contain many boulders. Many deposits are covered by a layer of topsoil a few centimeters thick. If till is present, it is usually processed with the underlying gravel. In contrast, some of the mined deposits in the Fort Langley Formation are overlain by laminated marine silts with a stripping ratio of almost 1:1. The gravel deposits of the Coquitlam Valley are geologically complex. They consist of sediments deposited during several major glacial advances and retreats as well as during nonglacial intervals in a glacier carved bedrock valley. Up to 150 m high cliffs expose deposits of Quadra sand, Highbury and Pre-Highbury sediments overlain by Vashon till with overburden ratios up to four times that of aggregate. The distribution of granular deposits throughout a major part of the Fraser Lowland has been documented by Hora & Basham (1981) as a three dimension block diagram.

Availability of the resource is influenced not only by the physical presence of the deposits and the economic viability of the product in the market area as a result of transportation costs, but also by conflicting interests that may sterilize existing deposits. For example, residential development favours areas underlain by gravel because of good drainage. Another limiting factor is public concern regarding noise, dust, water pollution and heavy traffic associated with aggregate extraction. Locally, even aesthetic aspects may play an important role in activating public pressure to eliminate existing production centres and to further restrict development of new deposits. Two Lower Mainland municipalities no longer allow privately operated

Years B.P.	Time Units	Geologic Climate Units	Lithostratigraphic Units	Comments	Important Sources of Mineral Aggregate
	Holocene	Post-glacial	Salish and Fraser River sediments	All post-glacial sediments	Alluvial gravels
10 000	Late Wisconsin	Post-glacial / Fraser glaciation	Capilano sediments; Sumas drift; Fort Langley Formation	Marine, deltaic, and fluvial deposits; Till, outwash, ice-contact deposits; Deltaic, ice-contact and outwash deposits, till	Deltaic and channel fill raised beach; Outwash and ice-contact; Deltaic, outwash and ice-contact
13 000	Late Wisconsin	Fraser glaciation	Vashon drift	Till, outwash, ice-contact deposits	Outwash
18 000	Late Wisconsin	Fraser glaciation	Quadra sand; Coquitlam drift	Proglacial sand, silt, gravel; Till	Gravel locally
26 000	Middle Wisconsin	Olympia non-glacial interval	Cowichan Head Formation	Fluvial, organic, colluvial	Gravel locally*
62 000	Early Wisconsin and Pre-Wisconsin	Semiahmoo glacial	Semiahmoo drift	Fill, glaciofluvial, glacio-marine	Gravelly outwash locally*
	Early Wisconsin and Pre-Wisconsin	Highbury non-glacial	Highbury sediments	Fluvial gravel, sand, and silt	Gravel locally*
	Early Wisconsin and Pre-Wisconsin	Westlynn glacial	Westlynn drift		Gravel locally*
	Early Wisconsin and Pre-Wisconsin		Older sediments		

*Exposed and mined in Coquitlam Valley as a complex of units.

Figure 2. Quaternary sediments of the Lower Mainland area, British Columbia, Canada (after Armstrong 1977).

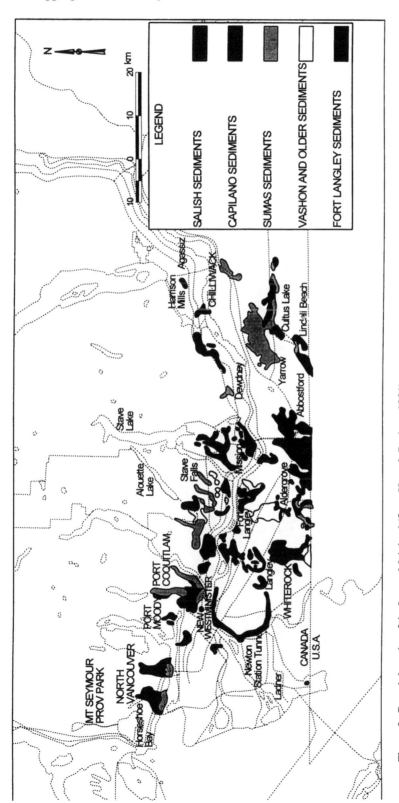

Figure 3. Gravel deposits of the Lower Mainland (from Hora & Basham 1981).

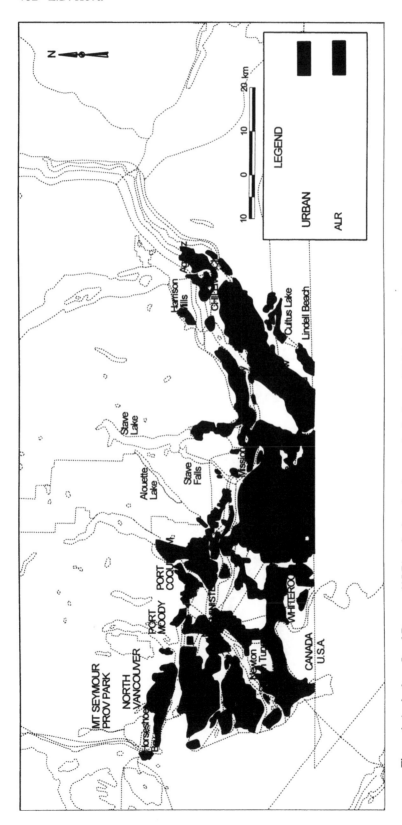

Figure 4. Agriculture Land Reserve (ALR) and urban development in the Lower Mainland.

Table 1. British Columbia consumption of aggregate in 1991 (thousand tonnes). Source: Statistics Canada (1993).

Fill	4,507
Road bed, surface	25,900
Ice control	39
Concrete aggregate	2,785
Asphalt aggregate	4,670
Railroad ballast	22
Backfill for mines	–
Mortar sand	75
Other purposes	3,994
Total	41,982

gravel pits and others restrict use of some heavy equipment like crushers to prohibit new gravel operations. Some of the deposits are several tens of meters thick and the gravel extends below the groundwater table. Municipal regulations, however, frequently limit gravel extraction to above the groundwater table. Another problem facing the aggregate industry is that most of the gravel deposits south of the Fraser River, and outside of the city limits, are located within the Agriculture Land Reserve (ALR) (Fig. 4). Application for exemption from the ALR to operate a gravel pit must be approved by local authorities and the Land Commission and can, therefore, become a political issue. In the end, abundant aggregate resources are reduced by the above pressures which sterilize resources needed for residential, commercial, industrial and transportation development.

In general terms, north of the Fraser River and along the coast, gravel availability is controlled primarily by geological factors and the physical presence of the deposits. South of the Fraser River, the limiting factors are availability of land and limitations of permitting procedures. The largest untapped resource, for example, is under the Abbotsford airport. The industry estimates that available resources on the south side of the Fraser River will be fully depleted in 5 to 10 years with no replacement reserves being planned (Irvine 1996). Since about 75% of aggregate consumption is used for public projects, it will be the taxpayer who carries the burden of significantly higher transportation costs to bring the aggregate from more distant sources.

Table 1 shows the most recent end use breakdown available as of 1996 (Table 1). Breakdowns published for 1977 and 1978, for both British Columbia and the Lower Mainland, show a similar share for individual end uses (Hora & Basham 1981).

6 QUARRIED AND CRUSHED AGGREGATE

Quarried crushed rock may substitute for natural sand and gravel in many applications. In the study area, crushed aggregate production at Pitt Lake, operated to blend its product with excess fines from Mary Hill pit, near Port Coquitlam, was phased out in the 1970's and the quarry at Watts Point, in Howe Sound, has been inactive for a number of years. In 1995, however, there are three major quarries opened in recent

Table 2. Processed mine waste (tonnes) from Texada Island quarries. (Thompson & Diggon personal communication 1995).

Year	Crushed/screened	Rip-rap
1994	935,000	57,000
1993	427,000	138,000
1992	535,000	179,000
1991	295,000	84,000
1990	167,000	93,000
1989	650,000	156,000
1988	685,000	153,000

Table 3. British Columbia aggregate production. Annual Lower Mainland consumption 30-60% of the B.C. total. Source: Mineral Policy, B.C. Ministry of Energy, Mines and Petroleum Resources.

Year	Quantity (10^3 tpa)	Value (10^3 $)	Total B.C. minerals value %
1994	41,837	146,790	5.6
1993	40,241	135,398	5.6
1992	39,883	128,024	5.1
1991	41,982	134,942	4.8
1990	41,278	140,585	4.7
1985	49,007	117,015	4.8
1980	45,278	98,666	4.5

years in the central part of the Fraser Valley and producing approximately two million tonnes annually.

Data published by the US Geological Survey (Langer 1988) indicate that production costs of crushed quarried aggregate are 25-30% higher than those for sand and gravel. This means that the two products cannot be competitive if they come from local sources. It is quite possible, however, that with increasing transportation costs for deposits more distant from the market, the price of crushed quarried rock in the Lower Mainland will again become competitive.

Limestone quarries on Texada Island, are producing large volumes of mine waste (Table 2). Granite dykes form a significant part of the limestone deposits, and for the lime and cement industry the dyke material is deleterious. Since selective mining of only limestone is frequently impractical, dykes are usually mined out and wasted. The limestone industry on Texada Island developed a crushed stone market for construction projects along the coast by processing such mine waste. This is in part in competition with coastal sand and gravel pits, but also supplements the market with otherwise unavailable products like rip rap and jetty stone (Table 3).

7 AGGREGATE PRODUCTION AND USE DISTRIBUTION PATTERNS

The distribution of production centres depends in general on the local market size and availability of the resource. As discovered during a 1980 survey (Hora & Basham 1981), the market appears to bear transportation costs up to approximately

50 km by truck and 150 km by barge. For 1995, those transportation limits remain the same. What has changed are local production volumes with the same transportation pattern. Transportation cost, therefore, seems to be the main limiting factor in the Lower Mainland by dictating the size of production from individual production centres. Only large deposits with large markets within economic transportation distances can afford several large producers concentrated in a relatively small area. Availability of transportation corridors is also an extremely important factor for marketability of aggregates in the lower mainland. The lack of available crossings on the Fraser River further constrains construction aggregate marketing from one side to the other.

For many years Greater Vancouver construction activities have relied on gravel imported from other areas. Since the major production centres in the Fraser Lowland and adjacent areas are distant from the urban core and trucking costs are not always competitive, the industry has developed production units along the coast and is barging aggregate to Greater Vancouver to supply the local construction industry. At the same time, the aggregate producers developed tidewater supply depots in the proximity of Greater Vancouver downtown areas. The shipping companies meanwhile developed self-unloading barges to serve the industry. Some deposits in the Howe Sound area have already been depleted, but about 50% of the deltaic deposits located along the shores of Jervis and Sechelt Inlets have not been explored or developed. An additional area of aggregate potential are the shores of Indian Arm and Pitt Lake. A final possibility includes dredging gravel from the Strait of Georgia.

Over the years, industry has shown a great level of flexibility in adjusting to changing the supply-demand situation with diminishing availability of aggregate from local sources (Figs 5-6, Tables 4-5). Since the 1960's, the major aggregate producers started to develop production centres along the coast to supply the Vancouver market. Capilano raised deltaic deposits are scattered along the mainland coast and, to a minor extent, on Vancouver Island. The thickness observed in some exposures has reached 65 m and the reserves are frequently in order of tens of million tonnes. Producing pits in Capilano sediment deposits that supply the Greater Vancouver

Table 4. Exports and imports of aggregate in tonnes, Lower Mainland, B.C. Source: Natural Resources Canada 1995.

Year	Exports	Imports
1994	327,042	381,524
1993	255,079	446,941
1992	297,402	428,318
1991	316,421	509,283
1985	225,090	864,675

Table 5. Aggregate prices in the Lower Mainland. FOB plant or supply depot source: Industry price lists.

	1978 ($/tonne)	1995 ($/tonne)
Pit run	0.63-1.83	3.25- 4.50
Screened and washed	2.14-4.06	8.50-11.50
Screened and crushed	1.85-3.66	8.00-11.95

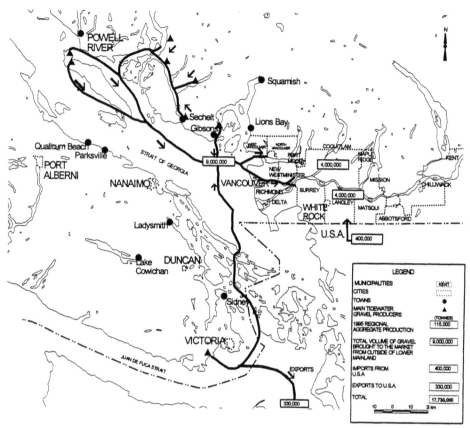

Figure 5. 1980 gravel production and shipments in the Lower Mainland, British Columbia.

market are along Howe Sound (partially depleted already), Jervis and Sechelt Inlets, and at Colwood, near Victoria, on Vancouver Island. Another coastal pit at Friday Harbour, on San Juan Island in Washington State, provided up to one million tonnes annually, most of it shipped to Vancouver until 1985, when this deposit was depleted. Closure of this production centre significantly reduced the aggregate imports from Washington State and opened some export markets for British Columbia producers, namely the Colwood pit on Vancouver Island. The industry sources indicate the transportation cost from Texada Island, for example, to the average downtown Vancouver supply depot in 1978 was $1.00 per tonne of aggregate, in 1995, it is $2.50 per tonne of aggregate. This is still only about 25% of the processed product price quoted for the Lower Mainland FOB plant.

Until 1985, most of British Columbia aggregate imports were shipped by barge, in 1995, imported gravel is shipped by trucks from pits just south of the 49th parallel.

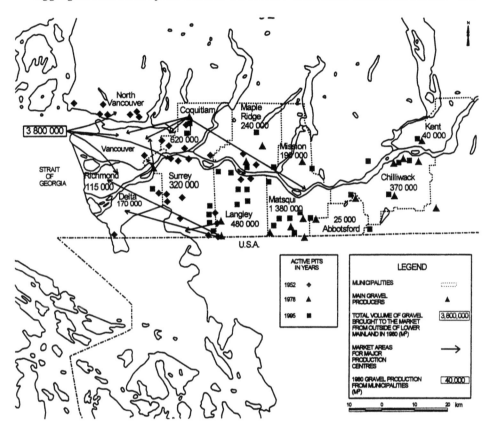

Figure 6. 1995 gravel production and shipments in the Lower Mainland, British Columbia.

8 CHALLENGES OF 1995

British Columbia does not have a comprehensive aggregate resources management policy. In total, ten government agencies from municipal to federal level have regulations which, in one way or another, affect the aggregate industry. In urban areas, particularly, a maze of regulatory requirements exists that operators of commercial gravel pits must potentially face. For example, there are as many as eight agencies that have potentially significant requirements that must be satisfied. Looking at the industry and regulation practices, it may be said that the aggregate production in British Columbia is over-administered and under-regulated. The main problem is that current focus is on the consequences of mining and processing aggregate instead of on the resource itself.

A major proportion of the province's population and population growth is in the Lower Mainland. The Greater Vancouver Regional District's (GVRD) growth plan with higher density housing is, however, not entirely coincident with municipal development scenarios. Neighbourhoods and communities usually see aggregate mining as a negative land use activity. Reclamation strategies integrated with long term growth plans can address this negative image by having community input defining the final land use. Such 'opportunity landscape' can have recreation (park, golf course), environmental (wetland), residential or industrial value.

A 'single window' regulatory process to deal with present overlap and fragmentation of responsibilities should be able to resolve the present problems and to ensure availability of aggregate at affordable prices in the Lower Mainland for many years to come.

REFERENCES

Armstrong, J.E. 1977. *Quaternary stratigraphy of Fraser Lowland*, Fieldtrip Guidebook, Geological Association of Canada, Annual Meeting.

Armstrong, J.E. 1980. Surficial Geology of Vancouver, New Westminster, Mission and Chilliwack Map Areas, British Columbia, Geological Survey, Canada, Maps 1484A, 1485A, 1486A, 1487A.

Baker, D. 1996. Planning for Aggregate Resource Extraction: Putting the Inventory into context (Abstract). In P.T. Bobrowsky, N.W.D. Massey & P.F. Matysek (eds), *Aggregate forum – developing an inventory that works for you!. Richmond, British Columbia. Mar. 30, 1995*. Victoria: British Columbia Ministry Energy Mines & Petroleum Resources, Inform. Circ. 1996-6.

Hora, Z.D. 1988. *Sand and Gravel Study 1985*, Transportation Corridors and Populated Areas, B.C. Ministry of Energy, Mines and Petroleum Resources, Geological Survey Branch, Open File 1988-27, 41 pp.

Hora, Z.D. & Basham, F.C. 1981. *Sand and Gravel Study (1980)*, British Columbia Lower Mainland, B.C. Ministry of Energy, Mines and Petroleum Resources, Geological Division, Paper 1980-10, 74 pp.

Irvine, B. 1996. Status Report for the Supply of Aggregates in British Columbia (Abstract). In P.T. Bobrowsky, N.W.D. Massey & P.F. Matysek (eds), *Aggregate forum – developing an inventory that works for you!. Richmond, British Columbia. Mar. 30, 1995*. Victoria: British Columbia Ministry Energy Mines & Petroleum Resources, Inform. Circ. 1996-6.

Langer, W. H. 1988. Natural Aggregates of the Conterminous United States, USGS Bulletin 1594, 33 pp.

Lee, S. 1996. Ensuring Ongoing Economic Sources of Highway Construction Aggregates Through a Gravel Resource Management Program (Abstract). In P.T. Bobrowsky, N.W.D. Massey & P.F. Matysek (eds), *Aggregate forum--developing an inventory that works for you!. Richmond, British Columbia. Mar. 30, 1995*. Victoria: British Columbia Ministry Energy Mines & Petroleum Resources, Inform. Circ. 1996-6.

Taylor, C.D. 1989. A Study of the Aggregate Industry in the Vancouver Area, unpublished thesis, Simon Fraser University, 74 pp.

Thurber Engineering Ltd. 1990. Sand and Gravel Industry of British Columbia, report to B.C. Ministry of Energy, Mines and Petroleum Resources, Thurber Engineering Ltd., 74 pp.

Statistics Canada 1993. Quarries and Sand Pits 1991, Statistics Canada, 30 pp.

Aggregate resources in Norway and their quality requirements

PEER-RICHARD NEEB
Geological Survey of Norway, Trondheim, Norway

1 INTRODUCTION

Aggregates used in Norway are mainly derived from Quaternary glaciofluvial sand and gravel deposits and crushed hard rock. The bedrock geology of Norway is complex being dominated by Precambrian gneisses, igneous rocks and metasediments and Cambro-Silurian metasedimentary and metavolcanic rocks. Most aggregates for building and construction purposes are composed of lithologies with good mechanical strength, and which do not have problems with mineralogical alteration (weathering).

Files on sand, gravel and hard rock aggregate deposits have, in the past, been held by a number of local and regional governmental bodies, arising from their differing needs for information on the location, volume, composition and quality of the deposits. The Civil Defence Force, the Road Authority, the State Railways and the Geological Survey (NGU) have all maintained separate files.

In 1978, the Norwegian Department for the Environment took an initiative to assemble all information into a central, complete, computer-based register at NGU. The information in these databases is available to users via the county map offices (part of the State Cartographic Service). Data has been collected, analysed and entered into the database by NGU. This work has been financed by the Ministries of the Environment, Industry and Energy, with contributions from county and local authorities and the road authorities.

The gross production value of Norway's most important mineral products in 1994 is NOK 1430 mill for aggregate and 1000 for sand and gravel. Sand, gravel and crushed hard rock aggregates are Norway's most important construction materials, with an annual market value of approx. NOK 2300 million. Annual consumption is ca. 50 million tonnes of sand, gravel and hard rock aggregate. This corresponds to approximately 12 t/capita/year of which 5.7 tonnes is sand/gravel and the remainder is hard rock aggregate.

Forty six percent of the production goes to road construction, 20% to concrete and the remainder to other uses (Figs 1-2). The export of aggregate in 1995 was 5.9 million tonnes to Europe with an export value of NOK 300 million. Norway is the fourth largest per capita producer of sand, gravel and aggregate behind Iceland (34 t/capita /year), Canada (16.5 t/capita/year) and Finland (14 t/capita/year).

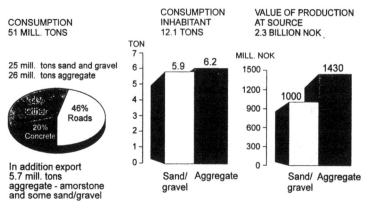

CONSUMPTION
51 MILL. TONS

CONSUMPTION
INHABITANT
12.1 TONS

VALUE OF PRODUCTION
AT SOURCE
2.3 BILLION NOK

25 mill. tons sand and gravel
26 mill. tons aggregate

46% Roads
20% Concrete

In addition export
5.7 mill. tons
aggregate - amorstone
and some sand/gravel

Figure 1. Consumption and production value of sand, gravel and hard rock aggregate 1994/1995.

Figure 2. Use of aggregates by society.

Gravel and Hard Rock Aggregate Databases are computerised databases which provide an inventory of all known deposits in the country along with information on their volume and quality. The Databases have made a significant contribution to the better management of non-renewable national resources and are an important part of the Norwegian Geological Survey's information system.

2 USE OF AGGREGATES BY SOCIETY

2.1 *Land-use conflicts and resource economics*

The construction industry and the State Road Authority are the most important consumers of sand, gravel and hard rock aggregate in Norway. However improvements in concrete technology and road construction, have led to increasingly stringent requirements regarding the quality of the construction materials. Hence the demand for knowledge about the location, volume, composition and properties of the deposits available has increased.

Sand, gravel and hard rock aggregate are relatively inexpensive resources measured in kroner/m^3. Because of the frequent need for large volumes of these materials the distance from source to consumer is an important economic factor. This leads to pressure on, and rapid exploitation of deposits found near the larger cities. This is most obvious in the Oslo region where the most conveniently located deposits are nearing exhaustion. Continued rapid growth in the construction of new housing, business developments, roads, railway lines and not least, a new major airport, will necessitate the use of raw materials from an increasingly wide catchment area. Similar problems are found in many towns where the only sand and gravel deposits are in built-up areas.

2.2 *Environmental factors*

Increasing environmental awareness limits industrial activity with the vicinity of housing and recreational areas. The quarrying of building materials leads to scars in the countryside, increased noise, dust and traffic of heavy goods vehicles. There is also an increasing interest in the protection of both typical and rarer land forms in the vicinity of the larger conurbations. Furthermore groundwater is becoming an increasingly important component in our water supply. Important groundwater sources were often located in sand and gravel deposits which leads to difficult strategic decision-making and increases the need for adequate attention to be given to natural resources in local planning.

Sand, gravel and hard rock aggregate have traditionally been regarded as more or less inexhaustible resources in Norway. That this is a myth is obvious from economic, technical and environmental factors. This has led to an increasing need for tools which give both an overview of, and detailed information on resources available. The Gravel and Hard Rock Aggregate Databases are such tools leading to better management of these resources and the environment.

2.3 *Hard rock aggregate or natural gravel*

Selection of hard rock aggregate or natural gravel as building material is determined by factors such as bulk quality, economics and the quantitative and qualitative particle distribution of the natural gravel. Hard rock aggregate is considered a natural substitute in areas with inadequate supplies of natural gravel, although this presupposes that the local source of aggregate would be more competitive than imported natural gravel. The same applies in areas in which the available resources of natural gravel do not comply with the general or specific quality criteria for construction purposes. In recent years hard rock aggregate has taken over most of the market for material used in road pavement and road metal and future consumption of hard rock aggregate is thus likely to increase more rapidly than that of natural sand and gravel.

3   USE OF THE DATABASE AND RESULTS

3.1   *The Database*

Existing information on deposits, in the area being considered, is evaluated along with information from Quaternary and Bedrock maps and reports. An aerial photo interpretation is made of the area and a field study with sampling is carried out on the deposits. The Databases are built up by systematic assessment of all available material – commune by commune and county by county. The relevant information is entered on to a standard data sheet.

Data collection is focused on areas within an acceptable distance from the existing communication network and population centres. Deposits are registered if the probable total volume above the water table, moraine, silt, clay or rock exceeds 50,000 m^3 and the average thickness exceeds two meters. These criteria are applied with discretion in areas where smaller or thinner deposits would have particular significance, for instance, along the coast where surficial deposits are very limited. The following information on individual sand, gravel and hard rock aggregate deposits is entered into the database:

– Co-ordinates,
– Volume (area x thickness), except for deposits of rock,
– Quality (see below),
– Present land use, except for deposits of rock,
– Land use conflicts in the event of exploitation,
– Production status and description of deposits where there already are established quarries,
– References for reports by the Geological Survey, the State Road Authority, consultants and others.

Analyses in the Gravel and Hard Rock database, rocks are divided qualitatively based on such criteria as flakiness, brittleness value, resistance to wear, abrasion value, ball mill value, Los Angeles value and for export polished stone value (PSV). The content of mica, schist, mafic rock and grains of other types in the sand fraction is also assessed. The Database also contains information on rock type(s) present, grain size, degree of compaction, quality classification, texture and petrography from thin-section descriptions, and possible alkali-reactive aggregates (more than 20% of counted grains).

Development of the Gravel and Hard Rock Aggregate databases began in 1978. The Gravel Database encompasses sand and gravel deposits with a combined volume of 12,000 million cubic meters. Ten percent of the area covered by the deposits is built-up and is therefore not available for production (Fig. 3). The Gravel Database contains information on 8790 deposits. These include 810 active quarries, 3865 in intermittent production and 1871 abandoned quarries for a total of 6546 quarries.

The Hard Rock Aggregate database will cover special parts of the country by 1999 and now contains information on 934 deposits where of 186 are in continual production and 149 in intermittent production. A number of quarries (237) have been in production but are now abandoned for a total of 572 quarries. The total numbers of sand and gravel extraction points and hard-rock quarries in Norway (both active and abandoned) is in the order of 7100. Potential deposits of hard rock aggregate near

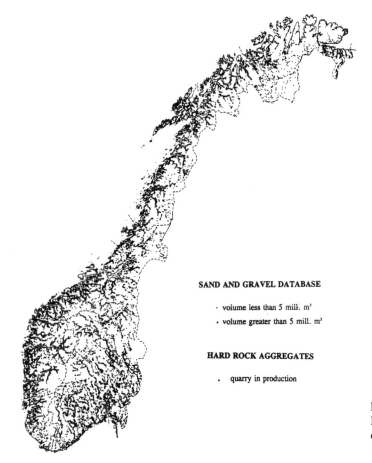

SAND AND GRAVEL DATABASE

· volume less than 5 mill. m³
· volume greater than 5 mill. m³

HARD ROCK AGGREGATES

. quarry in production

Figure 3. Gravel and
Hard Rock Aggregate
databases – status
1996.

major roads have been mapped in 10 of 18 counties for the local road authorities and county authorities.

A total of 636 maps of sand, gravel and hard rock aggregate resources, at a scale of 1:50,000 have been issued (01.01.96), some of them printed in colour (Fig. 4). The extent of the database allows us to define and rank the largest sand and gravel deposits in the country: most of them are in production. The largest deposits are located near Oslofjord, Buskerud county, Østfold county, Telemark county and in the counties of Finnmark and Hedmark.

3.2 *Sand and gravel for use in concrete*

The petrographic character of sand and gravel or hard rock aggregate deposits is important in terms of their usability for concrete and is dependent on the nature of the local bedrock. For example, material with a high content of mica can pose problems. Concrete with a high mica content in sand has a reduced workability and a higher water requirement. This, in turn, leads to reduced mortar strength after solidification. Deposits registered in Norway show considerable variations from deposits with al-

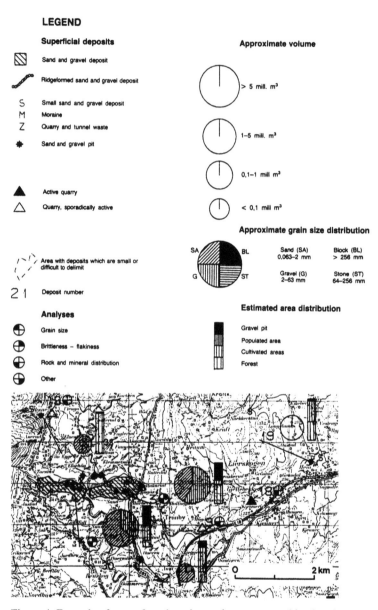

LEGEND

Superficial deposits

Sand and gravel deposit

Ridgeformed sand and gravel deposit

S   Small sand and gravel deposit
M   Moraine
Z   Quarry and tunnel waste
★   Sand and gravel pit

▲   Active quarry
△   Quarry, sporadically active

Area with deposits which are small or difficult to delimit

2 1   Deposit number

Analyses

⊕   Grain size
⊕   Brittleness – flakiness
⊕   Rock and mineral distribution
⊕   Other

Approximate volume

> 5 mill. m³

1–5 mill. m³

0,1–1 mill m³

< 0,1 mill. m³

Approximate grain size distribution

SA   BL
G   ST

Sand (SA)   Block (BL)
0,063–2 mm   > 256 mm

Gravel (G)   Stone (ST)
2–63 mm   64–256 mm

Estimated area distribution

Gravel pit
Populated area
Cultivated areas
Forest

Figure 4. Example of map of sand- and gravel resources and hard rock aggregates; part of the Lier map-sheet, 1:50,000 in Buskerud county.

most no mica in southernmost Norway to micaceous sands in Nordland county. The mineralogical compositions of aggregates are important because of relevance for potential alkali reactivity. Aggregates for concrete should since 1992 be tested in accordance with the procedures outlined by the Norwegian Concrete Society. The first step in testing involve testing by petrographic examination. If a higher quantity of reactive rock-types than 20% is present, the aggregate is classified as reactive.

3.3 *New quality requirements for aggregate for use in road construction*

Revised standards for the use of aggregates in road construction were implemented in Norway since 1992 with a publication of the handbook '018 Road construction', published by the Norwegian public roads administration. A number of the requirements have been made more stringent especially those with respect to brittleness, flakiness, abrasion and resistance to wear. Furthermore the Norwegian public roads administration and NGU have also used the Nordic ball mill test and the Los Angeles test after CEN standards.

It is important to use the correct quality of rock type for road construction in order to reduce maintenance costs. Stone materials in Norway are classified into five stone classes, according to brittleness and flakiness. The capacity of materials for resistance to wear is derived from their abrasion values and their brittleness. The resistance to wear value (Sa-value) equals the square root of the brittleness value multiplied by the abrasion value. These mechanical properties are all registered in NGU's Hard Rock Aggregate Database for existing and potential quarries.

Analyses of the mechanical properties from the Database have been used to set the new classification limits. The new requirements for aggregates are shown in (Figs 5-6) for brittleness, flakiness, abrasion value and Sa-value. The abrasion value and Sa value are important parameters in the evaluation of the wear of course material. Requirements for the wear of course material are set in relation to the annual-daily traffic. Samples of rocks which have very good resistance to wear have been collected at many sites. Registered rock-types which give crushed material of this quality are hornfels, quartzite, gabbro, mylonite, anorthosite, rhom porphyry, eclogite, diabase, dolerite, syenite, rhyolite, sandstone, conglomerate and augen gneiss.

There is a large variation in Sa-values for each rock-type and it is difficult to distinguish rock-types with low Sa-values. Localisation of bedrock with high PSV value will be of great importance in the future. Some anorthosites, gneisses and sandstones are of very good quality with polished stone value (PSV) over 60.

At present PSV-testing is considered to have little importance in Norway due to

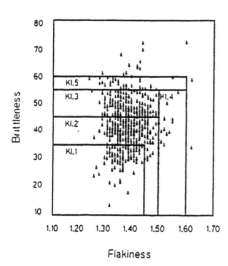

Figure 5. Database for Norwegian hard rock aggregates, brittleness and flakiness from 611 analyses.

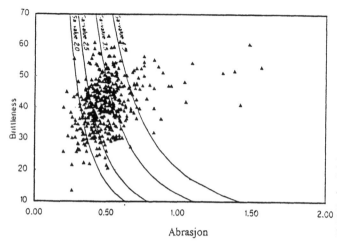

Figure 6. The resistance to wear value (Sa-value) from 472 analyses. The Sa-value is the square root of the brittleness value multiplied by the abrasion value.

the widespread use of studded-tires in the long winter months. However, the Norwegian Ministry of Environment is now considering the banning of such studded tires in major urban centres. If legislation to the effect is enacted, PSV-testing will be of rapidly increasing importance for road-surface aggregates. After the new CEN standard, we also use Los Angeles values with test results from 10.0 to 40.0 for Norwegian bedrock.

3.4 *Resource accounting*

Sand, gravel and hard rock aggregate are important non-renewable resources. Consumption is nevertheless rapid, especially in densely populated areas. A shortage of resources is already presenting problems in some parts of the country. NGU has developed a system for resource budgeting and accounting which gives a picture of the extraction and subsequent transport flow of construction materials. This provides information to reserve areas containing high quality deposits for future generations.

There is an acute shortage of construction materials in several of the local government areas in Norway at present, either because the originally available resources were very limited or because they have been exhausted. This leads to transport of the necessary materials over increasing distances with correspondingly increased costs. Long term national planning thus requires the mapping of our resources of building materials and a more sensible management of these non-renewable resources. NGU has now assessed the consumption and transport flows of building materials in 10 counties in Norway, based on the data in the Gravel and Hard Rock Aggregate Databases. This gives an overview of the availability of resources and of annual consumption in these areas and may reveal present or potential supply problems in relation to existing or planned construction activity. The interest for aggregates in the southern part of Norway for export to Europe has increased in recent years, especially Denmark, the Netherlands, Great Britain, Belgium, France and Germany.

Each local government area to be covered in the resource account is assessed with respect to availability of raw materials, production, import/export and consumption.

The resource account is presented in the form of diagrams for each local government area based on the data for any one year.

3.5 *Use of the Gravel and Hard Rock Aggregate Databases*

The Gravel and Hard Rock Aggregate Databases contain a range of data which allows different user groups to extract information for their separate purposes. Communal, county and national planning and development authorities can obtain information, maps and data relevant to land planning, communications and industrial developments from the databases.

An important aspect of annual resource accounting is the possibilities it provides for projection of the future situation and thus the ability to tackle future supply problems in a more effective manner. Sound information on consumption patterns enables prognoses to be made for future demand for sand, gravel and hard rock aggregates. This demand is heavily dependent on the level of activity in the building and construction industries, and on plans for the construction of new roads and the maintenance and improvements of the existing road network.

Good prognoses allow the prediction of the areas in which the demand for raw materials will be greatest in the years to come. This information, along with data on the availability of resources and on the transport of raw materials, allows local and regional authorities to delimit appropriate areas for production and to work out plans for how and when it is to take place.

The combination of data on sand and gravel, and hard-rock aggregate in the same database provides information on the total resource situation for the construction industry in any region. Users may evaluate alternative resources in relation to land-use conflicts and transport economics using a computer-based map, as in the Arc-Info Arc-View system.

4 COASTAL SUPERQUARRIES IN NORWAY

Increasing international demand for raw construction materials and stricter environmental requirements for aggregate production in Europe and Scandinavia may lead to the situation whereby Norway becomes a major exporter of aggregate based essentially on coastal quarrying.

There is also a somewhat disputed exploitation of raw construction materials from the sea-bed of the North Sea. In this case, large areas of sand and gravel are 'vacuumed' up. This destroys the breeding grounds for fish, and leaves large, submarine, desert-like areas. There is reason to believe that the North Sea countries will introduce restrictions in this activity in the relatively near future.

Development and maintenance of settlements, roads, airports, harbour areas and other constructions are totally dependent on supplies of good quality raw construction materials. New environmental restrictions in some areas will therefore increase the pressure on resources where the environmental consequences of exploitation are less. The search for more remote, but still accessible reserves has already begun. In this light, Norwegian stone stands out as an interesting possibility.

Aggregate companies have expanded internationally with the establishment of large, financially strong, industrial companies who themselves produce, sell and use construction raw materials. Such companies, amongst them Tarmac, ARC, Redland, Caemas and McAlpine in Europe and Vulcan in the USA, consider Norway as an especially interesting raw material supplier for the future. NGU have received enquiries from a number of European companies on the possibilities of establishing aggregate plants along the coast.

The export of aggregate in 1995 was 5.9 million tonnes to Europe with an export value of NOK 300 million, which is about 23% of Norwegian aggregate production (Fig. 7). In 1982, the value of 500,000 tonnes of exported aggregate was circa NOK 15 million. The present-day export is supplied by 17 large aggregate quarries in the south of Norway, and goes mainly to buyers in England, Denmark, Germany, the Netherlands and Belgium. These countries together have a total consumption of 280 million tonnes of aggregate. Estimate analysis shows that the growth in aggregate consumption may increase by 2.5% per year. In that case, the markets in the above-mentioned countries will increase to 460 million tonnes of aggregate by the year 2010.

With today's share of the market and stable prices, Norwegian raw material sources may be expected to reach an annual export of at least 10 million tonnes in the year 2010, and an export value of ca. NOK 460 million. If we assume that the companies who base their production on Norwegian raw materials are capable of conquering 5-10% of the market, then the consumer growth in Europe may form a basis

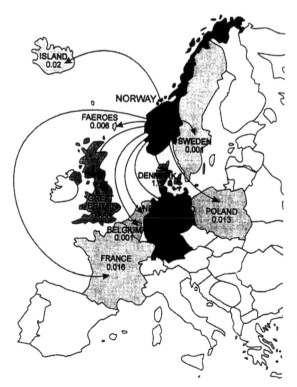

Figure 7. About 23% of Norwegian aggregate production, or 5.9 million tons, was exported to the continent in 1995.

for the establishment of several Norwegian superquarries with a total yearly production of 20-30 million tonnes of aggregate. In this case, this will represent an export value of NOK 1.5 billion. In addition, there is the employment and income created by the transport. International companies also consider Norway as an interesting future raw material supplier for markets on the North American east coast, a perspective which has not been included in this calculation.

It is important already today to clarify whether or not the natural, and physical conditions in Norway are appropriate for establishing superquarries, such that Norwegian companies daring to take a risk will have a realistic basis for a long-term market and investment strategy. NGU, in co-operation with the mining and quarrying industry, will work out how such resource mapping should best be carried out.

Norway has the natural resources to enable it to play a major role in the international aggregate market. There are several different rock-types which provide excellent opportunities for aggregate production. Parts of the Norwegian coastline are situated in a favourable manner so that transportation has a competitive advantage for the European market.

Superquarries will involve environmental inconveniences such as noise and dust emission, as well as creating large visible scars in the landscape. The major environmental challenge will be to establish production which will not be visible from the sea. This will entail the excavation of bedrock in mountainous terrain in quarries which are sheltered from view. The rock mass will be processed in an underground crushing/sifting plant and stored in silos inside the mountain itself. The transport of the aggregate will be done on conveyor belts through a tunnel out to the shipping area. In this way, the landscape maintains its form, noise is locked inside the mountain, and the dust can be dealt with in closed storage systems.

A major challenge for the aggregate industry will be to find a use for the fine fraction (material < 2 mm). Many aggregate quarries store this part of the production material today, without any means of disposing of it. With a production of the size of a giant aggregate quarry, the lack of a suitable market for this material could represent a considerable environmental problem. Possible areas of application could be industrial mineral products (quartz, feldspar, mica), fine material additives (asphalt, concrete, plastic, ore), rock-wool materials, etc.

The mapping programme that NGU has proposed will concentrate on the following:
– Topography and harbour conditions,
– Mapping of environmental conflicts,
– Analysis of existing infrastructure,
– Geological conditions.

5 CONCLUSIONS

1. Sand, gravel and crushed hard rock aggregates are Norway's most important construction materials, with an annual market value of approximately NOK 2300 million. Annual consumption is circa 50 million tonnes of sand, gravel and hard rock aggregate. 2. The Gravel Database encompasses sand and gravel deposits with a combined volume of 12,000 million cubic meters. 3. The total numbers of sand and

gravel extraction points and hard-rock quarries in Norway (both active and abandoned) is in the order of 7100. 4. A total of 636 maps of sand, gravel and hard rock aggregate resources at a scale of 1:50,000 have been issued. 5. A number of the requirements have been made more stringent especially those with respect to brittleness, flakiness, abrasion and resistance to wear. 6. In Norway there are many rocktypes present along the coastline which have interesting potential in terms of a range of aggregates for export. 7. Aggregate is being exported today from the south of Norway to England, Denmark, Germany, the Netherlands, Belgium, France, Island and Sweden. 8. Norway has 17 large aggregate quarries which exported in 1995 a total of circa 5.9 million tonnes valued at about NOK 300 million, exclusive of ship transport. The bedrock is dominated by gneiss, granite, anorthosite, mylonite, gabbro, syenite and quartz-diorite. 9. Demand for aggregate is expected to increase towards and beyond the year 2000 in the most populated areas providing an export potential for Norway. 10. There is reason to believe that coastal superquarries can provide the basis for Norway's most important mining industry in the future. 11. NGU, in collaboration with the aggregate industry, is planning to map the bedrock and rock-types along parts of the coast where environmental criteria, harbour conditions and topography are of potential interest for large scale aggregate extraction

REFERENCES

Erichsen, E. 1991. Holder norsk stein mål? Våre Veier, 18. årg. nr. 10/91, pp. 39.

Neeb, P.-R. et al. 1992. Byggeråstoffer. Kartlegging, undersøkelse og bruk. Tapir forlag, 374 pp.

Neeb, P.-R. 1994. *Aggregate recourses in Norway*. Report on the 2 nd International Aggregates Symposium in Erlangen, October , 1990.

Neeb, P.-R. 1995. Aggregate resources in Norway superquarries an important mining industry of the future. NGU Repport nr. 95.062.

Neeb, P.-R. 1995. Årsmelding for Grus- og Pukkregisteret 1994 med katalog over kart og rapporter. NGU Rapport 95.074

Sturt, B.A et al. 1995. *Opportunities in development of Mineral Resources in Norway*. NGU Report nr. 95.116.

Statens vegvesen. 1992. Håndbok 018, Vegbygging.

Ontario's aggregate resources inventory program

R.I. KELLY & D.J. ROWELL
Ontario Geological Survey, Sudbury, Ontario, Canada

H.M. ROBERTSON
Mineral Sector Analysis Branch, Sudbury, Ontario, Canada

1 INTRODUCTION

Situated in central Canada, the province of Ontario has a population of nearly eleven million people and a land mass of some 820 000 km^2 (Corpus Almanac and Canadian Source Book 1995) (Fig. 1). The majority of the people reside in southern Ontario, one of the major reasons why the aggregate industry is concentrated largely in that part of the province. The aggregate industry is important to the Ontario economy. In 1993 the value of all construction work purchased in the province was nearly 33 billion dollars (Ontario Ministry of Natural Resources 1995), with the value of aggregates being nearly one billion dollars (Burton 1993). The majority of aggregate products, which include bedrock-derived crushed stone as well as naturally formed sand and gravel, are consumed in the province's road-building and construction industries.

Very large quantities of aggregate are used each year in the province. For instance, in 1989 Ontario's mineral aggregate production peaked at a total of 197 million tonnes (Ontario Ministry of Natural Resources 1995) or nearly 13 tonnes per capita. By 1993 the tonnage of aggregate produced had decreased to 131 million tonnes, however current short term trends indicate a slight increase in production (Ontario Ministry of Natural Resources 1995). Despite the lower tonnages of aggregate produced in recent years the total is greater than that of any other metallic or non-metallic commodity mined in the province (Ontario Ministry of Northern development and Mines 1994).

Although mineral aggregate deposits are plentiful in Ontario, they are fixed-location, nonrenewable resources, which can be exploited only in those areas where they occur. Mineral aggregates are characterized by their high bulk and low unit value so that the economic value of a deposit is a function of its proximity to a market area as well as its quality and size. The potential for extractive development is often greatest in areas where land-use competition is extreme. For these reasons the availability of adequate resources for future development is now being threatened in many parts of the province and especially in urban areas where demand is greatest. A recently completed report on the state of aggregate resources in major market areas of southern Ontario identified numerous regions that are currently experiencing shortages of certain aggregate products or are expected to within the next few years (Planning Initiatives and Associates 1993).

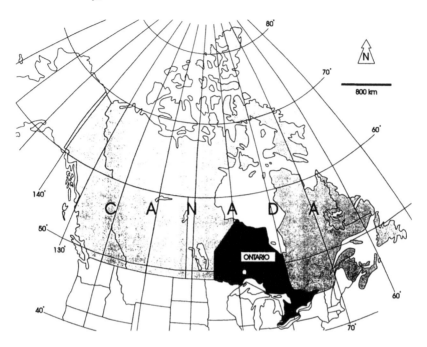

Figure 1. Location map of Ontario.

Comprehensive planning and resource management strategies are required to make the best use of available resources, especially in those areas experiencing rapid development. At the same time, an increased environmental awareness by the public and recently issued environmental and planning legislation by the government of Ontario places a much stronger emphasis on environmental considerations. In some cases, the best aggregate resources are found, in or near, environmentally sensitive areas, resulting in the requirement to balance the needs of other natural resources. Therefore planning strategies must be based on a sound knowledge of the total mineral aggregate resource base at both local and regional levels.

2 PLANNING FOR MINERAL RESOURCES

2.1 *How many dimensions in land use planning?*

Principles of land use planning and associated decision making processes, whether enshrined in enabling statutory legislation and regulations, policy or best practices, have been traditionally based on land as a horizontal two-dimensional plane. In recent years it has become more apparent that decisions about how land is to be used, what is to be used on that land, or, whether the land should be used at all, are no longer isolated in their impact. Decision-making in planning systems today has become more comprehensive and integrated. For example, ecosystem-based planning and watershed management planning embody integrated approaches to land use planning taking into consideration such factors as size, form and functions of natural ecosystems.

However, in the case of unknown, undiscovered or unmapped mineral deposits, decisions about the use of that land have failed to consider the subsurface. Some exceptions to this are the concern with potential impacts of decisions on groundwater quantity and quality or with the potential for soil contamination.

The depiction of geographic areas, particularly in land use planning and in survey plan documents, often reflects only the two dimensional plain regardless of the underlying mineral rights. On the other hand, in mineral land tenure both surface and subsurface rights are considered in 'land'. Moreover, legal interpretations of 'land' do not distinguish between land on the surface and land in the subsurface. It is all 'land'. The legal treatment of 'land' as a whole is exemplified by certain exemptions in Ontario's Planning Act (Section 50) amended by Bill 163, proclaimed on March 28, 1995. The exemptions effectively acknowledge the existence of the 'third dimension' or subsurface by restricting land division requirements of the Planning Act to a horizontal plane. Furthermore, land division documents, that is, plans of subdivision and consents, make no reference to underlying or subsurface land tenure. Consequently, a land use decision or a land division approval may have the potential impact of compromising the rights of the holder of the subsurface mining rights when the surface land rights are held in separate ownership. It should be noted that the Planning Act applies only to private land, which is approximately 14% of Ontario's land area and is mostly located in central and southern Ontario, also the location of most land use conflicts.

Mineral deposits, whether aggregate, industrial or metallic minerals are generally not visible on the land until the deposit is actually extracted. Even then, the extraction may be very subtle and largely unnoticed, particularly if the method of extraction is an underground mining operation with minimal surface buildings, infrastructure or land disturbance. However, the extraction of aggregates is in most cases conducted by open pit or quarrying methods, depending on the type of aggregate commodity. Pits and quarries, considered 'uses of land' under Ontario's Planning Act (Section 34) are more obvious on the landscape. More land is disturbed as the deposit is extracted. Under the Aggregate Resources Act progressive rehabilitation is encouraged to minimize this type of disturbance and ultimately return the land to its former use or a suitable alternative use.

2.2 *Minerals must be considered early in the planning process*

A fundamental difference between minerals in the ground and many land uses or activities on land is the fact that mineral deposits can be extracted or mined only where the deposit is located in the land. In a way, this could be compared to such complex ecosystems as Ontario's provincially significant wetlands. A wetland or wetland complex, while a renewable resource over the long term, cannot simply be relocated, readily replaced or reconstructed if destroyed.

Unlike residential, commercial, industrial or institutional land uses, that may have a choice of alternate sites before selection of the best or preferred site, mineral deposits cannot be relocated. They are also finite and non-renewable. Consequently, decisions about the land in or under which these deposits are located may, depending on site specific circumstances, facilitate, hinder, restrict or ultimately prohibit the ability to access and extract or mine the mineral deposit.

After land use decisions have been approved, the ability to protect a mineral deposit may be very constrained and the opportunity to reverse a planning decision may be difficult. For example, where formal decision appeal processes are in place, they have been traditionally cumbersome, lengthy and costly with no guarantee of the resultant outcome. The increasing risk of land use conflicts and consequent sterilization of subsurface mineral deposits develops when the presence of these 'third dimension' resources is not factored in at an early stage in the planning process; in other words, not planned. In order to properly plan for the protection of all types of mineral resources it is extremely important that these resources be identified in advance of planning decisions. The importance of identifying and mapping mineral deposits and areas for potential future extraction or mining cannot be overstated.

2.3 *A proactive option for protecting minerals*

Increased land use conflicts, competing interest in land, greater public demand for aggregate resources and a diminishing supply of the non-renewable resource prompted legislators, bureaucrats, planners and the aggregate industry to examine and develop mechanisms for proactive planning for mineral resources in an effort to reduce land use conflicts and to better protect the resource for the future. The *Aggregate Resources Act, 1990*, which replaced the former *Pits and Quarries Control Act, 1971,* provides the legislative instrument to administer and regulate aggregate operations in the province of Ontario. However, neither of these statutes was ever intended to deal with land use planning. As a result, a mechanism was sought under the Planning Act to protect the provincial interest in mineral aggregate resources where land use planning matters were involved.

2.4 *A mineral aggregate resources policy statement*

Section 3 of the *Planning Act* enables the Minister to issue provincial policy statements for matters of provincial interest where private land is concerned. The *Public Lands Act*, which deals with Crown Land, contains no similar provision for Crown Land planning.

The first policy statement ever approved under the Planning Act was the Mineral Aggregate Resources Policy Statement (MARPS), which came into effect in 1986 only after a lengthy period of testing, first as a 10-point policy for Official Plans in 1978 and later as a planning policy in 1982.

The goal of MARPS is 'to ensure all parts of Ontario possessing mineral aggregates, an essential non-renewable resource to the overall development of any areas, share a responsibility to identify and protect mineral aggregate resources and legally existing pits and quarries to ensure mineral aggregates are available at a reasonable cost and as close to markets as possible to meet future local, regional, and provincial needs'.

The policy statement contains 6 principle policies with the following underlying intent:

1. Identify and protect legally existing pits and quarries from incompatible uses,

2. Identify and protect mineral aggregate resource areas from land uses that may be incompatible with future extraction,

3. Official Plan policies may be established to permit non-aggregate land uses or developments in areas of mineral aggregate resources,

4. Clear and reasonable mechanisms should be included in Official Plan policies to allow for the establishment or expansion of pits and quarries,

5. Wayside (temporary) pits and quarries should be permitted without requiring an amendment to the Official Plan or Zoning By-law, except in areas of existing development or environmental sensitivity,

6. Rehabilitation should be compatible with long term uses.

More recently, the Ontario government has taken an integrated and comprehensive approach to providing planning policy direction where the provincial interest is of concern. The Commission on Planning and Development Reform in Ontario (Sewell Commission) studied Ontario's planning system for two years. In its Final Report, June 1993 (Commission on Planning and Development Reform in Ontario 1993), the commission recommended to the Ontario government that:

'Provincial policy must be clearly stated and developed in a manner credible to those affected. This is important for two reasons:

1. Provincial policy provides a context for provincial planning decisions,

2. It defines the framework within which municipal planning can occur'.

'To provide clarity and consistency in the definition of provincial interest in planning, the province adopt a comprehensive set of policy statements under Section 3 of the Planning Act'.

On March 28, 1995 a Comprehensive Set of Policy Statements (CSPS) came into effect. MARPS, as approved in 1986, was amalgamated with a combination of six other existing and new provincial policy statements approved under Section 3 of the Planning Act, as amended by Bill 163, proclaimed the same day. The CSPS contains a range of provincial policy statements providing a balance between environmental protection and economic development. The six approved provincial policy statements are: A-Natural Heritage, Environmental Protection and Hazard Policies; B-Economic, Community Development, and Infrastructure Policies; C-Housing Policies; D-Agricultural Land Policies; E-Conservation Policies; and, F-Mineral Aggregate, Mineral and Petroleum Resources Policies.

All planning decisions under the Planning Act, as amended by Bill 163, must be consistent with the Comprehensive Set of Policy Statements.

2.5 *New recognition of the provincial interest in all minerals*

With approval of the CSPS came the recognition that other non-aggregate minerals and petroleum resources required protection for their access and resource use in Ontario. The resulting policy statement for Mineral and Petroleum Resources has as its goal 'to protect mineral and petroleum resource operations, deposits of minerals and petroleum resources, and areas of potential mineral and petroleum resources for resource use' (Ontario Ministry of Municipal Affairs 1995).

The policy statement contains 4 policies that reflect the following planning intent:

1. Identify mineral and petroleum resources for resource use and protect from incompatible land uses,

2. Development that precludes or hinders future access to and use of these resources will be permitted only under special circumstances,

3. Incompatible development adjacent to resource operations and deposits will be discouraged and permitted only under special circumstances,

4. Rehabilitation will be required after extraction and other related activities have ceased.

It is yet too early to determine the effectiveness of the new policies for mineral and petroleum resources. However, the implementation of MARPS over a period of approximately 9 years has demonstrated its use as an effective planning tool to ensure that the provincial interest in mineral aggregates is taken into consideration in the development of Official Plan policies and in all planning decisions. Aggregate resources of southern Ontario, A state of the resource study (Planning Initiatives and Associates 1993) prepared for the Ontario Ministry of Natural Resources has some of the following comments about MARPS:

'An analysis of 24 selected Official Plans in the Market Areas indicate that the majority of the municipalities recognize and identify for the public, the locations of existing licensed operations in the Official Plan, and adopt either Option 4a or 5 of MARPS, thereby offering the highest levels of recognition and protection for aggregate resources'.

'Notwithstanding the above statements, there appears to be inconsistency and misunderstanding with respect to municipal implementation of MARPS policies in Official Plans'.

'Both MARPS and the Aggregate Resources Act embody the principles of sustainable development through the recognition of, and requirement for, an examination of environmental and economic factors'.

While it is known that land use conflicts will never be totally avoided or eliminated, it is hoped that through clear policy direction, effective and up-front introduction and implementation of provincial policy statements to the land use planning process, land use conflicts can be minimized to the greatest degree possible. At the same time it is hoped that these planning policies will allow for mineral resources in Ontario to be developed using effective, sound, responsible and environmentally sensitive resource management practices.

Author's Note: At the time of publication of this manuscript, the government of the Province of Ontario amended the Planning Act, known as Bill 20, the Land Use Planning and Protection Act. One of the significant changes in Bill 20 is the removal of the requirement for planning decisions to *be consistent with* provincial policy statements. Bill 20 returns the wording to the pre-Bill 163 Planning Act, which requires that planning decisions *shall have regard to* provincial policy statements. Concurrently, the Ontario government released a new Provincial Policy Statement, which replaced the CSPS with proclamation of Bill 20. The new Provincial Policy Statement still contains policies for Mineral Aggregate, Mineral and Petroleum Resources. The general intent of the relevant policies remains largely unchanged from the former CSPS.

3 GEOLOGICAL OVERVIEW OF THE PROVINCE

The province of Ontario is underlain by rocks of Precambrian, Paleozoic, Mesozoic and Quaternary age. Precambrian rocks underlie approximately 60% of the province (Fig. 2). While they provide the host rocks for all of Ontario's metallic mineral wealth and a range of non-metallic industrial minerals, the contribution of Precambrian strata to the provinces total production of crushed stone aggregate is small.

Paleozoic rocks occur in both the northern and southern parts of the province (Fig. 2). The strata consist of limestones, dolostones, shales, sandstones and evaporitic rocks, including salt and gypsum. Paleozoic rocks underlie most of the more densely populated areas of southern Ontario as well as sparsely populated areas in northern Ontario around Hudson and James Bays. In southern Ontario, the Paleozoic rocks provide the bulk of the provinces industrial mineral wealth in addition to supplying approximately 30% of the total provincial production of mineral aggregates (White 1983). At present, this total percentage may be closer to 40% (Ontario Ministry of Natural Resources 1995).

Mesozoic rocks are found in the Moose River Basin area of northeastern Ontario (Fig. 2). The rock deposits identified include calcareous clays, quartz sands, kaolinitic

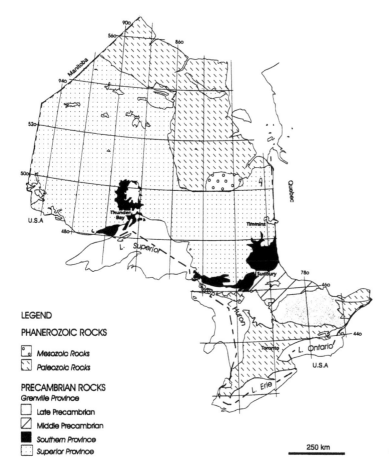

LEGEND

PHANEROZOIC ROCKS

▢ Mesozoic Rocks
▢ Paleozoic Rocks

PRECAMBRIAN ROCKS
Grenville Province
▢ Late Precambrian
▢ Middle Precambrian
■ Southern Province
▢ Superior Province

250 km

Figure 2. Generalized bedrock geology of Ontario.

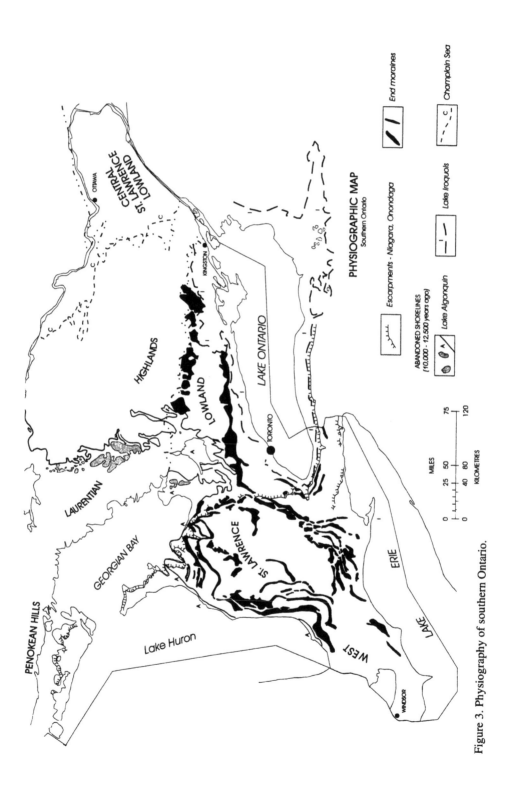

Figure 3. Physiography of southern Ontario.

clays and lignite. From a mineral aggregate perspective none of the strata have much importance.

Quaternary deposits cover a large part of the province. On several occasions during the Quaternary Period the entire province was completely glaciated. Of all materials deposited as a result of glaciation, the majority are attributed to the Wisconsinan or youngest glacial period.

The cover of glacial sediments is not continuous everywhere. In southern and northern Ontario drift thickness can exceed 200 m, but is generally between 30 and 60 m (Karrow 1989). There are large areas of the Canadian Shield with exposed bedrock, however, many areas of southern Ontario underlain by Paleozoic rock also have limited drift cover. The major physiographic features of southern Ontario are shown on Figure 3.

The Quaternary materials were deposited directly by glacial ice or meltwaters that were generated by, and discharged from the glaciers. Till is the most widespread of the glacial deposits and includes a variety of textures ranging from sand- to clay-rich (Barnett 1992). Till is generally not well suited for aggregate production, however, locally some coarser textured tills are utilized for fill.

As the glaciers melted, large volumes of meltwater were generated by, and discharged from the ice. Large volumes of material were carried by the meltwater and deposited beneath the glacier, along the ice margin or beyond the ice margin in rivers and streams, lakes and marine bodies. Materials deposited by meltwater range from sand and gravel rich, to silt and clay rich. In the province these glacial deposits supply approximately 70% of the total mineral aggregate production (White 1983).

4 AGGREGATE RESOURCES INVENTORY PROGRAM

4.1 *Office work*

Aggregate resource inventory studies in the province of Ontario fall under the jurisdiction of the Ontario Ministry of Northern Development and Mines. In those areas of the province that are organized into administrative districts individual projects are generally completed on a township or county basis. Where administrative boundaries are absent, project limits are defined by geographic boundaries such as longitude and latitude or UTM coordinates. Project mapping is normally completed on a 1:50,000 scale, however, projects entailing both larger and smaller scale mapping have been undertaken.

The methods used to prepare Aggregate Resources Inventory Papers (ARIPs) involve the interpretation of published geological data such as bedrock and surficial geology maps and reports, as well as field examination of potential resource areas.

Prior to starting field investigations for any project a considerable amount of background material, derived from a number of sources, is gathered and reviewed. The first step is to gather all relevant surficial and bedrock geology maps for a particular study area. As with the Aggregate Resources Inventory Program, the Ontario Ministry of Northern Development and Mines has the responsibility to map the geology of Ontario. Consequently, the bulk of geological reports and maps used in aggregate surveys have been generated, and are available, in-house. Additional map and report coverage for some areas of the province, has been produced by the Geological Survey of Canada, a Federal Government agency.

Review of available geological maps and documents provides a working knowledge of the geology of the study area as well as allowing potential aggregate sources to be targeted for further investigation. Regardless of whether an area has existing geological map coverage, black and white aerial photograph coverage for a study area is usually obtained so as to allow field stations to be located and geological observations noted. The information gathered during field work can then be used to verify existing map units or, if required, allow changes to be made to existing ones. If the study area lacks geological mapping then considerably more aerial photograph interpretation and field work will be required to identify, delineate and assess potential aggregate resources. In addition to using aerial photographs other remotely sensed images are sometimes employed to provide a regional geological overview. Remotely sensed images utilized include LANDSAT TM images, colour infrared imagery and radar imagery.

In Ontario, a number of different provincial government ministries have a direct interest in the aggregate industry. These ministries collect information that is extremely useful in assessing aggregate deposits. The Ontario Ministry of Transportation has the mandate to plan, design, construct and maintain the provinces highway system and as a result has an extensive database on many of the aggregate deposits that are currently being exploited as well as past producers. Data contained in these files include field estimates of the depth, composition, and 'workability' of deposits, as well as laboratory analyses of the physical properties and chemical suitability of the aggregate. Information concerning the development history of the pits and acceptable uses of the aggregate is also recorded. The provincial Ministry of Natural Resources has the responsibility, under the provincial legislation of the Aggregate Resources Act (1989), to license and oversee aggregate operations throughout much of the province. A range of information is collected under the Act and includes: the name and address of the owner/operator, location, size, and depth of extraction of licensed pits and quarries. Site plans outlining operational characteristics and rehabilitation are also required under the Act. The above information is maintained on file with regional offices of the Ontario Ministry of Natural Resources where the aggregate operation is located. The Ministry of Natural Resources also has the responsibility of regulating and overseeing the provinces petroleum industry under the provincial legislation of the Petroleum Resources Act. Information on file with the Ministry of Natural Resources, gathered during the drilling of wells, provides useful data on drift thickness and bedrock formation thickness.

Other background information useful to the project is obtained through a provincial water well database that is maintained by the provincial Ministry of the Environment and Energy. Water well information is used in some areas to corroborate deposit thickness estimates or to indicate the presence of buried granular material. These records are used in conjunction with other evidence.

Information that is deemed useful to a specific project may be utilized and published in a report with the permission of the Ministry that retains that information. Reports on geological testing for type, quantity and quality of aggregates, generated during property evaluations or licensing procedures, may also be obtained from specific individuals or companies with their permission.

4.2 *Field work*

Review of all relevant background data prior to field investigation allows the project geologist to identify areas that are known or likely to be potential aggregate sources. Field investigation is concentrated on these 'targets': Field study routinely involves the examination of granular materials in natural and man-made exposures such as quarries and sand and gravel pits. As well, deposits or potential deposits that have limited or no exposure are also investigated. Data collected at pit sites include: the depth of stripping required to access granular material, estimates of the total face height, the proportion of gravel- and sand-sized fragments, size, shape and lithology of the particles, deleterious lithologies and their percentages, geological environment and a sketch of the operation. This data is important in estimating the quality and quantity of the aggregate.

In quarries, data collected includes: depth of overburden, face height, bed thickness of strata, rock lithology(s), presence of any deleterious lithologies and their percentage, and notable structural features if present. A sketch map of the operation is also made.

In areas of limited exposure, various techniques may be employed to assess subsurface materials. Generally the most common techniques for investigating surficial deposits involves hand augering. Test pitting, using a tractor mounted backhoe is also widely utilized. Occasionally, test hole drilling using hollow stem augers or rotasonic drilling is employed to provide deeper subsurface data. Geophysical surveys including hammer seismic or electromagnetic conductivity (EM) are sometimes employed to supplement data collected by other subsurface investigations.

In the investigation of bedrock resources, analysis of existing drillcore, if available, may be undertaken, or in some cases diamond drilling may be employed to obtain data where information is scarce. Deposits with potential for further extractive development or those where existing data are scarce, are studied in greater detail.

Representative sand and gravel layers or bedrock strata in these deposits are characterized by collecting 11 to 45 kg bulk samples from existing pit faces, test pits, quarry faces, bedrock outcrop or drillcore. Both sand and gravel and bedrock samples are routinely tested for absorption, resistance to abrasion, soundness quality and undergo petrographic analyses. Sand and gravel samples are analyzed for grain size distribution. Other tests that are sometimes conducted for both sand and gravel and bedrock samples include abrasion resistance of sand particles, analysis for potential alkali reactivity problems (Rogers 1985), bulk density, and resistance to polishing. Analyses are performed by the laboratories of the Ontario Ministry of Transportation, or in private testing facilities.

5 MAPS: FORMAT AND GENERATION

After completion of field investigation, the preparation of maps and an accompanying report is undertaken. Two maps are generated for a specific study area. One map outlines all surficial and known buried sand and gravel resources while the second map deals with bedrock resources.

Topographic maps of the National Topographic System, at a scale of 1:50,000 are used as a base for the field and office data. All data relevant to the project is plotted on a series of acetate overlays that are registered to the map base. Once all information is plotted, the map base is scanned to form a digital base. Information from the acetate overlays is then digitized in a series of layers and tied to the digital map base. Currently MicroStation TM software is used for cartographic production. Field data is entered into a computer spreadsheet program, where it is manipulated and made ready to be incorporated into tables and the body of the report.

5.1 *Sand and gravel resources map*

The sand and gravel resources map shows the extent and quality of sand and gravel deposits within the study area and the present level of potential extractive activity. The map is derived from existing surficial geology maps, field investigation and mapping, if required, and from interpretation of aerial photographs. The present level of potential extractive activity is indicated by those properties that are licensed for extraction under the Aggregate Resources Act and unlicensed pits (abandoned pits or pits operating on demand under authority of a wayside permit).

The sand and gravel resources map also presents a summary of available information related to the quality of aggregate contained in all the known aggregate deposits in the study area. Much of this information is contained in the symbols that are found on the map. The Deposit Symbol appears for each mapped deposit and summarizes important genetic and textural data. The Texture Symbol is a circular proportional diagram, which displays the grain size distribution of the aggregate in areas where the bulk samples were taken.

5.2 *Bedrock resources map*

The bedrock resources map is compiled using information from bedrock geology, drift thickness and bedrock topography maps, water well data from the Ontario Ministry of the Environment and Energy, oil and gas well data from the Ontario Ministry of Natural Resources and from geotechnical test hole data from various sources.

The geological boundaries of the bedrock units are shown by a dashed line, as are three sets of contour lines. The contour lines delineate drift thickness of 1, 8 and 15 m. Quarrying may be considered in areas where drift cover is up to 8 m, however, areas where bedrock outcrops or is within 1 m of the ground surface constitute the most favourable potential resource areas because of their easy access. Depending on the amount and type of other locally available aggregate resources, however, quarrying may not be economical in areas with more than 5 m of overburden. Bedrock areas overlain by 8 to 15 m of overburden constitute resources that have extractive value only in specific circumstances, such as for some specialized industrial mineral applications. Outside of these delineated areas, the bedrock can be assumed to be covered by more than 15 m of overburden, a depth generally considered to be too great to allow economic extraction.

6 RESOURCE TONNAGE CALCULATION TECHNIQUES

6.1 *Sand and gravel resources*

Once the interpretative boundaries of the aggregate units have been established, tonnage estimates of the possible resources available can be made. First, the area of the deposit, as outlined on the final base map is calculated. Deposit thickness is estimated from the face heights of pits developed in the deposit, or on subsurface data such as test holes and water well logs. Original tonnage values can then be calculated by multiplying the volume of the deposit by 17,700 (the density factor). This factor is the approximate number of tonnes in a 1.0 m thick layer of sand and gravel, 1.0 ha in extent, assuming an average density of 1770 kg per cubic meter

$$\text{Tonnage} = \text{Area} \times \text{Thickness} \times \text{Density Factor}$$

Tonnage calculated in this manner must be considered only as an estimate. Furthermore, such tonnages represent amounts that existed prior to any extraction of material (i.e. original tonnage).

The possible resources available in selected sand and gravel deposits are calculated in the following way. Two successive subtractions are made from the total area: 1. The number of hectares unavailable because of the presence of permanent cultural features and their associated setback requirements; and 2. Those areas that have previously been extracted (e.g. wayside (short-term license) and abandoned pits are included in this category). The remaining figure is the area of the deposit that, without any other land-use restrictions, is potentially available for extraction. The available area is then multiplied by the estimated deposit thickness and the density factor to give an estimate of the sand and gravel tonnage possibly available for extractive development and/or resource protection. It should be noted, however, that a recent study in Ontario (Planning Initiatives and Associates 1993) showed that anywhere from 15 to 85% of this tonnage figure in any resource area may be further constrained or not accessible because of such things as environmental considerations, resident opposition, or other matters. Since the purpose of the Aggregate Resources Inventory Program is to provide basic geological information these additional constraints are not considered as part of the aggregate assessment.

6.2 *Bedrock resources*

The method used to calculate resources of bedrock-derived aggregate is much the same as that described above. The areal extent of bedrock formations overlain by less than 15 m of unconsolidated overburden is determined from bedrock geology maps, drift thickness and bedrock topography maps, and from the interpretation of water and petroleum well records. The measured extent of such areas is then multiplied by the estimated quarriable thickness of the formation, based on stratigraphic analyses and on estimates of existing quarry faces in the unit. In some cases a standardized estimate of 18 m is used for thickness. Volume estimates are then multiplied by the density factor (the estimated weight in tonnes of a 1.0 m thick section of rock 1.0 ha in extent).

Because bedrock resources in Ontario are derived primarily from Paleozoic rocks the majority of density calculations that are used relate to rocks found in those strata. The most commonly used density factors are dolostone, 2649 kg per cubic meter; sandstone, 2344 kg per cubic meter; shale resources, 2408 kg per cubic meter; and limestone, 2554 kg per cubic meter (Telford et al. 1976).

7 SELECTION OF RESOURCE AREAS

7.1 *Selected sand and gravel resource areas*

The process by which deposits are evaluated and selected involves the consideration of two sets of criteria: site specific and regional. The main selection criteria are site specific, related to the characteristics of individual deposits. A second set of criteria involves assessment of local aggregate resources in relation to total resources of the region.

All the selected sand and gravel resource deposits are first delineated and then classified into one of three levels of significance: primary, secondary and tertiary. Each area of primary significance is assessed as to its probable relative value as a resource in the study area and is given a deposit number. Resource areas classed at the primary level of significance represent areas in which a major resource is known to exist, and may be reserved wholly or partially for extractive development and/or resource protection.

Deposits of secondary and tertiary significance are also outlined on the sand and gravel resources map. Although deposits of secondary significance are not considered to be the 'best' resources in the report area, they may contain large quantities of sand and gravel and should be considered as part of the aggregate supply of the area. Deposits classified at the tertiary level of significance contain either poor quality aggregate material or very limited quantities of good aggregate.

7.1.1 *Site specific criteria*

Deposit size
Ideally, selected deposits should contain available sand and gravel resources large enough to support a commercial pit operation, using a stationary or portable processing plant, for a significant number of years. In practice, much smaller deposits may be of significant value depending on reserves in the rest of the study area.

Aggregate quality
The limitations of natural aggregates for various uses result from variations in the lithology of the particles composing the deposit, and from variations in the size distribution of these particles. Four indicators of the quality of aggregate may be included in the deposit symbols shown on the sand and gravel resources map. They are: gravel content, percent fines, percent oversize and lithology.

Three of the quality indicators deal with grain size distribution. The gravel content indicates the suitability of aggregate for various uses. Deposits containing at least 35% gravel, in addition to a minimum of 20% material greater than the 26.5 mm sieve, are considered to be the most favourable extractive sites, since this content is the minimum from which crushed products can be economically produced. For the purposes of the aggregate program the classification of materials is based on the system used by the Ontario Ministry of Transportation. In this classification gravel is considered to consist of those particles between 5 mm and 75 mm in size.

Excess fines (high silt and clay content) may severely limit the potential use of a deposit. Fines content in excess of 10% may impede drainage in road sub-base aggregate and render it more susceptible to the effects of frost action. In asphalt aggregate, excess fines hinder the bonding of particles.

Deposits containing more than 20% oversize material (greater than 10 cm in diameter) may also have use limitations. In Ontario the oversize component is unacceptable for uncrushed road-base use, so it must either be crushed or removed during processing. In certain application, such as high strength Portland cement concrete, and some asphalt binder courses, oversize material can have a beneficial effect.

Another indicator of the quality of an aggregate is lithology. Just as the unique physical and chemical properties of bedrock types determine their value for use as crushed rock, so do various lithologies of particles in a sand and gravel deposit determine its suitability for various uses. The presence of objectionable lithologies such as chert, siltstone, and shale, even in relatively small amounts, can reduce the quality of an aggregate, especially for high quality uses such as concrete and asphalt. Similarly, highly weathered, very porous and friable rock can restrict the quality of an aggregate. Alternatively, rock types such as diabase (traprock) provide aggregate for high quality end uses such as asphalt for high traffic density highways.

Analyses of unprocessed samples obtained from test holes, pits or sample sites are plotted on grain size distribution graphs and included in the text of an Aggregate Resources Inventory Paper. By plotting the gradation curves with respect to the specification envelopes, it can be determined how well the unprocessed sampled material meets specified criteria for each product.

Location and setting
The location and setting of a deposit has a direct influence on its value for possible extraction. The evaluation of a deposit's setting is made on the basis of natural and man-made features which may limit or prohibit extractive development.

First, the physical context of the deposit is considered. Deposits with some physical constraint on extractive development, such as thick overburden or high water table, are less valuable because of the difficulties involved in resource recovery. Second, permanent man-made features, such as roads, railways, powerlines and housing developments, which are built on a deposit, may prohibit its extraction. The constraining effect of legally required setbacks surrounding such features are also included in the evaluation. A quantitative assessment of these constraints can be made by measurement of their areal extent directly from the topographic maps.

7.1.2 Regional considerations
In selecting deposits for resource development, it is important to assess both the local and the regional resource base, and to forecast future production and demand patterns.

Some appreciation of future aggregate requirements in a region may be gained by assessing present production levels and by forecasting future production trends. Such an approach is based on the assumptions that production levels in a region closely reflect the demand and that the present production 'market share' of an area will remain roughly at the same level.

The aggregate resources in the region surrounding a municipality should be assessed in order to properly evaluate specific resource areas and to adopt optimum resource management plans.

7.2 Selected bedrock resource areas

7.2.1 Selection criteria

Criteria equivalent to those used for sand and gravel deposits are used to select bedrock areas most favourable for extractive development. The evaluation of bedrock resources is made primarily on the basis of performance and suitability data established by laboratory testing at the Ontario Ministry of Transportation.

Deposit 'size' is related directly to the areal extent of thin drift cover overlying favourable bedrock formations. Since vertical and lateral variations in bedrock units are much more gradual than in sand and gravel deposits, the quality and quantity of the resource are usually consistent over large areas.

Quality of the aggregate derived from specific bedrock units is established by the performance standards previously mentioned. Location and setting criteria and regional considerations are identical to those for sand and gravel deposits.

Selection of bedrock resource areas is restricted to a single level of significance. Three factors support this approach. First, quality and quantity variations are gradual. Second, the areal extent of a given quarry operation is generally smaller than that of a sand and gravel pit producing an equivalent tonnage of material. Third, since some crushed bedrock products have a higher unit value than sand and gravel, longer haul distances can be considered. These factors allow the identification of alternative sites having similar development potential.

8 REPORT FORMAT

Reports generated to accompany maps are divided into a number of parts. The first two parts of the report introduce and explain the purpose of the Aggregate Resources Inventory Program, explain the methology used, and defines the more relevant terms used throughout the report (e.g. Texture Symbol).

The third part of the report details information about the specific study area. Such topics as: location and population, physiography, extractive sand and gravel activities, selected sand and gravel resource areas, extractive bedrock activities and selected bedrock resources are clearly defined and detailed. This section is then followed by a series of tables which summarize much of the data and information, aggregate grading curves, and a reference listing for that particular report.

Finally, a series of appendices follow in which many other pertinent geological and aggregate industry terms are defined, including product specifications.

9 PROGRAM STATUS

To date over 150 aggregate reports with accompanying maps have been produced. Many of these reports cover townships or counties within the southern part of the province, simply due to the higher resource demands, population and land-use pressures in that part of the province. A number of reports have also been generated for many of the major urban areas in northern Ontario, as well as less populated regions where aggregate information was required for specific purposes.

Traditionally aggregate reports were produced in a hard copy format only. Recently, however, the Ministry of Northern Development and Mines undertook and completed a project to convert all existing paper maps and reports to digital format.

10 CLIENT UTILIZATION

The Aggregate Resources Inventory Program has proven to be very popular with a wide variety of clients. Typical client users include: other provincial and federal government ministries, municipal and regional governments, consulting firms involved in many areas related to the aggregate industry, the aggregate industry itself, real estate agents and the general public. Information on aggregate resources gleaned from the reports and maps is routinely implemented into planning documents such as municipal or regional government official plans.

11 FUTURE DIRECTIONS

With the recent advances in computer technology related to geological mapping and handling of spatial data, new directions may be charted regarding generation of aggregate maps and reports. These advances should lead to the generation of maps at a variety of scales for specific regions (i.e. 1:10,000 or 1:20,000 as compared to the traditional 1:50,000), or to the production of maps that have Geographic Information System (GIS) capabilities. Since other government ministries rely on this information, computer linkages should be established with these agencies and other clients.

Much of Ontario has yet to be mapped by the Aggregate Resources Inventory Program, so there is still a demand to complete the coverage of the province, at least, in areas of extreme land use competition. Many of the earlier reports (circa 1980) are becoming outdated, since deposits are depleted over time. Client input has indicated that it is important to update the older reports, especially those in high aggregate demand areas, i.e. the Greater Toronto Area of southern Ontario, so that the information is as current as possible. Currently, reports are being updated for a number of counties surrounding the Greater Toronto Area. Finally, the program will try to evolve to address legislative changes and to meet the requirements of our clients in a more effective and efficient manner.

SUMMARY

The Aggregate Resource Inventory Program is designed to provide a database of geological information that allows the identification of potential mineral aggregate resource areas in planning strategies. The Aggregate Resources Inventory Papers should form the basis for discussion on those areas best suited for possible extraction. The aim is to assist decision makers in protecting the public well-being by ensuring that adequate resources of mineral aggregate remain available for future use.

Aggregate Resources Inventory Papers are technical background documents, based for the most part on geological information and interpretation. They have been designed as a component of the total planning process and should be used in conjunction with other planning considerations, to ensure the best use of the resources.

REFERENCES

Barnett, P.J. 1992. Quaternary geology of Ontario. In *Geology of Ontario*. Ontario Geological Survey, Special Volume 4(2):1011-1088.

Burton, B. 1993. Aggregate resources - A Canadian perspective. *Canadian Aggregates* 7(7): 5-10.

Commission on Planning and Development Reform in Ontario 1993. New Planning for Ontario, Final Report 1993. *Commission on Planning and Development Reform in Ontario*. Queen's Printer for Ontario 1993, Toronto.

Corpus Almanac and Canadian Source Book 1995. Editor: Barbara Law, Southam Inc. Don Mills, Ontario: 17-303.

Karrow, P.F. 1989. Quaternary geology of the Great Lakes subregion. In *Quaternary Geology of Canada and Greenland*. Geological Survey of Canada, Geology of Canada, No. 1: 326-350.

Ontario Ministry of Municipal Affairs. 1995. Comprehensive Set of Policy Statements 1995. *Ontario Ministry of Municipal Affairs*. Queen's Printer for Ontario, Toronto.

Ontario Ministry of Natural Resources. 1995. Mineral aggregates in Ontario, overview and statistical update 1993. *Ontario Ministry of Natural Resources*, Queen's Printer for Ontario, Toronto, 51 pp.

Ontario Ministry of Northern Development and Mines. 1994. Ontario mineral SCORE 1992. *Ontario Ministry of Northern Development and Mines*, Queen's Printer for Ontario, Toronto, 199 pp.

Ontario Mineral Aggregate Working Party. 1977. A policy for mineral aggregate resource management in Ontario; *Ontario Ministry of Natural Resources*, Queen's Printer for Ontario, Toronto.

Planning Initiatives and Associates. 1993. *Aggregate resources of southern Ontario. A state of the resource study*. Ministry of Natural Resources. Queen's Printer for Ontario, Toronto.

Rogers, C.A. 1985. *Evaluation of the potential for expansion and cracking due to the alkali-carbonate reaction*. Ontario Ministry of Transportation and Communications.

Telford, W.M., L.P. Geldart, R.E. Sheriff & D.A. Keys. 1976. *Applied Geophysics*: 25-27.

White, O.W. 1983. Geology of Ontario. *19th Forum on the Geology of Industrial Minerals*. Ministry of Natural Resources MP114: 1-8.

Geologic and geographic aspects of sand and gravel production in Louisiana

JOANN MOSSA
Department of Geography, University of Florida, Gainesville, FL., USA

WHITNEY J. AUTIN
Institute for Environmental Studies, Louisiana State University, Baton Rouge, LA., USA

1 INTRODUCTION

Naturally-deposited sands and gravels are used in nearly all residential, industrial, and commercial building construction and in most public-works projects. In the US, about 96% of the sand and gravel has been consumed by the construction industry with the remainder being used for other industrial purposes (Yeend 1973, Langer 1988). Because of such common and diverse uses, the sand and gravel mining is the largest non-fuel mineral industry in terms of volume in the United States. However, in many locations, especially where development is removing land from future mineral production, it has been suggested that available domestic resources will barely meet demand by the year 2000 (Yeend 1973).

Although the United States as a whole is faced with a need for managing aggregate resources, each state has its own particular prospects and problems. In Louisiana, a low-lying state on the Gulf coastal plain, one such problem is that there is limited availability of rock or stone (Langer 1988) because virtually the entire state consists of Tertiary and Quaternary sediments. Despite the fact that rock and stone are lacking, Louisiana has not been, nor will likely be, a top producer of sand and gravel in the United States. In 1938, Louisiana ranked thirtieth among the sand and gravel producing states with about 2 million metric tonnes annually (Woodward & Gueno 1941). At its peak during the mid-1970s, production in Louisiana was about 6% of the 650 to 900 million metric tonnes consumed in the US (Evans 1978). However, more recent production figures from the early 1990s for Louisiana are well below 2% of the national production of 900 million metric tonnes in 1990 (Langer & Glanzman 1993) and the estimated production of 1000 million metric tonnes for this decade (Tepordei 1982).

Even though Louisiana is still near or below the national average in terms of production rates, the sand and gravel mining industry faces a unique set of issues. Most gravel and sand resources in Louisiana are located below the water table, differing from other areas where estimates of gravel quantities often explicitly exclude quantities below the water table (Bliss 1993). Because some of the largest gravel and sand reserves occur in river floodplains, there are serious environmental and engineering consequences associated with mining, which include channel instability, hydrologic alterations, biological impacts, and bridge failures. Yet, at present, no regulations directly related to mining and reclamation are in place, in contrast with most other states (Banks et al. 1981, Mossa 1985). Thus, Louisiana is not just an 'average' state but constitutes an interesting case study of the issues associated with mining in an unregulated, low-lying setting deficient in rock and stone.

The objectives of this paper are: 1. To overview history and technology of aggregate mining in Louisiana, 2. To examine geologic and geographic aspects of aggregate production in the state, and 3. To evaluate how current issues, especially environmental and economic constraints, might affect the future of mining. This synthesis can be used to assist in resource and land use planning in Louisiana, and provides some recommendations on best approaches for minimizing problems associated with mining in this and similar settings.

2   TECHNOLOGY AND HISTORY OF SAND AND GRAVEL MINING IN LOUISIANA

2.1   *Technology of sand and gravel mining in Louisiana*

Changing technologies have affected both spatial and temporal patterns of gravel and sand mining in Louisiana. Mining in Louisiana generally began in upland settings where the water table is located below the base of the pit. This is referred to as open pit or dry mining because water is generally not utilized in mining and subsequent sorting. Initially, gravel and sand were removed by hoe, pick, shovel and hand (Woodward & Gueno 1941). More recently, it has shifted to use of bulldozers, cranes and other heavy equipment.

Some years later, in the mid-1930s, the hydraulic dredge was introduced for mining in areas with high water tables, a process known as wet mining (Fig. 1). Since its

Figure 1. Aerial photograph of the middle Amite River, southeastern Louisiana, showing the pits, ponds, and piles created by mining of river floodplains. During flooding, the flow may be diverted into the pits on the lower floodplain, as there is very little buffer between these pits and the channel.

introduction, wet mining has comprised an increasing percentage of total production. In the early years, as observed in aerial photographs taken in the 1950s and 1960s, direct mining of the point bars and channel bottom were the most common practices. Most of the operations established since have been located on adjacent floodplain surfaces and low terraces, where practices have included removal of vegetation and fine-grained overburden, and digging of pits and ponds. Because sand occurs in much higher proportions than gravel but demand has been nearly equal, excess sand extracted by mining operations typically has been stored in conical tailings piles on the floodplain surface (Fig. 1).

2.2 *History of aggregate production rates*

The history of gravel and sand production rates in Louisiana can be characterized by data compiled by the US Geological Survey (USGS), and the Louisiana Department of Revenue and Taxation (LDRT). The first published estimates, compiled by the USGS, indicate that gravel and sand production were each less than 300,000 metric tonnes in 1905 (Woodward & Gueno 1941). Since 1920 when statewide data became available from the LDRT (with some exceptions in 1940, 1941, 1942, 1946), production has experienced several major and minor rises and falls. Production showed an initial peak in the mid-1920s followed by a drop associated with the Depression (Fig. 2). Through the early and middle part of the century, the general trend in aggregate production was upward. Production reached a peak in the mid-1970s, and has generally declined since to about one-third of the maxima (Fig. 2). This represents a different trend that the United States as a whole, where production rates have been rather consistent in recent two decades (Langer & Glanzman 1993). Gravel produc-

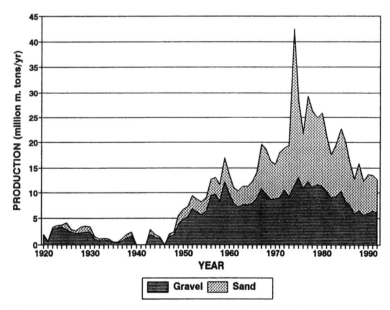

Figure 2. Historic mining production of gravel and sand in Louisiana, 1920 to 1992. Data were acquired from Woodward & Gueno (1941) and the Louisiana Dept. of Revenue and Taxation.

tion has been relatively steady in Louisiana, with annual production ranging from about 5-13 million metric t/yr since 1950. Sand production has been somewhat variable, ranging from 1-30 million metric tonnes annually since 1950.

Little is known about causes of temporal variations in production, although there are likely several physical and human explanations. Improvements in technology have allowed for increased supply, especially since the 1930s. Depletion of areas where supply is insufficient may be one reason for the recent decline in production rates. There may be a number of outside influences including the nation's economics, such as periods of growth and recession, and transportation of material from out-of-state. However, the most important influence in Louisiana has likely been the petroleum industry and associated changes in population and prosperity, especially in the 1970s. Construction of several interstate highways occurred during the 1970s, coinciding again with the greatest production rates. Further research is necessary to evaluate the role of these factors and other influences on supply and demand in a temporal context.

3 GEOLOGIC ASPECTS OF AGGREGATE PRODUCTION

3.1 *Aggregate suitability mapping in Louisiana*

General categories that are often considered in studies of aggregate potential include: 1. Deposit volume and geometry, 2. Size distribution of the material in the deposits, 3. Physical characteristics of the material, and 4. Chemical composition and chemical reactivity of the material. Our suitability ranking for Louisiana is based primarily on the first two categories, yet we recognize that there are potential consequences if certain physical and chemical standards are not met (Mather & Mather 1991).

In general, the quantity of sand and gravel in Louisiana is strongly related to the distribution of geologic units. Thus, the geologic-geomorphic units in Louisiana were divided into areas of relative suitability for aggregate exploration (Fig. 3). Regional mapping of aggregate suitability was accomplished by using the Geologic Map of Louisiana (Snead & McCulloh 1984) with some modifications of terminology (Autin et al. 1991, Autin & Snead 1993). Areas where sand is limited and gravel is lacking are characterized as not suitable for aggregate exploration. The remaining areas were categorized as possibly suitable, suitable with limitations, and highly suitable for aggregate exploration. Because gravel is more limited than sand, but in equal demand, our comments refer primarily to the availability of gravel. Due to variability within various geologic units, the classification only provides information on production potential and is inadequate for prospecting purposes at the present map scale.

As elsewhere in the United States (Bliss 1993), nearly all gravel and sand deposits mined in Louisiana are Quaternary. In sediments that are Pleistocene and older, thickness of overburden generally increases close to the Mississippi Valley where loess thickness is greatest (Miller et al. 1985, Mossa & Miller 1995). The most common minerals of the gravel deposits are chert, quartz, and pseudomorphous quartz such as petrified wood, and occasional quartzite and novaculite (Woodward & Gueno 1941, Donellan & Ferrell 1978, Self 1986). Some additional information on

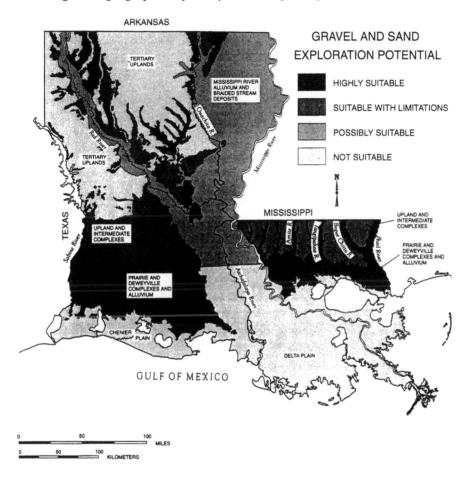

Figure 3. Suitability classification and aggregate-bearing environments of Louisiana (base information from Snead & McCulloh 1984).

the physical and chemical characteristics of aggregate material can be found in these studies.

3.2 *Areas not suitable for aggregate production*

Geologic units considered not suitable for gravel and sand mining include Tertiary upland deposits and the Holocene deltaic and chenier plains (Fig. 3). Although such areas may have localized zones where sand and shell mining occurs, evidence from geologic investigations and historical production rates suggests that these areas lack aggregate reserves of sufficient quantity and quality for commercial production. Instead, it is considered fortuitous if such areas meet some local needs for roadfill.

3.2.1 *The Tertiary uplands*

This area occurs in north Louisiana between the Mississippi and Red river valleys and between the Red and Sabine river valleys (Fig. 3). The Tertiary upland deposits largely consist of interbedded sands, silts, clays, limestones, lignites, and volcanic ash (Snead & McCulloh 1984). Chert gravel is rare, but has been locally reported in lenses of the Miocene Williamson Creek Member of the Fleming Group. Other than sand and gravel, potential sources of aggregate include crushed sandstone from the Oligocene Catahoula Formation, scattered ironstone in the Eocene Claiborne Group, and salt dome caprock. Some of this material may be suitable for roadfill, and there is localized mining activity.

3.2.2 *The deltaic plain*

The delta plain in southeastern Louisiana (Fig. 3) primarily consists of Holocene deposits of abandoned and active channels and distributaries of the Mississippi River and interdistributary backswamp and marsh deposits (Frazier 1967, Saucier 1994). This area lies between the southern limit of Pleistocene deposits and the Gulf of Mexico. The deposits generally lack gravel, and areas with sand and shell are localized and not abundant. Coastal beach ridges along erosional headlands contain local borrow pits (Harper 1977) and constitute one of the better onshore sources of sand. Sand has been produced locally from point bars along the Mississippi River, although the size and quality of material decreases appreciably downstream. Shells have also been mined from lakes and bays, especially in the lower Atchafalaya basin and Lake Pontchartrain.

Because of the serious coastal land loss problem in Louisiana (Gagliano et al. 1981, May & Britsch 1987, Walker et al. 1987), much sand will likely be required in the future for coastal restoration and storm protection projects (i.e. Nakashima & Mossa 1991, Penland et al. 1990, Suter et al. 1989). Local sources of material for coastal restoration have dominantly come from marine waters including inlets and their ebb and flood-tidal deltas, shoals, and inner shelf sands.

3.2.3 *The chenier plain*

The chenier plain in southwestern Louisiana (Fig. 3) is comprised of alternating or coalescing ridges and mudflat or marsh deposits of late Holocene age (Byrne et al. 1959, Gould & McFarlan 1959), largely tied to historical changes in the course of the ancestral Mississippi. Cheniers are isolated linear ridges of sand and shell upon which a substantial oak tree population typically occurs.

Coastal erosion is also a problem in the chenier plain, although it is less serious than in the delta plain (Walker et al. 1987). In some places, offshore sand resources may be used for coastal restoration or shoreline protection. Stratigraphic studies of sediments on the continental shelf in southwest Louisiana show that there are extensive offshore sand resources, which include fluvial channels and transgressive shoal sands (Suter & Berryhill 1985, Suter et al. 1987).

3.3 *Areas possibly suitable for aggregate exploration*

Alluvium of the Mississippi and Red Rivers and the braided stream deposits of northeast Louisiana are considered possibly suitable based on reports of aggregate in geologic investigations (Fig. 3, Saucier 1968, 1994). Available evidence suggests that production from these areas has been sparse, and that depth to aggregate and limited gravel resources may keep extraction from being economically viable. However, there are generally sufficient quantities of sand for roadfill and other construction needs.

3.3.1 *Mississippi and Red River alluvium*

Historically, the Mississippi and Red River valleys of northeast and northwest Louisiana have not been heavily used for sand and gravel production. In addition to the modern course, both rivers have several abandoned meander belts (Saucier 1974, Saucier 1994). Distinctive landforms and deposits of active and abandoned courses include natural levees, crevasses and crevasse splays, ridge and swale topography, oxbow lakes and backswamp deposits (Russ 1975, Smith & Russ 1974, Saucier 1994).

In both valleys, sand-dominated alluvium is typically overlain by fine-grained overburden (Smith & Russ 1974, Saucier 1994). The fine-grained material was largely produced by floodplain deposition of meandering rivers during the Holocene. Thickness of overburden varies and in places makes aggregate extraction uneconomical. Gravel distribution is not well-defined, but most favorable deposits would include active and abandoned point bars and alluvial fans where tributary streams merge with these valleys.

3.3.2 *Braided stream deposits*

This area of northeast Louisiana is associated with Pleistocene-age glacial outwash carried from the midwestern United States by the Mississippi River. The deposits are known to contain gravel (Smith & Saucier 1971, Saucier 1968) and are capped by relatively thin overburden of silty loess deposits. Few investigations have been conducted on the distribution, concentration, and depth to gravel of these deposits, therefore we tentatively classify these in the possibly suitable category.

3.4 *Areas suitable with limitations for aggregate exploration*

The Upland and Intermediate Complexes have gravel-rich exposures (Cullinan 1969, Campbell 1971, Snead & McCulloh 1984, Self 1986, 1993, Mossa & Autin 1989) and are suitable for aggregate exploration. A map which includes the southeastern Louisiana uplands, an important area for gravel production, shows the distribution of these geologic deposits (Fig. 4). Both local (Woodward & Gueno 1941) and regional sources (Langer 1988) consider the area where these deposits are mapped as having potential for sand and gravel production. However, neither source mentions limitations, especially the matrix clays and iron oxides often associated with these deposits. We consider this important as sand and gravel mined from these geologic units may require washing before use in some construction projects. Also, because of low water tables, dry mining is required.

3.4.1 *The Upland Complex*

The Upland Complex corresponds closely to areas mapped by others as the Citronelle Formation (Matson 1916). The deposits forming the Upland Complex consist predominantly of coarse-grained sediments, the source of which has been regarded to be the continental interior (Fisk 1939, Woodward & Gueno 1941), an eastern Gulf or

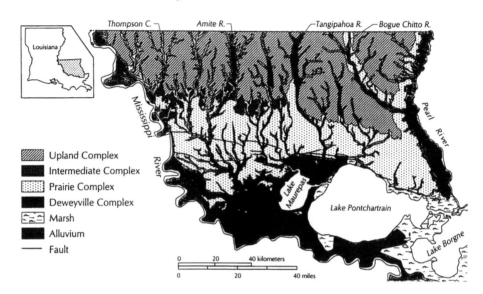

Figure 4. Generalized geologic map of southeastern Louisiana, showing geologic units comprising the aggregate suitability map (modified from Snead & McCulloh 1984). Alluvium, and the Deweyville and Prairie complexes are considered to be highly suitable for exploration, with the former being the most environmentally vulnerable. The Upland and Intermediate complexes are considered to be suitable with limitations, because of increased consolidation, clay and iron oxides, and lower water tables.

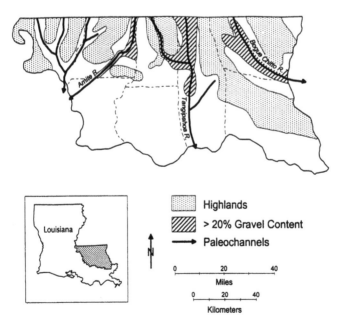

Figure 5. Paleogeography of southeastern Louisiana during the early Pleistocene to late Pliocene. Mapping of gravel content of the Upland Complex deposits was used to make interpretations of paleochannels and paleohighlands (modified from Self 1986).

Appalachian provenance (Rosen 1969, Cullinan 1969) or, more likely, a combination of these and possibly other sources. Most recent studies suggest that these gravels in southeast Louisiana were deposited by braided streams, possibly associated with small alluvial fans (Self 1993). The age of these deposits has been a subject of contention due in part to the scarcity of paleontological data and in part because these gravels overlie Tertiary deposits of varying ages. Pleistocene, Pliocene, and Miocene ages have been suggested (Mossa & Autin 1989, Self 1993). Gravel in younger deposits is dominantly reworked from this source.

The predominance of coarse-grained deposits has been attributed to stream rejuvenation caused by uplift of the continental interior (Doering 1956, Durham et al. 1967, Rosen 1969), deposition by a larger river system such as an ancestral Tennessee River (Brown 1967, Campbell 1971), or a more arid climatic regime (Alt 1974), although none of these ideas has been fully substantiated. Although the original geomorphic expression of the surface has been obliterated because of dissection and structural influence, some studies suggest where best to prospect. Brown (1967) suggested that gravel-rich deposits often occur beneath basin divides, selectively preserved from erosional processes. Other areas likely to be rich in gravel include paleochannels, several of which have been mapped by Self (1986) (Fig. 5).

3.4.2 *The Intermediate Complex*
The Intermediate Complex is not well-defined, but is considered to have lower dissection, slope, topographic position, and less soil development, than the Upland Complex. The sediments forming the Intermediate Complex have not been well-equated with a distinctive suite of depositional environments or lithologies. In southwestern Louisiana, the deposits have generally been interpreted as fluvial-deltaic (Fisk 1944, Saucier 1994). In southeastern Louisiana, the Intermediate Complex deposits may be derived from hillslope and alluvial reworking of loess and the sediments making up the Upland Complex (Mossa & Autin 1989, Mossa 1991). Estimates of the age of these deposits are also controversial (Fisk 1944, Fisk & McFarlan 1955, Saucier 1974, Alford & Holmes 1985).

Where sediments are fluvial or reworked from the Upland Complex, there is potential for appreciable sand and gravel production. Given that its nature and distribution is controversial, the production potential is highly variable and difficult to evaluate. In general, the geology is more poorly documented and the gravel production has been lower in quantity than the Upland Complex.

3.5 *Areas highly suitable for aggregate exploration*

The Prairie and Deweyville complexes, and alluvium of streams draining these and older complexes (Upland and Intermediate) are considered highly suitable for aggregate exploration (Figs 3 and 4). These units have water tables near the surface and, in places, their deposits contain abundant gravel. Best reserves are associated with modern valleys and paleochannels of late Quaternary age. Reworking of gravel from the older Pleistocene deposits may improve sorting and certainly reduces limitations associated with clay and iron oxides.

3.5.1 *The Prairie Complex*

The Prairie Complex in south Louisiana replicates the spectrum of modern environments. The age of the deposits has been regarded as either mid-Wisconsinan (Farmdalian) or Sangamonian or some range within this span (Mossa & Autin 1989, Autin et al. 1991, Saucier 1994). Although prospecting of Prairie Complex for gravel and sand is more challenging than in confined alluvial valleys, preliminary studies suggest it may have a large amount of untapped reserves. Water tables are high and adequate for hydraulic mining, and environmental impacts would be less pronounced than mining alluvium.

The parts of the Prairie Complex with greatest sand and gravel content include late Pleistocene paleochannels, relict barrier islands (Graf 1966) and beach ridges. In general, most paleochannels are located within modern alluvial valleys and flank the modern streams. However, toward the south, some of these paleochannels diverge from modern valleys, with multiple paleochannels being evident in Louisiana (Mossa & Miller 1995), as well as nearby Texas (Galloway 1981). Some of these deposits occur along sand ridges and hills, which occur at the highest landscape positions along east-to-west transects (Mossa & Miller 1995) (Fig. 6). Because studies of paleochannels are limited and fairly recent, these areas are not included in syntheses of mining in the United States (Langer 1988).

3.5.2 *The Deweyville Complex*

The Deweyville Complex in south Louisiana is considered to be a river-trending deposit of fluvial origin. It is found along streams of intermediate size and is topographically higher than the Holocene alluvium but lower than the Prairie Complex (Snead & McCulloh 1984). It also may be present along smaller streams although detailed studies are lacking. As with the other surfaces, it is an enigma because sublevels may be present (Saucier & Fleetwood 1970). Spatial relationships across the region and with other deposits are poorly understood (Kolb et al. 1975, Saucier 1977), and temporal and causal relationships are varied or undetermined (Bernard 1950, Gagliano & Thom 1967, Alford & Holmes 1985). Soils that develop on these deposits are locally the same as on the Prairie Complex (e.g. Trahan et al. 1991).

The Deweyville Complex has large meander scars on its surface, with wavelengths three to six times larger than those of modern rivers. Given the seemingly larger size of ancestral rivers, there is tremendous potential for sand and gravel deposits. Most locations have coarse sediments, high water tables, and are located far enough from rivers that environmental impacts would not be severe if certain precautions were taken. Prospecting efforts could be confined to a limited number of locations because the known distribution of this deposit is spatially limited.

3.5.3 *Alluvium*

Alluvial valleys and Holocene floodplains are the most common and prolific sources of gravel and sand deposits. At present, as well as in the past (Woodward & Gueno 1941), many active mines were located along the Amite and Bogue Chitto rivers. There also have been and still are a number of mines along floodplains of several smaller rivers, such as the Tangipahoa (Fig. 6).

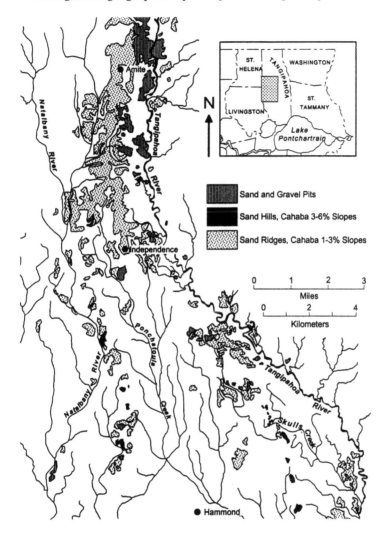

Figure 6. Sand and gravel operations occur in paleochannels of the late Pleistocene Prairie Complex (see Mossa & Miller 1995 for further discussion) and in alluvium, located adjacent to the modern Tangipahoa River. Paleochannels in part correspond with sand hills and ridges, mapped as the Cahaba (Typic Hapludult) series, 3-6% slopes and 1-3% slopes (data from McDaniel et al. 1990). Paleochannels are also associated with other soils, and are buried in places, especially toward the southern part of the map. Located away from the floodplain, the area of sand ridges between Amite and Independence is a good location for prospecting sand and gravel.

Modern channels, as well as paleochannels, show appreciable variability in the gravel content and size distribution of bottom sediments. One cause of variation in grain size is the depositional environment, with point bars and channel lag deposits being sand and gravel-rich, and overbank deposits being dominantly fine-grained. Because sediment size generally decreases downstream and deposit volume, manifest by the size of the overall channel and the individual bars, generally increases downstream, the middle segment of valleys often have the best combination of quantity and content.

Across the state, there is also considerable variability between rivers in terms of the quantities of gravel and sand resources. Variability between channels is in part a function of the gravel and sand content of the source deposits, which is in turn affected by historical geology. Gravel content is likely related to hydraulic factors, including stream gradient and discharge, although no studies have established such relationships.

Even though we recommend avoiding mining floodplains, where possible, because of environmental impacts, studies of the bed materials and depositional environments of modern channels could be useful for developing analogs. Stratigraphic and sedimentologic studies can be used to document relationships and develop models regarding occurrences of sand and gravel content. These models can then be tested in paleochannels of varying age, and applied in prospecting.

4   GEOGRAPHIC ASPECTS OF AGGREGATE PRODUCTION IN LOUISIANA

4.1   *Historic sand and gravel mining localities*

To acquire some understanding of the geographic aspects of aggregate production in Louisiana, background political and transportation information is necessary. The LDRT compiles gravel and sand production into political units known as parishes (counties) (Fig. 7). Major cities and highways represent important localities of ag-

Figure 7. Location of parishes and interstate highways in Louisiana.

gregate consumption (Fig. 7). One of the more important regions of aggregate production in the state is known as the Florida Parishes (West Feliciana, East Feliciana,
East Baton Rouge, St. Helena, Livingston, Tangipahoa, Washington, and St. Tammany), an area that was formerly an extension of the Florida panhandle from the
mid-1500s to early 1800s. This portion of southeastern Louisiana has become important because it constitutes the primary source of aggregate for New Orleans and Baton Rouge, the two largest cities in the state.

A previous study (Woodward & Gueno 1941) was used to map historic sand and
gravel mining activity in Louisiana (Fig. 8). In southeastern Louisiana, historic mining activity was concentrated in the Amite River valley and adjacent areas in Livingston, St. Helena, and West Feliciana Parishes, the Tangipahoa River valley and
adjacent areas in Tangipahoa Parish, and the Bogue Chitto and lower Pearl River
valleys and adjacent areas in Washington and St. Tammany Parishes. Other notable
concentrations in the Florida Parishes include Bayou Sara in East Feliciana Parish,
and Thompson Creek in East and West Feliciana Parishes. In the southwestern part
of the state, the largest areas of active and abandoned mines are located in Vernon
and Beauregard parishes, and secondly in Jefferson Davis and Beauregard parishes.

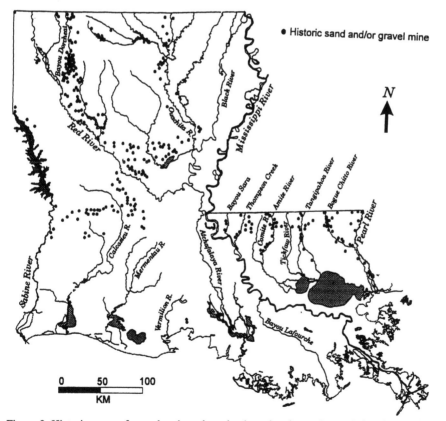

Figure 8. Historic areas of gravel and sand production, showing active and abandoned mines prior
to 1941 (data source is Woodward & Gueno 1941). Some abandoned mines have since been reacti-
vated, and several new mines are now in operation.

Heaviest concentrations of historic mining activity in northwestern Louisiana occur in the Bayou Dorcheat drainage and adjacent Pleistocene surfaces near Minden, Webster Parish. Notable concentrations in northeastern Louisiana occur near West Monroe in Ouachita Parish. In central Louisiana, heavily mined areas include the Pleistocene deposits flanking the Red River north and south of Alexandria in Grant and Rapides Parishes. In east-central Louisiana, appreciable gravel also occurs on an erosional remnant known as Sicily Island, located in the Mississippi Valley in Catahoula Parish.

From our knowledge of the state, there is considerable similarity between areas of historic and modern mining concentrations. However, earlier studies (Woodward & Gueno 1941) map fewer than 300 mines, whereas our preliminary investigations based on aerial photographs and large-scale topographic maps have documented over 1200 active and abandoned sand and gravel mines. Follow-up studies are recommended to produce maps which include mines established since the 1940s, and to classify sites based on geologic units. These geologic units have been revised and remapped considerably (Snead & McCulloh 1984) since earlier studies (Woodward & Gueno 1941), and will be remapped at a larger scale more suited to such assessments in coming years.

4.2 *Mapping of gravel and sand production*

Historic, recent, and average production data from the LDRT can be used to provide additional perspectives on spatial variations in gravel and sand production. Unfortunately, such data are collected and compiled based on political, rather than geologic, units. These quantities are expressed per unit area as it otherwise would create bias to compare production rates of small and large parishes. It is not known whether the quantities of sand and gravel produced includes dredging of river bottoms and coastal waters for navigational maintenance, beach nourishment, and construction of artificial levees and other flood control projects. Given that these projects are often conducted using local materials by federal, state, and local agencies, they may not be taxed or included in figures compiled by the LDRT.

Gravel production shows considerable spatial variation, because it occurs only in certain geologic units. Early data (1932-1939) shows that the most productive parish, based on tonnage per unit area per year was East Feliciana (Fig. 9), which borders the western boundary of the upper Amite River and includes some mines along Thompson Creek. Other areas that were historically important for gravel production include Livingston, Tangipahoa, Washington, Webster, and Ouachita parishes (Fig. 9). Recent data (1985-1992) show that St. Helena Parish, a rural area bordering the eastern side of the upper Amite River, is by far the largest producer of gravel, producing close to 2000 metric $t/km^2/yr$ (Fig. 9). Washington Parish is second, exceeding 500 metric $t/km^2/yr$. Other significant producers of recent years include most of the Florida Parishes, especially East and West Feliciana, Livingston, and St. Tammany, and scattered locations throughout the state, including Webster, Ouachita, Red River, Rapides, and Beauregard parishes (Fig. 9). Average data representing the production rates over many years (1932-1992) show that St. Helena Parish tops the list overall (Fig. 9). Several of the Florida parishes have been highly productive, with East and West Feliciana, East Baton Rouge, Tangipahoa, Washington and St. Tam-

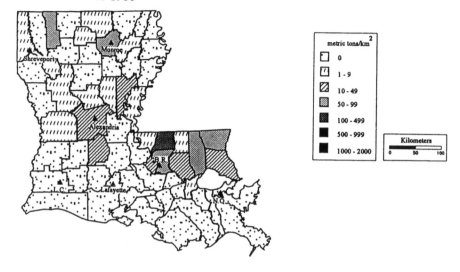

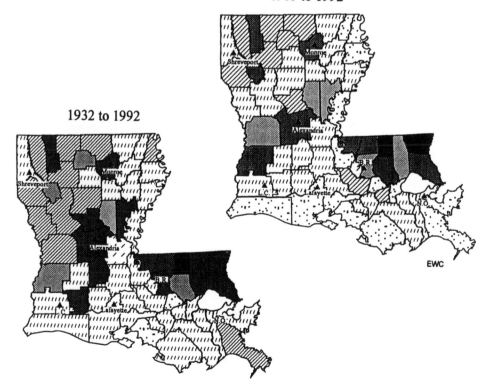

Figure 9. Early (1932-1939), recent (1985-1992), and average (1932-1992) gravel production rates by parish in metric tonnes per year. Data were acquired from Woodward & Gueno (1941) and the Louisiana Dept. of Revenue and Taxation.

1932 to 1939

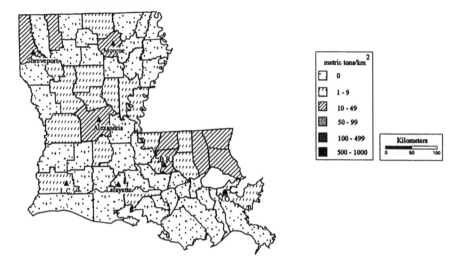

1985 to 1992

1932 to 1992

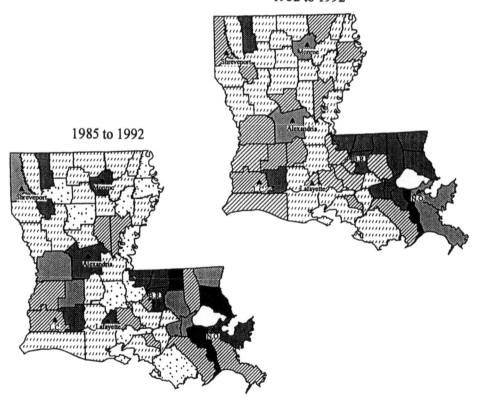

Figure 10. Early (1932-1939), recent (1985-1992), and average (1932-1992) sand production rates by parish in metric tonnes per year. Data were acquired from Woodward & Gueno (1941) and the Louisiana Dept. of Revenue and Taxation.

many producing more than 100 metric t/km^2/yr. Other parishes that have achieved this level of production include Webster, Ouachita, Catahoula, Grant, Rapides, Evangeline, and Jefferson Davis parishes (Fig. 9). In most cases, with the possible exception of Plaquemines Parish, where gravel is rare and one year showed unusually large production, the data corroborates our knowledge of mining in the state.

In terms of sand production, the early years (1932-1939) show about nine parishes more productive than the rest, five of which are in the southeast (Fig. 10). The most productive included East Feliciana, East Baton Rouge, Tangipahoa, Washington and St. Tammany parishes in the southeast and Caddo, Webster, Ouachita, and Rapides parishes in northern and central Louisiana. Each produced more than 10 metric t/km^2/yr. More recent data (1985-1992) show new areas of activity, with three parishes in the southeast, St. Helena, St. Tammany, and Jefferson parishes, exceeding 500 metric t/km^2/yr in sand production (Fig. 10). Over many years (1932-1992), only around New Orleans in Jefferson Parish did sand production exceed 500 metric t/km^2/yr (Fig. 10). However several other parishes in the southeast had sand production rates exceeding 100 metric t/km^2/yr. These included all of the Florida parishes except Livingston, St. James, St. John the Baptist, and St. Charles parishes in southeastern Louisiana, Jefferson Davis Parish in the southwest and Webster Parish in the northwest.

4.3 *Interpretation of spatial variations and changes*

The surface distribution of geologic deposits has a major influence on mining patterns in Louisiana. Pits rarely are dug through multiple units, because of the expense of removing overburden and the limitations of the dredging machinery associated with increased depths. This in part explains why the Florida parishes, an area dominated by highly suitable and suitable geologic units, has been, and still is, an area of very active production. The outward shifting over the study period from the southern (East Baton Rouge and Livingston) to northern (West Feliciana, St. Helena, Washington) Florida Parishes suggests that some areas closer to markets are experiencing declining reserves. Similarly, several parishes that were not initially mined around Shreveport, Alexandria, and Lake Charles are presently active areas of production, either because of prospecting discoveries and/or inadequate resources close to market.

From a geologic perspective, there are several areas of prospective study for those interested in analysis of spatial variations and exploration issues. Based on the classification in this paper, a preliminary production potential index could be developed for each parish by classifying geologic deposits according to their production potential, multiplying the area of each class by the weighing and computing a weighed total. Also, if the records of various mining operations could be acquired, and if more detailed assessments were made from maps, it would be desirable to relate production and size more directly to the geologic units. Although such approaches would provide a regional overview, local geologic studies targeted at mapping paleochannels and the distribution of coarse-grained deposits are also lacking. Such studies would be useful to miners interested in prospecting new areas, especially away from alluvial valleys.

Demographic and economic factors also affect spatial patterns of mining in Louisiana and their changes. Transportation costs are a significant consideration, since sand and gravel are high-bulk, low-cost commodities. The prospecting area is typically confined by an upper distance, determined by shipping costs, and a lower distance, imposed by higher land values, competing uses, and zoning constraints close to urban centers. Gravel for New Orleans and Baton Rouge, the two largest cities, has been supplied largely by the eight Florida parishes in varying proportions. Transport distances to Baton Rouge are reasonable, but somewhat longer for New Orleans, and may become a more serious issue as local resources are tapped. Shreveport, the next largest city, has been supplied dominantly by Webster Parish (Figs 7, 9 and 10). Other cities which exceed 100,000 people, including outlying areas, are Lafayette, Lake Charles, Alexandria, and Monroe (Fig. 7). Alexandria appears to have sufficient production within Rapides Parish, and similarly Ouachita Parish appears to have adequate supply for the city of Monroe. However, Lafayette and Calcasieu parishes, the site of Lafayette and Lake Charles, have not been productive recently or historically, and aggregates have likely come from nearby production in Evangeline, Jefferson Davis, and Beauregard parishes.

Public works projects also influence the spatial patterns of mining, with several recent interstate highway projects (I-10, I-12, I-20, I-55, and I-49) being one example (Fig. 7). Langer (1988) estimated that for an average six-lane and four-lane interstate highway with bridges and interchanges, about 59,000 metric t/km and 48,000 metric t/km aggregate are used. However, because transportation corridors often pass through or connect population centers, it is often difficult to separate their effects. More detailed studies could focus on the consumption by various facets of the construction industry and more comprehensive analysis of the influence and role of human factors on spatial variations in production.

5   THE FUTURE OF MINING IN LOUISIANA

Earlier portions of this paper have described the historic expansion of mining and the concentration of mining in alluvium. The sand and gravel industry has a precarious future in Louisiana because of the environmental effects associated with floodplain mining, more awareness of liability, the growing involvement of regulatory agencies, and the increasing operating costs. The ability of miners to adjust to new challenges, and take proactive and interactive approaches to reclamation and land use planning, may well affect their longevity in the business.

5.1   *Environmental effects*

Alluvial mining practices include removal of vegetation and overburden from the banks and floodplain, dredging of pits and ponds, creation of floodplain tailings piles, direct mining of point bars, and removal of gravel armor from the channel perimeter (Figs 1 and 11). These practices have resulted in geomorphic, hydrologic, and biologic impacts, only a few of which have been documented in Louisiana. During low flow, direct impacts are primarily associated with increased sediment loads associated with in-stream mining and occasional breaches between the channel

Figure 11. Aerial photograph of the middle Amite River showing direct mining of the channel bottom and point bar surface. Channel modification includes the creation of pits and mounds, and removal of gravel armor, which makes the underlying sandy bed material more susceptible to movement and increases turbidity. Such channel changes also affect aquatic organisms.

and turbid floodplain dredge ponds (Figs 1 and 11). During floods, geomorphic changes occur when flow spreads across bare sand areas, tailings piles, and mine ponds (Mossa 1995), and these changes, in turn, have a number of hydrologic and biologic consequences.

Some types of geomorphic changes along, downstream, and upstream of mined reaches in Louisiana include riverbank erosion, cutoffs, avulsions, aggradation and degradation (Mossa 1995). Implications to humans include loss of property, roads, and buildings, disputes between landowners over riparian property, problems with access, and complications of ownership and taxation. In aggrading reaches there is reduced channel capacity, and in degrading mined reaches, there is also increased potential for floods to undermine bridges, pipelines and other structures. Several bridges on the main stem and tributaries of mined rivers in the Florida Parishes have been undermined during floods; these are costly to repair and represent serious concerns for public safety.

Changes in hydrology often occur along and downstream of mined reaches, often in conjunction with geomorphic changes. Within mined reaches, degradation results in reduced flooding, and deterioration of wetlands. Downstream of mined reaches, sediments eroded upstream may be deposited on the channel bed (Mossa 1995). The decreased channel capacity, even with a constant discharge regime, results in more common, severe, and widespread flooding. Thus, flooding problems previously attributed to other human activities including urbanization (Turner & Bond 1980) and canalization, may also be aggravated by mining.

Flooplain mining also has potential consequences to terrestrial and aquatic ecosystems. Some effects are direct, including removal of riparian forest, which has declined from 42% to 21% of the total area in Louisiana (Turner et al. 1981). Floodplain mining, while not the dominant cause, is a major contributor; it is the primary cause of disturbance of 4276 ha in the Amite River basin alone (State of Louisiana 1992). Less direct effects take much more effort to quantify and document, and need further study. For example, decreases in the area and quality of bottomland forest often result in decreased diversity; even after many years of natural recovery, xeric vegetation such as grasses and pine trees are often the only flora found on abandoned mines. The wildlife that use river and floodplain ecosystems are likely to decrease in quantity and diversity as well.

Aquatic biologists are increasingly concerned about the influence of mining and other land use activities on threatened and endangered species. Direct and indirect changes in mined reaches, including increased turbidity and sedimentation, accelerated erosion, decreased depth, loss of habitat diversity, substrate instability, and decreased stream canopy cover, decrease the quality and quantity of aquatic life (Berkman & Rabeni 1987, Chutter 1969, Hartfield 1993, Houp 1993, Murphy et al. 1981). Perhaps most severely impacted are freshwater mussels, which have been depleted, extirpated, or otherwise affected in a number of rivers in Louisiana (Hartfield 1993).

Although far less common in Louisiana, analogous problems may also occur because of mining in coastal waters. The creation of borrow pits in coastal waters affects wave refraction, which in turn may affect coastal erosion and accretion rates and patterns (Suter et al. 1989), and possibly nearshore currents. As in the river environment, there are concerns about aquatic organisms, especially non-migratory organisms and those vulnerable to increases in turbidity.

5.2 *Economic constraints*

Economic constraints on the mining industry are increasing, and may be a contributing reason for decreasing production in the past few decades. Economic constraints of mining gravel and sand include equipment costs, resource depletion, environmental regulation and permitting, liability issues, and prospecting. Over time, as resources become depleted, miners will require better technology to acquire gravel from deeper pits. Environmental regulations are becoming more costly; in the past, only rarely were state regulatory agencies involved in enforcement, with small fines given for water quality violations associated with sediment releases. Recently, federal agencies are asserting that any modification of wetlands now requires environmental permitting, and operators need to hire consultants to assist them in this process. Because there is more knowledge of geomorphic and environmental impacts than previously, and more inclination toward legal action than the past, there is also more potential for liability between various parties involved. The costs of prospecting also may increase as known resources are depleted and more exploration is conducted elsewhere.

Because of the increasing costs, sand and gravel mining is being conducted by fewer producers with larger operations. The costs may put all but the largest and most efficient miners out of business, and increase the costs of aggregates for consumers. As transportation costs increase, different areas may be affected in different ways. For instance, places such as New Orleans, which lack abundant supply locally and are already at considerable distance from the source, will likely pay proportionately more in coming years.

5.3 *Management resolution*

Given the necessity of sand and gravel in a state which lacks other types of aggregate material, it would be unfeasible to ban mining. Thus, to resolve problems, considerable thought should be given regarding how to balance the economic return and environmental effects. Management, research, and education can be used to direct future exploration, and to improve operations of existing mines. Potential benefits include reducing environmental costs and increasing land values following mining.

In terms of future exploration, one recommended approach is to direct more attention and resources toward finding aggregates on late Pleistocene complexes. Although prospecting in such areas is less predictable and more costly, mining these areas will minimize environmental effects and reduce the amount of regulatory control on obtaining wetlands permits. State agencies with a vested interest in obtaining gravel and sand could direct funding toward geologic investigations.

In currently mined areas, more emphasis could and should be placed on best management practices for mining and reclamation of previously mined areas (Mossa 1985). Specifically, many potential problems can be avoided by prohibiting certain practices entirely, establishing vegetated buffer zones along the channel edges, construction of fences, and sediment traps, as well as requiring the mining industry to take a proactive stance in the reclamation of pits and ponds. Contouring, replacement of topsoil, and re-establishment of riparian vegetation would facilitate floodplain recovery. Other uses of reclaimed mines in floodplains include recreation, storm-water management, and farmland; on uplands, mines also may be reclaimed for residential developments and landfills. As of 1981, fifteen states including Louisiana lacked state and local laws requiring reclamation of aggregate mining operations (Banks et al. 1981). Perhaps through taxation, some portion of the revenue could be set aside for reclamation. Clearly, some legislation is desperately needed in Louisiana. Some discussions regarding management and reclamation have begun (State of Louisiana 1992), although no serious action has yet been taken.

River restoration and other approaches to mitigate past and future impacts is a future consideration, but unlikely to occur any time soon in Louisiana. In some other locations where it has been realized that extensive damage was done by channel modification, river restoration has been performed by government agencies. Although some facets of the river system may never recover, the goal of restoration often is to restore the natural function and physical environment as best possible, such that ecological recovery can occur slowly. Another strategy sometimes used elsewhere in an effort to mitigate future impacts is government purchase of sensitive lands, including those along river corridors. Although considerable damage has already been done in a number of rivers in Louisiana, such approaches, if taken, could prevent destruction of additional waterways.

Although politicians must ultimately pass various forms of legislation, scientists and others can assist by being a part of the process with potential roles in education, research, and planning. More awareness is necessary and through environmental education, a number of miners might willingly adopt best management practices and prospect in new areas if given adequate information. Geologists and geographers can assist prospectors and map paleochannels, especially where they occur away from river floodplains. Botanists and others could conduct applied studies to examine recolonization in environments without much topsoil. Social scientists could examine economic aspects of the industry, identify new issues, and suggest areas of potential compromise. Planners can provide ideas and plans to individual miners interested in reclaiming their property. A variety of individuals have potential roles in further assessment of impacts, in developing pilot reclamation projects, and in suggesting strategies for mitigation.

6 SUMMARY AND CONCLUSIONS

Sand and gravel production in Louisiana has gone through a boom and bust, likely related to the expanding petroleum industry and interstate construction, followed by declines in these activities. Both physical and human factors affect spatial patterns of mining, as most production occurs in coarse-grained alluvium within reasonable proximity of urban areas. More prospecting of the Deweyville and Prairie complexes is recommended because they are highly suitable for exploration but are less environmentally sensitive than floodplains. Such areas are likely to face increasing regulatory and liability pressures because of their geomorphic and biologic sensitivity.

Although more research is recommended, this synthesis can be used to assist in resource and land use planning in Louisiana, and serves as an example of some unique problems created or aggravated by mining floodplains. In the future, additional problems can be averted by land use planning and proper reclamation procedures. If both the scientific community and mine operators take steps to consider future needs and problems, there will likely be sufficient gravel and sand to supply Louisiana for the coming years and restoration of lands to alternative uses once local resources are consumed.

ACKNOWLEDGMENTS

Maps and figures were produced and revised from prior works by Lisa Ann Walsh, Ed Carter, Mark McLean, Tim Vitzenty, and Jeff Lower of the Department of Geography, University of Florida. Our research on aggregate suitability mapping was supported by the Louisiana Transportation Research Center (contract #87-2GT). Additional support was provided by our present employers, and our former employer – the Louisiana Geological Survey.

REFERENCES

Alford, J.J. & Holmes, J.C. 1985. Meander scars as evidence of major climate change in southwest Louisiana. *Annals of the Association of American Geographers* 75: 395-403.
Alt, D. 1974. Arid climate control of Miocene sedimentation and origin of modern drainage, southeastern United States. In R.Q. Oaks, Jr & J.R. DuBar (eds), *Post-Miocene Stratigraphy, Central and Southern Atlantic Coastal Plain*: 21-29. Logan, Utah: Utah State University Press.
Autin, W.J., Burns, S.F., Miller, B.J., Saucier, R.T. & Snead, J.I. 1991. Quaternary geology of the Lower Mississippi Valley. In R.B. Morrison (ed.), *Quaternary Non-glacial Geology, Contermi-*

nous US: *The Geology of North America, v. K-2:* 547-82. Boulder, Colorado: Geological Society of America.

Autin, W.J. & Snead, J. (eds) 1993. *Quaternary Geology and Geoarcheology of the Lower Red River Valley.* Baton Rouge: Friends of the Pleistocene, South Central Cell.

Banks, P.T., Nickel, R.E. & Blome, D.A. 1981. *Reclamation and Pollution Control: Planning Guide for Small Sand and Gravel Mines.* Washington, D.C.: Department of the Interior, Bureau of Mines.

Berkman, H.E. & Rabeni, C.F. 1987. Effect of siltation on stream fish communities. *Environmental Biology of Fishes* 18: 285-294.

Bernard, H.A. 1950. Quaternary Geology of Southeast Texas (Ph.D. dissertation). Baton Rouge: Louisiana State University.

Bliss, J.D. 1993. *Modeling Sand and Gravel Deposits – Initial Strategy and Preliminary Examples. Open File Report 93-200.* Tuscon, Arizona: US Geological Survey.

Brown, B.W. 1967. A Pliocene Tennessee River hypothesis for Mississippi. *Southeastern Geology* 8: 81-84.

Byrne, J.V., LeRoy, D.O. & Riley, C.M. 1959. The chenier plain and its stratigraphy, southwestern Louisiana. *Transactions of the Gulf Coast Association of Geological Societies* 9: 237-59.

Campbell, C.L. 1971. The Gravel Deposits of St. Helena and Tangipahoa Parishes, Louisiana (Ph.D. Dissertation). New Orleans, LA: Tulane University.

Chutter, F.M. 1969. The effects of silt and sand on the invertebrate fauna of streams and rivers. *Hydrobiologia* 34: 57-76.

Cullinan, T.A. 1969. Contributions to the Geology of Washington and St. Tammany Parishes, Louisiana (Ph.D. Dissertation). New Orleans: Tulane University.

Doering, J.A. 1956. Review of the Quaternary surface formations of the Gulf Coast region. *American Association of Petroleum Geologists Bulletin* 40: 1816-1862.

Donellan, M.S. & Ferrell, R.E., Jr. 1978. Mineralogy of Pleistocene terrace deposits, Louisiana. *Transactions of the Gulf Coast Association of Geological Societies* 28: 123-130.

Durham, C.O., Jr, Moore, C.H. & Parsons, B.E. 1967. An agnostic view of the terraces, Mississippi alluvial valley and terraces. *Geological Society of America Field Trip Guidebook, Part E.* New Orleans: Geological Society of America.

Evans, J.R. 1978. Sand and gravel. *US Bureau of Mines Commodity Profile MPC-23.* Washington, D.C.: US Department of the Interior.

Fisk, H.N. 1939. Igneous and metamorphic rocks from Pleistocene gravels of central Louisiana. *Journal of Sedimentary Petrology* 9: 20-27.

Fisk, H.N. 1944. *Geological Investigation of the Alluvial Valley of the Lower Mississippi River.* Vicksburg, Mississippi: Mississippi River Commission, US Army Corps of Engineers.

Frazier, D.E. 1967. Recent deltaic deposits of the Mississippi River: Their chronology and development. *Transactions of the Gulf Coast Association of Geological Societies* 17: 287-315.

Gagliano, S.M. & Thom B.J. 1967. Deweyville Terrace, Gulf and Atlantic coasts. *Coastal Studies Institute, Technical Report No. 39.* Baton Rouge: Louisiana State University.

Gagliano, S.M., Meyer-Arendt, K.J. & Wicker, K.M. 1981. Land loss in the Mississippi River deltaic plain. *Transactions of the Gulf Coast Association of Geological Societies* 31: 295-300.

Galloway, W.E. 1981. Depositional architecture of Cenozoic Gulf coastal plain fluvial systems. In F.G. Ethridge & R.M. Flores (eds), *Recent and Ancient Nonmarine Depositional Environments: Models for Exploration, SEPM Special Publication 31*: 127-55, Tulsa, Oklahoma: Society of Economic Paleontologists and Mineralogists.

Goldman, H.B. & Reining, D. 1983. Sand and gravel. In S.J. Lefond (ed.), *Industrial Minerals and Rocks*: 2, 1151-66. Littleton, Colorado: Society of Mining Engineers.

Gould, H.R. & McFarlan, E., Jr. 1959. Geologic history of the chenier plain, southwestern Louisiana. *Transactions of the Gulf Coast Association of Geological Societies* 9: 261-270.

Graf, C.H. 1966. The Late Pleistocene Ingleside Barrier Trend, Texas and Louisiana (M.S. thesis). Houston, Texas: Rice University.

Harper, J. 1977. Sediment dispersal trends of the Caminada-Moreau beach-ridge system. *Transactions of the Gulf Coast Association of Geological Societies* 27: 283-289.

Hartfield, P.D. 1993. Headcuts and their effect on freshwater mussels. In K.S. Cummings, A.C. Buchanan & L.M. Koch (eds), *Conservation and Management of Freshwater Mussels, Proceedings of the Upper Mississippi River Conservation Committee Symposium, October 1992*: 131-41. St. Louis, Missouri: Upper Mississippi River Conservation Committee.

Houp, R.E. 1993. Observations on long-term effects of sedimentation on freshwater mussels (Mollusca: Unionidae) in the North Fork of Red River, Kentucky. *Transactions, Kentucky Academy of Sciences* 54: 93-97.

Langer, W.H. 1988. *Natural Aggregates of the Conterminous United States. US Geological Survey Bulletin 1594*. Denver: US Government Printing Office.

Langer, W.H. & Glanzman, V.M. 1993. *Natural Aggregate – Building America's future: US Geological Survey Circular 1110*. Denver: US Government Printing Office.

Mather, K. & Mather, B. 1991. Aggregates. In G.A. Kiersch (ed.), *The Heritage of Engineering Geology: The First Hundred Years. Centennial Special Volume 3*: 323-332. Boulder, Colorado: Geological Society of America.

Matson, G.C. 1916. *The Pliocene Citronelle Formation of the Gulf Coastal Plain. United States Geological Survey Professional Paper 98*. Washington, D.C.: US Government Printing Office.

May, J.R. & Britsch, L.D. 1987. *Geological Investigation of the Mississippi River deltaic plain: Land loss and Accretion, Technical Report GL-87-13*. Vicksburg, Mississippi: US Army Engineer Waterways Experiment Station.

McDaniel, D.M., Daugereaux, D., Stephens, W., Fleming, B. & Seeling, P. 1990. *Soil Survey of Tangipahoa Parish*. Fort Worth, Texas: US Department of Agriculture.

Miller, B.J., Lewis, G.C., Alford, J.J. & Day. W.J. 1985. *Loesses in Louisiana and at Vicksburg Mississippi. Friends of the Pleistocene Field Trip Guidebook, South-Central Cell*. Baton Rouge: Louisiana State University.

Mossa, J. 1985. Management of floodplain sand and gravel mining. *Flood Hazard Management in the Government and Private Sector: Proceedings of the Ninth Annual Meeting of the National Association of State Floodplain Managers*: 321-328. Madison, Wisconsin: The Association of State Floodplain Managers.

Mossa, J. 1995. Sand and gravel mining in the Amite River floodplain. In C.J. John & W.J. Autin (eds), *Guidebook of Geological Excursions, The Geological Society of America Annual Meeting, New Orleans*: 325-360. Baton Rouge: Louisiana State University.

Mossa, J. & Autin, W.J. (eds) 1989. *Quaternary Geomorphology and Stratigraphy of the Florida Parishes, Southeastern Louisiana*. Field Trip Guidebook Number 5. Baton Rouge, Louisiana: Louisiana Geological Survey.

Mossa, J. & Miller, B.J. 1995. Geomorphic development and paleoenvironments of late Pleistocene sand hills, southeastern Louisiana. *Southeastern Geology* 35: 79-92.

Murphy, M.L., Hawkins, C.P. & Anderson, N.H. 1981. Effects of canopy modification and accumulated sediment on stream communities. *Transactions of the American Fisheries Society* 110: 469-478.

Nakashima, L.D. & Mossa, J. 1991. Responses of natural and seawall-backed beaches to recent hurricanes on the Bayou Lafourche headland, Louisiana. *Zeitschrift für Geomorphologie* 35: 239-56.

Penland, S., Mossa, J., McBride, R.A., Ramsey, K.E., Suter, J.R., Groat, C.G. & Williams, S.J. 1990. Offshore and onshore sediment resource delineation and usage for coastal erosion control in Louisiana: The Isles Dernieres and Plaquemines Barrier Systems. In M.C. Hunt, S. Doenges, & G.S. Stubbs, *Proceedings of the Second Symposium on Studies Related to Continental Margins – US Minerals Management Service*: 74-86. Austin, Texas: Minerals Management Service, US Department of the Interior.

Rosen, N.C. 1969. Heavy mineral and size analysis of the Citronelle Formation of the Gulf Coastal Plain. *Journal of Sedimentary Petrology* 39: 1552-1565.

Russ, D.P. 1975. The Quaternary Geomorphology of the Lower Red River Valley, Louisiana (Ph.D. Dissertation). University Park, PA: Pennsylvania State University.

Saucier, R.T. 1968. A new chronology for braided stream surface formation in the Lower Mississippi Valley. *Southeastern Geology* 9: 65-76.

Saucier, R.T. 1977. *The Northern Gulf coast during the Farmdale Substage, A search for evidence. Miscellaneous paper Y-77-1.* Vicksburg, Mississippi: US Army Corps of Engineers Waterways Experiment Station,

Saucier, R.T. 1994. *Geomorphology and Quaternary Geologic History of the Lower Mississippi Valley.* Vicksburg, MS: US Army Waterways Experiment Station.

Saucier, R.T. & Fleetwood, A.R. 1970. Origin and chronologic significance of late Quaternary terraces, Ouachita River, Arkansas and Louisiana. *Geological Society of America Bulletin* 81: 869-890.

Self, R.P. 1986. Depositional environment and gravel distribution in the Plio-Pleistocene Citronelle Formation of southeastern Louisiana. *Transactions of the Gulf Coast Association of Geological Societies* 36: 561-573.

Self, R.P. 1993. Late Tertiary to early Quaternary sedimentation in the Gulf Coastal Plain and lower Mississippi Valley. *Southeastern Geology* 33: 99-110.

Smith, F.L. & Russ, D.P. 1974. *Geological Investigation of the Lower Red River-Atchafalaya Basin Area, Technical Report S-74-5.* Vicksburg, Mississippi: US Army Corps of Engineers Waterways Experiment Station.

Smith, F.L. & Saucier, R.T. 1971. *Geological Investigation of the Western Lowlands Areas, Lower Mississippi Valley. Technical Report S-71-5.* Vicksburg, Mississippi: United States Army Corps of Engineers Waterways Experiment Station.

Snead, J.I. & McCulloh, R.P. (compilers) 1984. *Geologic Map of Louisiana,* scale 1:500,000. Baton Rouge: Louisiana Geological Survey.

State of Louisiana. 1992. *Governor's Interagency Task Force on Flood Prevention and Mitigation: Amite River Sand and Gravel Committee.* Baton Rouge: State of Louisiana.

Suter, J.R. & Berryhill, H.L., Jr. 1985. Late Quaternary shelf-margin deltas, northwest Gulf of Mexico. *American Association of Petroleum Geologists Bulletin* 69: 77-91.

Suter, J.R., Berryhill, H.L. & Penland, S. 1987. Late Quaternary shelf-level fluctuations and depositional sequences, southwest Louisiana continental shelf: of the Gulf coastal plain. *Journal of Sedimentary Petrology* 57: 198-219.

Suter, J.R., Mossa, J. & Penland, S. 1989. Preliminary assessments of the occurrence and effects of utilization of sand and aggregate resources of the Louisiana Inner Shelf. *Marine Geology* 90: 31-37.

Tepordei, V.V. 1982. Sand and gravel and crushed stone in the midwest. In C.H. Ault & G.S. Woodard (eds), *Proceedings of the 18th forum on Geology of Industrial minerals: Occasional Paper 37:* 173-81. Bloomington: Indiana Geological Survey, Department of Natural Resources.

Trahan, L., Bradley, J.J., Morris, L. & Nolde, R. 1991. *Soil Survey of St. Tammany Parish.* Fort Worth, Texas: US Department of Agriculture.

Turner, R.E. & Bond, J. 1980. Urbanization, peak river flow and estuarine hydrology near Baton Rouge, Louisiana. *Proceedings of the Louisiana Academy of Sciences* 43: 111-118.

Turner, R.E., Forsythe, S.W. & Craig, N.J. 1981. Bottomland hardwood forest land resources of the southeastern United States. In J.R. Clark & J. Benforado (eds), *Wetland of Bottom Hardwood Forests:* 13-28. Amsterdam: Elsevier.

Walker, H.J., Coleman, J.M, Roberts, H.H. & Tye, R.S. 1987, Wetland loss in Louisiana. *Geografiska Annaler* 69A: 189-200.

Woodward, T.P. & Gueno, A.J., Jr. 1941. *The Sand and Gravel Deposits of Louisiana,* Geological Bulletin No. 19. Baton Rouge: Louisiana Geological Survey.

Yeend, W. 1973. Sand and gravel. In D.A. Brobst & W.P. Pratt (eds). *United States Mineral Resources, US Geological Survey Professional Paper 820:* 561-65. Washington, D.C.: US Government Printing Office.

Subject index